The Collapse of Darwinism

The Collapse of Darwinism

or

The Rise of a Realist Theory of Life

Graeme Donald Snooks

LEXINGTON BOOKS
Lanham • Boulder • New York • Oxford

LEXINGTON BOOKS

Published in the United States of America
by Lexington Books
A Member of the Rowman & Littlefield Publishing Group
4501 Forbes Boulevard, Suite 200, Lanham, Maryland 20706

PO Box 317
Oxford
OX2 9RU, UK

British Library Cataloguing in Publication Information Available

Library of Congress Cataloging-in-Publication Data

Snooks, G. D. (Graeme Donald)
    The collapse of Darwinism, or, The rise of a realist theory of life / Graeme Donald
Snooks.
        p. cm.
    Includes bibliographical references and index.
    ISBN 0-7391-0613-9 (cloth: alk. paper)
    1. Evolution (Biology) 2. Natural selection. 3. Species. I. Title: Rise of a realist
theory of life. II. Title.

QH366.2.S56 2003
576.8'2—dc21   2003047689

Printed in the United States of America

♾™ The paper used in this publication meets the minimum requirements of American
National Standard for Information Sciences—Permanence of Paper for Printed Library
Materials, ANSI/NISO Z39.48–1992.

To Ronald Charles Williams

My maternal uncle, who nourished an early interest in
the big questions about life.

Contents

Figures

Tables

Abbreviations

A.	*Australopithecus*
ADP	adenosine diphosphate
ATP	adenosine triphosphate
BP	before the present
Descent	C. Darwin (1948), *The Origin of Species by Means of Natural Selection, or, The Preservation of Favored Races in the Struggle for Life; and The Descent of Man and Selection in Relation to Sex*, New York, Random House.
Domestication	C. Darwin (1868), *The Variation of Animals and Plants Under Domestication*, London, John Murray.
DNA	deoxyribonucleic acid
GDP	gross domestic product
H.	*Homo*
IMF	International Monetary Fund
myrs	million years
OED	*Oxford English Dictionary*
Origin	C. Darwin (1979), *The Origin of Species by Means of Natural Selection or the Preservation of Favoured Races in the Struggle for Life*, New York, Avenel Books.
WTO	World Trade Organization

Preface

The Darwinists are confident that they have won the war of words about life. They assert that "Today, no biologist questions the reality of evolution or that its mechanism is natural selection" (Stenseth 1999: 1490). They have even convinced many others that this is so. The Royal Swedish Academy of Sciences, for example, recognized their claim by awarding the 1999 Crafoord Prize—"The Nobel Prize in fields for which no Nobel Prize is awarded"—to "three giants in the field of evolutionary biology," namely Ernst Mayr, John Maynard Smith, and George C. Williams. They are three prominent representatives of a discipline of thought usually called neo-Darwinism. An earlier Crafoord Prize had gone to William D. Hamilton for his work on kin selection, which is the bridge between the "three giants" and the "selfish gene" argument popularized by Richard Dawkins.

Things, however, are not what they seem. Just as neo-Darwinism has achieved a peak of acceptability with the biology profession, the wider scientific establishment, and the general public—if we exclude the creationists who appear to be making a comeback in the U.S. states of Kansas, Alabama, Oklahoma, Kentucky, and New Mexico—the whole structure has begun crashing down. A great chasm is opening up between the appearances and the reality of the Darwinian tradition in evolutionary biology. It is the purpose of this book not only to explain the inevitable collapse of Darwinism but to provide an alternative theory of life that, unlike Darwinism, emerges from a systematic analysis of reality. This new realist theory of life is called the dynamic-strategy theory.

What are my credentials for this daunting task? While I am not a geneticist, neither was Charles Darwin. Not only did he guess wrongly about the biology of inheritance but his guesses were Lamarckian in nature. And Lamarckism is a heresy almost as great as creationism to the neo-Darwinist. The point is that a total lack of understanding of genetics did not prevent Darwin from proposing a theory of the dynamics of life that has survived in one form or another for the past 150 years. This is because the essential skill required to explore the dynamics of life and of human society is the ability to analyze not the way organic structures replicate themselves but rather how dynamic systems operate. And this essential skill, which is not part of the tool kit of neo-Darwinists, can only be developed through the close and systematic observation of dynamic systems in the natural and human worlds.

Further, there are two excellent reasons for claiming that an understanding of dynamic systems can best be achieved by focusing initially on human society over the past two million years. First, we know a great deal more about our own species than about any other. Consequently we are often willing to believe arguments about other species that we would not countenance for an instant about our own. And second, the most critical test for any theory of life—a test that Darwinists of all descriptions consistently fail—is whether it can explain the dynamics of human society. For both reasons it is no liability to be a social scientist rather than a natural scientist when exploring the dynamics of life. Even geneticists realize that it is not possible to construct a general and universal theory about the dynamics of life from the building blocks of organic structures.

The dynamic theory of life presented in this book is the outcome of a lifetime of systematically observing the way human societies—and more recently the "societies" of other species—change over time. This new realist theory, unlike those of the various Darwinian traditions, can explain the dynamics of *both* nature and human society in a nonevolutionary framework. It involves, therefore, a rejection not only of Darwin's theory of natural selection in favor of my theory of **strategic selection** but also of evolution in favor of the **strategic pursuit**. For this reason I avoid using the term "evolution" except in reference to the bankrupt theories of Darwinism. Indeed, I suggest that the term **biotransition** be substituted for that of evolution.

New theories generate new visions. The neo-Darwinist vision of life is that of the "selfish gene" which holds individual organisms, including human beings, on a tight "genetic leash." In contrast, the dynamic-strategic vision is of individual organisms engaged in a strategic pursuit. In this pursuit they employ **dynamic strategies**, of which genetic change is merely one, to achieve their common objective of survival and prosperity. It is a vision of human society that rejects genetic determinism and embraces humanism.

New theories also generate new terms and concepts. These are bolded in the text the first time they are mentioned in each chapter and are included with explanations in a glossary at the end of the book.

Treading new paths is a difficult and dangerous activity. Few colleagues are willing or able to accompany one on the way. The exceptions, who I wish to thank but not implicate for their generous encouragement, are Gary Magee at Queen Mary College, University of London, Bertrand Roehner at the University of Paris, Angus Maddison, formerly at the University of Groningen, and Mark Elvin at the Australian National University. Tim Farmiloe, formerly my publisher at Macmillan but now retired, has been a wonderful source of encouragement and wise counsel about publication possibilities. He has my continuing gratitude.

As always I wish to thank my old colleague Barry Howarth for his much appreciated research assistance and copyediting. Also I thank Wayne Naughton for his expert work on the figures, and Debbie Phillips of DP Plus for her excellent word processing and formatting.

The text was completed in this form on 28 February 2001, although a few references have been added since then.

Graeme Donald Snooks
Sevenoaks
Canberra

Prologue:

"The Grandest Pile of Ruins I Ever Saw"

The great Church . . . stood on one side of the Plaza: it was of considerable size and the walls very thick, 4 to 6 ft and built entirely of brick:
the front which faced the NE forms **the grandest pile of ruins I ever
saw**; great masses of brick-work being rolled into the square as fragments of rock are seen at the base of mountains. Neither of the side
walls are entirely down, but exceedingly fractured; they are supported
by immense buttresses, the inutility of which is exemplified by their
having been cut off smooth from the wall, as if done by a chisel.

[A description of the collapsed
cathedral at Concepción,
March 1835.]

Charles Darwin,
Beagle *Diary*

*In a forest clearing stands a simple but elegant chapel. It adjoins a busy town
preoccupied with materialist concerns. The dissenting chapel, a product of rebellion against the old orthodox religion of the land, is dedicated to nature. Its
stained-glass windows depict plants and animals in profusion, struggling to
reach the light. Some succeed but most fall by the wayside. In prismatic translu-
cence a progression of life forms from the simplest to the most complex parades
before us. The central window, which overlooks a plain altar, depicts a transcendent being raised up on the shoulders of a host of other life forms. As one's
eye travels upward from the altar, the colored-glass image becomes increasingly fragmented and blurred. It is as if the artist's vision failed him in the final
act of creativity.*

*The chapel excited considerable interest from the townspeople, and many
visited out of curiosity. Most were puzzled by the large central window. After a
time, however, their numbers dwindled until only a few visitors, mainly from
distant towns, ventured into the chapel's quiet darkness. And of these only one*

or two noticed the flaws that had appeared in the foundations or the fine stress fractures in the central window.

This peaceful obscurity, however, was not to last. The day came when the forest echoed to a recurring metallic ring and the occasional crash of falling trees. The forest's new owners—an intellectual order that claimed to be inspired by the original chapel builder—were clearing the trees right back to the edge of the village. They had plans to build a great cathedral where the chapel stood, and to take the word of their master to the people. They intended to throw this new structure like a mantle over the simple chapel. It was to be a great Gothic structure. A place of boundless space, subtly filtered light, and uplifting spirituality. It was to be based on abstract rather than empirical principles, using advanced mathematics rather than scale models. The townspeople were particularly interested in this new venture as their taxes were being employed to partly fund it.

The cathedral builders made an impressive start, with a floor plan that was bold and imaginative. Yet despite their rhetoric to the contrary, the original chapel did not form the center of their new structure but was tucked away to one side, far from the grand altar. As the walls rose and the impressively vaulted ceilings enveloped them the extent of the ambition of this new sect became clear. The townspeople were dazzled. Only when the stained-glass windows were installed did they begin to feel apprehensive. And this apprehension increased with the installation of each group of windows designed by different factions in the new order. The first group of windows contained brilliantly colored images of trees and lesser plants being pollinated by a variety of insects and birds. While most onlookers were reassured, there were some who detected ominous overtones in the way certain plants appeared to be exercising a coercive influence over the attendant animals. But it was the next group of windows, containing images of mating animals of all types, that shocked most of the townspeople. Few were reassured by the explanation that physical love was better than the war of nature depicted in the windows of the original chapel. Most thought it unnecessarily voyeuristic.

Most disquiet, however, was created by the last group of windows to be installed. They depicted animals and humans as slaves to small alien beings. The animals had the appearance of sleepwalkers—some said of "lumbering robots"—who were directed by alien beings that appeared to inhabit their hosts. Some said that these images created an impression of science fiction rather than the reality they claimed to convey. But it was the last of these windows, which depicted a grotesque transcendent being, that disturbed the townspeople most. The humanist leader of the townspeople said that it reminded him of William Blake's depiction of a humanoid flea. When they asked for an explanation, the high priests simply told the townspeople that this final window depicted the force controlling life on Earth.

With the great cathedral approaching completion, even members of the new order began to feel uneasy. Not with the cathedral's underlying philosophy but with the widening cracks that had appeared in some of the load-bearing walls. To the objective observer it had become quite clear that the abstract mathemat-

ics of the purists had not been able to account for the interaction of all the dynamic forces involved. Accordingly, the walls and columns could not tame the outward thrust of their towering structure.

The cathedral builders, however, could not agree among themselves as to the causes and the seriousness of the problem. Members of the order split into two groups. Those calling themselves "the purists," who were more arrogant than the rest, dismissed these concerns as not worthy of their high calling. They were matters merely of grubby empirical detail. In contrast, those known as "the pragmatists" insisted that it was a very real problem, which had arisen because of the unrealistically abstract approach that the purists had taken to cathedral design.

Not surprisingly a dispute—dignified by the title of "the great debate"—broke out between these two groups about the cathedral's design and construction methods. The pragmatists maintained that in view of the growing evidence of flaws in the original design, it was essential to shore up the foundations and provide external buttressing to the walls. The purists were outraged. They had complete faith in both the assumptions and logic of their design and refused to compromise its purity.

Both groups took their cause to the growing crowds of townspeople attracted by the vigor of the dispute. Because the purists, who made their case from the high pulpit, were skilled and charismatic speakers they were able to persuade many of the townspeople to their point of view. They were the "stars" of the new order. In contrast the pragmatists were less able to dazzle the masses. Instead they patiently gathered all the available evidence about the real state of the cathedral. Only then did they begin to speak to the townspeople from a low pulpit just outside the master's old chapel. Their detailed and convoluted arguments, which were shaped by the master's very real presence, always brought a glazed look to their listeners' eyes. The only lively moments were when the pragmatists suddenly burst into attacks on the purists for departing from the master's teaching. Some even said that the master, when visiting a far distant land, had a vision of the collapse of a great cathedral. Despite such moments their audiences dwindled. Only a leading humanist from the town seriously challenged the purists.

In the end the pragmatists, unable to win this war of words, took matters into their own hands. They began to shore up the foundations and to build a series of flying buttresses against the external walls. To do so they employed stone from a nearby quarry famous for its fossilized images of frozen time. But, because the pragmatists were not skilled in the construction of flying buttresses, these ad hoc supporting structures developed flaws of their own.

Needless to say the purists were horrified. Yet in the face of aggressive solidarity from the pragmatists they decided to ignore the grubby business of foundation-shoring and buttress-building and instead to focus on the abstract purity of their faith. They were determined to employ the forces of reason to take their metaphysical design to its logical conclusion. This final stage in cathedral construction led to the creation of internal details and images that greatly offended the common sense of the townspeople.

Just when the purists had put the finishing touches to the great cathedral, something completely unexpected happened. In their anger that their hard-won taxes had been employed to finance this "metaphysical madness" the townspeople, led by the disputatious humanist, decided to retrieve their funds by dismantling the flying buttresses to obtain stone for the construction of a meeting place of their own—a people's palace dedicated to individualism rather than individual enslavement to either gods or aliens. It would appear that the townspeople had finally seen through the brilliant but empty words and images of the high priests. Finally they realized that they had been misled about the true nature of life by this arcane sect. They realized that it was necessary to go back to reality and begin again. Anew.

Early the next morning, before members of the new order had begun to attend to their duties in the great cathedral, there was a sudden reverberating rumble of rupturing masonry followed by a brief but profound silence that struck fear in the hearts of all those assembled. Then in chaotic cacophony the cathedral walls collapsed outwards and the roof came crashing down. One of the observers, a white-bearded visitor from a foreign land, exclaimed that it was "the grandest pile of ruins I ever saw."

When the dust had settled it became clear even to the high priests that their work, together with the master's original chapel, had been destroyed. They finally realized that the structure they had built was unable to control the forces that they had created. Once the buttresses were removed—they were so badly cracked that they would have collapsed anyway—the whole crumbling structure fell apart. Totally demoralized, the religious community disintegrated and quickly left the region forever.

With their departure the townspeople resumed removing the fossil-bearing stones and within a relatively short time a towering new edifice rose within the town's center. They called it La Pedrera. On this occasion the building's designer adopted an empirical rather than an abstract approach. Using an innovative dynamic method to calculate the real structural stresses, the architect meticulously made large-scale models of the building. In this way he was able to see how all the real forces in the building interacted to produce a dynamic whole. Hence there was no need for supporting buttresses, as the correctly inclined columns carried the loads directly down into the earth, thereby eliminating the outward thrust that had destroyed the great cathedral.

Those who organized the construction of the new secular structure told a different and persuasive story of life based on careful observation of what had gone before. In this way they were also able to see what was to come. The great cathedral of metaphysics had collapsed but a new realist structure dedicated to humanism had risen from the ruins.

Chapter 1

The Rise and Fall of Darwinism

Most intellectual disciplines are dominated by "the big idea." It defines and justifies a discipline and provides a sense of identity, context, and security for its members. Big ideas, however, are often wrong as human ambition usually outruns our intellectual abilities. But what is fascinating is that being wrong does not matter. Big ideas are a matter of faith, not of science.

Two of the biggest ideas in the life sciences—biology and social sciences—during the twentieth century were Darwinism and Marxism. Both emerged from mid-Victorian England, one to explain the dynamics of life and the other the dynamics of human society. At the core of Darwinism is a mechanical formula that purports to explain the inexorable "descent with modification" of life forms, and of Marxism it is a metaphysical dialectic to explain the inevitable emergence of the ideal state of communism.

Both concepts are fatally flawed. And for similar reasons. Both are based on naive, even nonscientific, methodologies; both require totally unrealistic assumptions to generate the desired outcomes; and the predictions of both are completely contradicted by real-world evidence. What is not usually realized is that if either one of these theories had been true it could have been employed to explain the dynamics of *both* life and human society. But neither can do either. Mid-Victorian intellectuals had no real idea about dynamic systems.

Big ideas can survive despite being wrong for as long as they are not seriously challenged by reality. For as long, that is, as they are not essential to mankind's struggle for survival and prosperity. Only when knowledge in a particular discipline is required in society's **strategic pursuit** will the flawed big ideas—the old comforting faith—be rejected, and more realistic and relevant ideas be adopted. Marxism's claim that a moribund capitalism would be replaced via dialectical materialism by a vital communism was challenged by the Cold War and found seriously wanting. The same, it is argued here, is true of Darwinism.

Superficially, Darwinism appears to be going from strength to strength. Despite an inability to explain the dynamics of human civilization—or, as will be revealed later, even nature—Darwinism has recently become more influential in the social sciences and humanities. In part, this is due to the poverty of deductive theory in both intellectual disciplines, but in larger part to the recent decline in Marxism. Former Marxist (and potential Marxist) thinkers are increasingly adopting that other mid-Victorian idea, Darwinism, often, after a shift from socialism to environmentalism.[1] The loss of one faith requires the search for another. Ironically, social scientists and humanists are expressing greater interest in the big idea of Darwinism just at a time when biotechnologists have realized that "natural selection" is being displaced by "social choice" as the determinant of genetic change. What neither group realizes is that natural selection has never played this role.

The popularity of Darwinism will be short-lived. In the near future Darwinism will be severely challenged by a reality that cannot be ignored. If through biotechnology we are able to control our genetic structure so that we can change even human nature, a realistic theory of life will be essential. The old Victorian fantasies will no longer suffice. If we continue to fail to understand the dynamics of life and the role of human nature in it, we will run the very real risk of undermining the forces driving the survival and prosperity of our species. For the first time in the history of life big ideas really matter. Not in a metaphysical way as in the past, but in a very practical way concerning the future of our species. From now on the most important characteristic of big ideas will be their truth rather than their myth-making quality. It is the strategic pursuit that will finally open our eyes to the poverty of Darwinism—to its state of collapse. The twenty-first century demands modern rather than mid-Victorian ideas.[2]

What Is Darwinism?

Essentially Charles Darwin (1809–1882) wanted to persuade the scientific community in the mid-Victorian period that all life as they knew it had "descended with modification" from a common ancestor hundreds of millions of years ago. At the time it was widely believed by academics at Oxford and Cambridge that all existing life had been divinely created either at the same time about six thousand years ago or in a number of stages since then.

To change this conventional view Darwin had to advance a convincing mechanism by which "evolution"—or "transmutation" as it was widely called—might have occurred. Although the idea of evolution had been in the air for generations—even Charles's paternal grandfather Erasmus had written a book on the subject—and was employed by some of his contemporaries—such as Herbert Spencer, a *social scientist*—no one had succeeded in constructing a persuasive evolutionary mechanism. Darwin initially called this discovery "natural selection," although Alfred Wallace (1823–1913), who also stumbled on the same concept, persuaded him to adopt the term "survival of the fittest" (coined by Spencer) to avoid the obvious question: "Selected by whom?" As it turned out this is the key question that must be asked about life, and our modern answer is very different to that of the Victorians. The Darwinists, therefore, had good cause to suppress it.

Natural selection is a simple idea that struck a chord in the mid-Victorian mind. British society in the mid-nineteenth century was a product of three hundred years of successfully pursuing the commerce (1550–1750) and the technological (1750–1850s) dynamic strategies (Snooks 1997: ch. 10). Owing to this success they accepted as the natural order of things that Britain, as the "fittest" society on Earth, should emerge as the dominant power in the struggle between "civilized" nations. Adopting this triumphal outlook the worldly philosophers—including Adam Smith (1723–1790), David Ricardo (1772–1823), and Thomas Robert Malthus (1776–1834)—advocated a policy of unfettered competition at home and abroad. Of course, under the prevailing conditions of intense competition the strategically strong nations, such as Britain, would survive and prosper while the less "fit" would be overwhelmed and absorbed.

Darwin's idea of natural selection, which had much in common with the ideas of the worldly philosophers, fell on fertile ground in 1859 when *The Origin of Species* was first published. It is probably fortunate that Darwin did not ask himself what was the basis for the "selection" of successful countries like Britain. Had he done so, and arrived at the correct answer, the concept of natural selection would never have emerged.

Despite the compatibility of natural selection with mid-Victorian aspirations and values, it could never have been more than a stopgap measure. No one at the time, least of all Charles Darwin, understood anything about "social dynamics." Not even John Stuart Mill (1806–1873), who coined the term. A general dynamic theory of either nature or human society was too complex for the Victorians. Simplicity rather than complexity was needed to win the initial great battle against the creationists. What was required was an uncomplicated—even homely—yet persuasive explanation, which Darwin was the ideal person to provide.

Darwin arrived at the concept of natural selection through what I have called the "farmyard" analogy. By assuming that organisms at all times and in all places are driven to maximize their number of offspring—what he called the "doctrine of Malthus"—Darwin was able to replace artificial selection in the farmyard with "natural selection," which he projected onto nature. This provided him with the following simple formula (or algorithm) to "explain" evolution:

- nature generates random variation in physical and instinctual characteristics;
- all organisms are somehow programmed to produced as many offspring as possible;
- this leads to intense competition for scarce resources—"the struggle for existence"—in which only those with a physical or instinctual advantage, no matter how slight, will survive;
- surviving individuals pass on these advantages to their offspring;
- the slow but continuous accumulation of these slight variations leads, over millions of years, to the emergence of new species.

This formula was simple, logical, and had the ring of authenticity about it for those of the landowning class familiar with experiments in domestic breeding.

Despite his brilliance as a naturalist and his vast practical knowledge of the selective breeding of domestic animals, Darwin knew nothing about the properties of dynamic models or the histories of dynamic systems. There is just no way he could have been expected to understand the dynamic reality underlying the society of which he was a part, let alone that underlying nature of which his main experience was in the barnyard.

Even J. S. Mill, a brilliant contemporary who declared his intention in *A System of Logic* (1843) to develop a theory of "social dynamics" (in contrast to Ricardian "social statics"), had to admit failure in this respect. If Mill, the best prepared and most finely tuned analytical mind of his generation, was unable to develop a general dynamic model, what hope did Darwin, who was educated to become a country clergyman, have of doing so. Darwin did his best under these unfavorable circumstances, but his best merely produced a mechanical formula (based on convenient but incorrect assumptions) to "explain" transmutation rather than a general dynamic model capable of explaining life (and, hence, transmutation) as well as human society. The only way to explain transmutation (what I will call **biotransition** arising from genetic change) is as part of the wider process of the dynamics of life. While Darwin helped to divorce science from religion, he was unable to construct a workable dynamic model of life. He merely replaced one faith with another.

With the passing of Darwin's generation the controversy excited by *The Origin of Species* (1859) died down. Indeed, by the beginning of the twentieth century Darwinism was on the retreat. At that time, biologists in the new discipline of genetics, who had rediscovered the forgotten pioneering work of Gregor Mendel (1822–1884), thought they could explain "descent with modification" without the assistance of Darwin's concept of natural selection. Not only were they critical of Darwin's failure to understand the hereditary process—he knew nothing about genetics—they also thought they could construct a dynamic theory of life from the building blocks of genetics.[3]

Darwinism, in name if not in substance, was rescued in the 1930s from early extinction by a later generation of geneticists. The so-called neo-Darwinists—Ronald Fisher (1930), J. B. S. Haldane (1932), and Sewall Wright (1931; 1932)—realized that it was not possible to explain evolution from the study of genetics alone. Accordingly they made an attempt, which was not completed until the 1950s, to reconcile natural selection with genetics (Gould 1982: 382). But in the process they radically reworked Darwin's concept of natural selection to suit their genetic interests. It became known as the "modern synthesis." In effect the neo-Darwinists hijacked Darwin's evolutionary mantle and surreptitiously transformed his theory to suit their purposes. They have used Darwinism as a cloak, which they have completely retailored, to disguise their attempt to develop a new dynamic theory of life from genetic building blocks.

The neo-Darwinists have transformed natural selection from an economic concept involving the struggle for scarce resources into a sociological concept involving the struggle for "reproductive success." When taken to its logical conclusion by modern neo-Darwinists such as Edward Wilson (1975) and Richard Dawkins (1976; 1982), the Darwinian focus shifts from organisms pursuing reproductive success to "selfish genes" attempting to maximize their presence in the gene pool by manipulating not only their host organism (or "survival machines") but also other organisms (or "extended phenotypes") as well. In doing so they have reduced Darwin's fatally flawed but serious concept into a subject of farce.

Some pragmatic Darwinists, such as Niles Eldredge (1995) and Stephen Gould and Richard Lewontin (1979), who are embarrassed by this recent expansion of neo-Darwinism, have attempted to distance themselves by labeling it "ultra-Darwinism" and distinguishing it from their own version of neo-Darwinism. Unfortunately for them this is not possible because ultra-Darwinism is merely a logical extension of neo-Darwinism. Which is why I only employ the latter term in this book. In any case, neo-Darwinism of any type is no more capable than is Darwin's original version of explaining the historical patterns of life, of deriving the laws that govern them, or of making predictions about the future.

Why Has Darwinism Collapsed?

Darwinism has collapsed because despite all the buttress-building that the true believers have been engaged in over the past half-century the cracks in the great structure have been widening inexorably. To continue believing in Darwinism the faithful have been forced to go into denial. They have attempted to distract themselves and the growing crowd of onlookers by generating a series of popular rationalizations of Darwinism. It is no coincidence that popularizers—the "media stars" of the profession, as Ernst Mayr has called them—such as Richard Dawkins, Edward Wilson, and Stephen Gould, among many others, have never been so active as in the final days of this great intellectual dynasty. Such massive public relations is not required when a dynasty is in its prime, only when it is in the process of collapse.

What has accounted for the widening cracks in the great cathedral of Darwinism? The fundamental problem is that the very foundation stone has crumbled. Natural selection is built on the totally untenable assumption that all organisms at all times and in all places attempt to maximize the number of their offspring. This is the exogenous (or external) force—the "doctrine of Malthus"—required to bring natural selection into operation. The problem for Darwinism is that it has no counterpart in reality.

But this is only the initial flaw. The concept of natural selection is not a general dynamic theory, only an enabling device, and a faulty one at that. To begin with, its focus is all wrong. It is only possible to understand the dynamic process of genetic change at the individual, species, or dynasty level by developing a general dynamic model of life in which it is but a component. Albeit an important component. This misfocus—encouraged by neo-Darwinism's exclusive concern with genetics—lies at the center of the failure of Darwinism. Although "evolution" is not the same as the dynamics of life, it is treated by the

Darwinists as if it were. It is similar to developing a (faulty) model of technological change and then claiming that it can explain the dynamics of human society. There is no escaping the fact that the Darwinian model has no endogenous
(or internal) driving force and no general dynamic mechanisms that can account
for macrobiological change.

Owing to this fundamental flaw it is not surprising that Darwinism is unable
to convincingly explain the historical patterns in either life or human society.
Any successful dynamic model must be able to do both. While our remoteness
from earlier life forms makes it possible to fudge the issue by introducing spuri-
ous assumptions, the inability of Darwinism to explain the changing fortunes of
human society, about which we know a great deal more, exposes the Darwinist
sleight of hand. To be able to explain our own species is the critical test that
Darwinism, from the great man himself to the whiz kids of today, has failed
totally.

A flawed theory cannot make realistic predictions. As is well-known, Darwin predicted quite incorrectly that "descent with modification" takes place in a
very slow but continuous fashion and that new species displace older species:
that "the extinction of old forms is the almost inevitable consequence of the production of new forms" (*Origin*: 342). The unrealistic nature of this prediction
has been highlighted by paleontologists in both Darwin's time (Falconer assisted
by Walker 1859) and our own (Eldredge and Gould 1972). It is now clear from
the fossil record that systematic genetic change occurs much more rapidly than
Darwin's model predicted, that it is associated largely with the initial period of
"speciation" (the emergence of a new species), that it is followed by much
longer periods of genetic stability, and that old species usually go extinct before
the rise of new species. Genetic change, therefore, is discontinuous, being confined to restricted periods of relatively rapid change in the history of species and
dynasties. This has been called—misleadingly as we shall dis-
cover—"punctuated equilibria" by Stephen Gould and Niles Eldredge.

The Darwinian model makes other predictions that are also falsified by reality. Darwin predicted that systematic transmutation (or genetic change) would
occur—in fact could only occur—under conditions of intense competition for
scarce natural resources. In reality genetic change associated with speciation
(resulting in greater access to natural resources) only occurs when the diametrically opposite conditions prevail—when competition is minimal and natural
resources are abundant. To compound their problem, the Darwinian scenario of
intense competition and scarce resources leads not to speciation but to wars
within and between species, and to their extinction. Darwin and his followers
could not be more wrong if they tried.

Darwinism has been unable to derive useful laws of life or society. Such
laws can only arise from a viable general dynamic model. With these laws it
should be possible to develop more specialized models to analyze detailed aspects of life and society. In the case of human society the laws of history that
have been generated by my dynamic-strategy theory have been used to construct
formal economic and political models that in turn are the basis for prediction

and policy formation (Snooks 1998b; 1999; 2000). The same can be done for the detailed study of nature.

Darwinism by contrast is unable to formally predict what will happen in the future. This is particularly serious today because of the control we are gaining over the genetic structure of life through biotechnology and the genome project. Even neo-Darwinists such as Edward Wilson (1998) acknowledge that the Darwinian model will have no predictive power in the future because "natural selection" will be replaced by "social choice" and Darwinian evolution will give way to "volitional evolution." In reality the predictive failure of Darwinism is the result not of biotechnology but of its inability to develop a general dynamic model of life. The point that needs to be recognized is that had the Darwinists been able to develop such a theory it would have been possible not only to explain the past but also to predict the future, including the role and implications of biotechnology. It is the purpose of this book to show not only that Darwinism has already collapsed but also that a new and realist general dynamic model has risen to take its place.

The Rise of a Realist Theory of Life

How can a realist theory of life be constructed? We must begin by clearing the entire ground before us. This can only be done by rejecting Darwinism completely. It is essential to discard both the analogical method that Darwin used to create his artificial farmyard theory of life and the inappropriate deductive method employed by the neo-Darwinists to create their fantasies about the "selfish gene." Instead we must adopt the realistic historical (or inductive) method, which involves creating theory from systematic observation of the way both life and human society change over time. Those natural scientists approaching this method most closely are the paleontologists, who are rightly concerned with the patterns of life found in the fossil record. But their insights, as valuable as they are, have been distorted by their refusal to abandon Darwin's unrealistic and unworkable theory of natural selection. We must, therefore, begin at the beginning. There is no other way. Not if we wish to succeed.

The inductive method involves identifying the historical patterns—or **timescapes** as I call them—constructing the general dynamic model that underlies these patterns, and using both to identify the specific mechanisms that have been operating in the past in both life and human society. From the general model and the specific dynamic mechanisms—but *not* the timescapes—it is possible to derive the laws of both life and history and to make sensible predictions about their futures. It is even possible to say something useful about life elsewhere in the universe. This method is made clear in part III.

The outcome of this realist approach to the dynamics of life and human society is my dynamic-strategy theory (Snooks 1996; 1997; 1998a; 1998b; 1999; 2000). This theory treats life as a strategic pursuit in which organisms adopt one of four **dynamic strategies** in order to achieve the universal objective of survival and prosperity. These dynamic strategies, which operate in both life and human society, include genetic/technological change, family multiplication (procreation *and* migration), symbiosis/commerce, and conquest. The all-important driving force in this dynamic system is the **materialist organism** (or **materialist man**), striving at all times, irrespective of the degree of competition, to increase its access to natural resources. It is this most basic force in life—which I have called **strategic desire**—that accounts for the dynamic-strategy theory's self-starting and self-maintaining nature. More intense competition merely raises the stakes of the strategic pursuit.

The development path taken by a society/species/dynasty is determined by the unfolding dynamic strategy and the sequence of dynamic strategies adopted by organisms. There is nothing teleological about this unfolding process, which is an outcome of organisms exploring their strategic opportunities on a day-to-day basis in order to gain better access to natural resources. There is no preordained outcome. Individual strategies for survival and prosperity become the dynamic strategies of societies, species, and even dynasties through the process of what I call **strategic imitation**, whereby the conspicuously successful strategic pioneers are imitated by the vast mass of strategic followers. This is the way in which individual "choice" and action is incorporated into my macrobiological theory. The development path of life, therefore, is an outcome of the exploitation and exhaustion of a dynamic strategy or sequence of strategies. Once replacement strategies are no longer available the society, species, or dynasty collapses. Hence, the rise and fall of groups of organisms at all levels is the outcome of the strategic pursuit of the individuals they contain.

Organisms pursue the dynamic strategy that provides them with the most effective access to natural resources. This is a matter of trial and error. Organisms invest time and energy in exploring the potential of a dynamic strategy. This involves developing the physical (or technological) and instinctual characteristics required to gain better access to natural resources in order to generate the energy needed to survive and prosper. As these physical/technological and instinctual characteristics are a response to the **strategic demand** generated by an unfolding dynamic strategy, they will vary with the type of strategy pursued and the stage reached in the unfolding process.

Under the dynamic strategy of genetic change the physical and instinctual characteristics of organisms vary in order to use existing natural resources more intensively or to gain access to previously unobtainable resources. The outcome of this dynamic strategy is the emergence of new species. The family-multiplication strategy on the other hand generates a demand for physical and instinctual characteristics that increase fertility and mobility; the commerce strategy demands characteristics that provide a monopoly over certain resources that can be exchanged for mutual benefit; and the conquest strategy demands weapons of offense and defense. The way in which these physical changes in organisms are achieved brings us to the centrally important concept of **strategic selection**.

Strategic selection distinguishes this theory from all others. It displaces the "divine selection" of the creationists and the "natural selection" of the Darwinists. Strategic selection empowers the organism and removes it from the clutches of gods, genes, and blind chance. It formally recognizes the dignity and power that all organisms clearly possess and in particular reinstates the humanism of mankind that the neo-Darwinists have done their best to demolish.

While strategic selection is examined in detail in chapter 12, a brief outline is required here. Organisms respond to strategic demand for a variety of inputs required in the strategic pursuit, such as skills, infrastructure, institutions (rules), organizations, and physical/technological and instinctual characteristics. Those organisms possessing the physical and instinctual characteristics required by the prevailing dynamic strategy will be, on average, conspicuously more successful in gaining access to natural resources than those who do not possess them. This success will attract the attention of other organisms with similar characteristics. Through cooperative activity these similarly gifted organisms will maximize their individual as well as group success. If of different gender they will mate together and pass on their successful characteristics to some of their offspring. They may even cull out—or allow their stronger offspring to cull out—those offspring that do not share these successful characteristics. This is done by animal and human society alike to increase their survival prospects.

The point of strategic selection is that individual organisms—rather than gods, genes, or blind chance—are responsible for selecting comrades, mates, and siblings that have the necessary characteristics to successfully pursue the prevailing dynamic strategy. It is important to realize that strategic selection operates under varying degrees of competition, not just under intense Darwinian competition, and that it responds to each of the four dynamic strategies, not just the genetic strategy. Also it is associated with the welfare of the self and not that of future generations.

If the prevailing dynamic strategy is genetic change, organisms will seek out associates that have the physical characteristics required to reinforce their own in order to gain greater access to natural resources. If and when they mate, these advantages will be passed on to at least some of their offspring. The others are usually culled out by the parents or more fortunate siblings. In this way new species will gradually emerge. There is no need here for either a divine "selector" or a mechanical natural "selector." The selection is undertaken by the organisms themselves in their strategic pursuit. It is revealing that this type of genetic change, which is associated with speciation, only occurs in reality when competition is minimal and resources are abundant: a situation in which Darwinian natural selection is totally unoperational. The reason is that the genetic strategy takes time and the guarantee of long-run monopoly profits to be successful. Darwinian "survival of the fittest," therefore, is a fiction.

Under the genetic strategy, organisms will only seek out associates that possess the correct characteristics. Hence those characteristics that would assist nongenetic strategies are, at this time, rejected. In other words, mutations that do not contribute to the success of the prevailing dynamic strategy are ignored. Individuals possessing them are regarded as "freaks" or "mutants," are isolated, and are often destroyed.

Once the genetic strategy has been exhausted and new species have emerged, organisms will pursue either the family-multiplication or commerce strategies. As these nongenetic strategies require only slight modifications to physical structure, the genetic profile of the species involved will, after the initial phase of relatively rapid change, approximate the horizontal. This explains the "punctuated equilibria" that paleontologists have detected in the fossil record. With the exhaustion of those strategies, competition becomes intense and resources scarce—the Darwinian scenario at last—but instead of pursuing the genetic strategy that results in speciation (resource-accessing technology), this mature species pursues the conquest strategy that requires only add-on "technology"—body armor, club tails, slashing teeth and claws—useful in war. This requires less time and only short-term returns on the investment of energy. In these circumstances organisms select their associates and mates on the basis of war skills and reject those that could, in the much longer run, lead to the development of new species. The outcome of Darwinian intense competition, therefore, is not speciation but war and, eventually, extinction.

The dynamic-strategy theory, which is explored in greater depth and context in part III, invests individual organisms with greater control over their welfare and with greater dignity. Organisms are no longer the playthings of gods, the victims of arbitrary mechanisms like natural selection, nor are they the robotic "survival machines" of the selfish gene. Indeed, organisms are not only largely responsible for their own fate but collectively they determine, through the process of strategic imitation, the great patterns of life and history together with their unfolding mechanisms.

But what of exogenous events? Life and history, as we shall see, are certainly not driven by the catastrophes, climatic change, or other natural forces preferred by most natural scientists. If they were there would be no laws of life or history, only the random outcomes of a great cosmic lottery. Exogenous forces merely provide the context within which the game of life is played (Snooks 1996: 2). The rise and fall of societies, species, and dynasties are the outcome of a host of individual organisms engaged in the strategic pursuit. Exactly how is discussed, together with many other associated issues, in part III. This introduction is merely to provide a taste of what is to come.

What Is to Come?

The book is divided into three parts. In part I, entitled "Darwin's Dissenting Chapel," I discuss Darwin's concept of natural selection, showing that it is just a farmyard theory that has been extrapolated onto nature, highlight the flaws in Darwin's method of analogy, and demonstrate that natural selection fails the critical test of explaining human society. In part II, entitled "The Collapsing Cathedral of Darwinism," I discuss the modern paleontologists, or "buttress-builders," to show how their historicism has been distorted by their refusal to abandon Darwin's theory despite its inability to explain the fossil record; I outline the absurdities and antihumanism of the neo-Darwinists, or "cathedral-reconstructionists," and I show how the sociobiologists, or "image-makers and evangelists," have lost their way.

In part III, entitled "Rising from the Ruins," I detail my realist dynamic-strategy theory. To chart a new course I identify the main patterns, or time-scapes, in the history of life and human society. On this empirical foundation I construct my general dynamic theory of life and then focus in greater detail on the main elements of that theory. This involves additional chapters on the driving force, strategic selection, the dynamic mechanisms of life, the emergence and role of social organizations, the laws of life, and the future of life on Earth and in the rest of the universe. By the end of this book I hope to have established not only that Darwinism as a system of thought has collapsed, but that a new realist theory of life has risen to take its place.

Part I

Darwin's Dissenting Chapel

The dissenting chapel, a product of rebellion against the old orthodox religion of the land, is dedicated to nature. Its stained-glass windows depict plants and animals in profusion, struggling to reach the light. . . . The central window, which overlooks a plain altar, depicts a transcendent being raised up on the shoulders of a host of other life forms. As one's eye travels upward from the altar the colored-glass image becomes increasingly fragmented and blurred. It is as if the artist's vision failed him in the final act of creativity.

Prologue

Chapter 2

The Chapel of Evolution: Darwin's Theory of Natural Selection

It is well-known that Charles Darwin (1809–1882) was not the first to discuss the momentous idea that the great diversity of life today has its distant origin in one ancestral organism. Evolution had been discussed by many earlier authors. Even the idea, if not the name, of "natural selection" can be found in a number of works before Darwin's. And, owing to his prevarication over publishing the theory, Darwin was almost upstaged by his younger Welsh contemporary Alfred Russel Wallace (1823–1913).

Why then is Darwin regarded as the father of evolution? While there are a number of reasons, the chief of these concerns the analogical method he employed in *The Origin of Species* (1859) not only to construct his theory of life but also to convince his readers of its truth. I will argue that while it was a brilliant method of persuasion it actually distorted our view of the dynamics of life on Earth. Darwin became, in effect, a prisoner of persuasion. Yet so effective was *Origin* as a means of persuasion that natural selection has formed the foundations of thinking about evolution down to the present day. It is on these foundations that the great cathedral of Darwinism has been built.

To understand Darwin's theory of natural selection, as opposed to that of his self-styled followers, we need to go back to *The Origin of Species*. The neo-Darwinists, as we shall see in the following chapter, have fundamentally distorted the master's theory, while at the same time posing as defenders of the faith. By revisiting this, the most influential book of the modern era, we can also see for ourselves why it was such a popular success, going through six editions in the last twenty-three years of its author's life. We discover that *Origin* is a masterpiece of gentle persuasion owing to its careful construction, its logical exposition, its fair and balanced judgment, but ultimately to its analogical method. It is, in a word, seductive. Sadly it is also wrong.

Darwin's Big Idea

Darwin's theory of evolution—of "descent with modification through natural selection"—is quite simple and straightforward. As Thomas Huxley is reported to have said when hearing of it for the first time: "How extremely stupid not to have thought of that!" But then most revolutionary ideas are relatively simple. Nevertheless, it takes great minds to invent them and good minds to embrace them when first presented. Only mediocre minds fail to appreciate the obvious.

Darwin's theory of natural selection can be encompassed in the following five points.

1   Nature is responsible for generating random variations in the physical and instinctual characteristics within species, providing individuals with slightly different abilities with which to cope with life.

2   Every organism in the plant and animal kingdoms is somehow programmed to produce as many offspring as possible in all places and at all times.

3   The resulting struggle for existence over scarce resources is always extremely severe—called by Darwin (*Origin*: 459) "the war of nature"—forcing individuals to employ any slight physical or instinctual advantage as an instrument of survival. While the successful survive, all others perish.

4   The surviving individuals pass on their physical and instinctual advantages—called "profitable variations"—no matter how slight, to their progeny. In this way the profitable variations are preserved and accumulated in successful species. This is the alleged power of natural selection, which Darwin also called "the principle of preservation" (*Origin*: 170).

5   The steady accumulation of very slight variations over vast periods of time—thought by Darwin to run into hundreds of millions of years (in fact he underestimated it by a factor of ten)—through this continuous process of natural selection led, Darwin claimed, to the emergence of new varieties, species, groups of species, genera, families, orders, and classes of life. All from one ancestral organism.

This persuasive theory of life led to the collapse of the case of the creationists, which stated that God had created all existing life forms, including mankind, either at the one time some six thousand years ago or in stages since then. More sophisticated religious thinkers, who accepted that old species had died out and had been replaced by new species since the first act of creation, argued for a continuing creative involvement by God. To appreciate Darwin's argument we need to examine each of these five points more closely.

1   Variation in Nature

Darwin begins *Origin* by outlining his analogy of farmyard selection, which he called "variation under domestication," comparing and contrasting this with "variation under nature." This appeal to what was familiar to his landowning audience set the stage for his method of both persuasion and analysis. Yet while

Darwin was able to illustrate the variation in both the farmyard and nature he was unable to account for it. He candidly admitted his "ignorance of the cause of each particular variation" (*Origin*: 173).

While there can be no "descent with modification through natural selection" if there is no physical or instinctual variation between individuals in a species, Darwin's generation had no understanding of what produced these differences. Indeed, Darwin's guesses about the source of mutation—his theory of "pangenesis" revealed in *The Variation of Animals and Plants Under Domestication* (1868)—could not be verified at the time, and ever since has been an embarrassment to neo-Darwinians because of their Lamarckian basis (by which lifetime acquired characteristics are said to be inherited by the subsequent generation). Darwin's concept of "pangenesis" states that if an organism experiences a physical change in adapting to the environment, the body cells of the affected organ will be excited and emit "gemmules" (or "pangenes")—minute representations of the organ—which will enter the bloodstream and, ultimately, the reproductive cells, and be passed on to the next generation. Neo-Darwinists, as we shall see in part II, have ignored this and other Lamarckian aspects of Darwin's theory, claiming that genetic information cannot pass from body (or somatic) cells to reproductive (or germline) cells, owing to Weismann's barrier.

It was only with the development of the discipline of genetics from the turn of the twentieth century—finally following up the pioneering work of Gregor Mendel (1866)—that biologists began to understand the source of Darwin's variations. At first these early geneticists thought they could dispense with Darwin's theory of natural selection—few biologists liked its close association with the political economy of Thomas Robert Malthus (1766–1834)—but, as they soon discovered, a knowledge of genetics tells us little about the dynamics of life.[1]

Conversely, even a total lack of understanding of genetics is no barrier to understanding the dynamics of life, as Darwin demonstrated. The essential and little understood point is that the dynamics of both life and society are the outcome of what I call the "strategic pursuit" and not of the means, whether genetic or technological, by which this pursuit is translated into the individual's primary objective of survival and prosperity. Hence it is not because of Darwin's total ignorance of genetics that his theory of evolution must be rejected. What we need, as I show in part III, is an entirely new dynamic model derived not from Darwin's flawed method of analogy but from the more robust and relevant historical, or inductive, method.

Darwin's views about the "laws of variation" can be briefly stated. While Darwin thought that variations arose randomly, in the sense that they occurred "in all directions," he did not hold that they were without causes, just that these causes were imperfectly understood:

> I have hitherto sometimes spoken as if variations—so common and multiform in organic beings under domestication, and in a lesser degree in those in a state of nature—had been due to chance. This, of course, is a wholly incorrect expression, but it serves to acknowledge plainly our ignorance of the cause of each particular variation. (*Origin*: 173)

In general he attributes these causes to changes in "the external conditions of life" such as "climate and food, etc.," habit in "producing constitutional differences," and use/disuse in strengthening/weakening the organs of individuals, which would eventually be passed on to future generations. This is a further embarrassment to neo-Darwinians, because of its Lamarckian aspects.

For Darwin, however, the bottom line is that it really does not matter how variations are generated, because it is natural selection that is responsible for accumulating the "profitable" ones. In Darwin's words:

> Whatever the cause may be of each slight difference in the offspring from their parents—and a cause for each must exist—it is the steady accumulation, through natural selection, of such differences, when beneficial to the individual, that gives rise to all the more important modifications of structure, by which the innumerable beings on the face of this earth are enabled to struggle with each other, and the best adapted to survive. (*Origin*: 203–4)

2 and 3   Procreation and Struggle

In *Origin* steps 2 and 3 are brought together (in chapter 3 entitled "Struggle for Existence") by Darwin, because his idea of the struggle for existence depends critically on the assumption of a geometric increase in the progeny of all organisms. To generate the slow accumulation of beneficial variations over vast periods of time, it is essential that the struggle for life be ever present. This is only possible in Darwin's theory if population presses urgently and continuously on natural resources as an outcome of the "principle of geometric increase."

How does Darwin attempt to support this very strong and critically important assumption? When introducing less important issues Darwin always bombards the reader with persuasive evidence. But in the case of this population assumption, which is critical to the entire theory of natural selection, Darwin the naturalist turns to Malthus the political economist for support.

Malthus, it is well-known, had earlier written his famous *Essay on the Principle of Population* (1798), in which he claims that, while food production increases at an arithmetic rate, it is outstripped by population which, he *asserts*, grows at a geometric rate. Only a struggle for access to scarce resources would, he argues, slow this growth of population owing to an increase in the incidence of disease, famine, and wars. In later editions Malthus concedes that, through education, celibacy, and emigration, a less wasteful outcome would be achieved. This, as it turned out, flawed economic thesis (Snooks 1999: ch. 13) rather than the observation of undisturbed ecosystems in the wild became the linchpin of Darwin's theory of evolution. As discussed in the next chapter, the reason Darwin adopted this population assumption was not because he was a Malthusian but because it was the only way of projecting the farmyard model onto nature.

Darwin discusses his use of the Malthusian population principle as follows:

> A struggle for existence inevitably follows from the high rate at which all organic beings tend to increase. . . . It is the doctrine of Malthus applied with

manifold force to the whole animal and vegetable kingdoms; for in this case there can be no artificial increase of food, and no prudential restraint from marriage. (*Origin*: 116–17)

To convince his readers that this was not just a rhetorical flourish he added that *"there is no exception to the rule* that every organic being naturally increases at so high a rate, that if not destroyed, the earth would soon be covered by the progeny of a single pair" (my emphasis).

Needless to say this is a very strong expression of his central assumption, because it can be disproved with just one exception. And in reality there are many exceptions in human society—where birth control has been practiced not only in "civilized" societies but in all preagricultural societies such as those formed by the first Australians, the Bushmen of the Kalahari, and the Inuit of Alaska—as well as in nature—where the self-regulation of population size has been widely observed.[2] In a recent book, *Global Transition* (Snooks 1999: 239–50), I demonstrate statistically that in human society population, rather than being a force out of control, is employed as an instrument in the strategic pursuit.

In the face of such a strongly stated assumption, the marshalling of evidence can be a dangerous exercise. To avoid stumbling across exceptions, particularly if there are many of them, it is better to accept the issue as self-evident and merely to discuss it in general terms. This is the tactic, as can be seen in the following extract, that Darwin employed. He claims that

> In a state of nature almost every plant produces seed, and amongst animals there are very few which do not annually pair. Hence we may confidently assert, that all plants and animals are tending to increase at a geometrical ratio, that all would most rapidly stock every station in which they could any how exist, and that the geometrical tendency to increase must be checked by destruction at some period of life. (*Origin*: 118)

Hence, Darwin asks us to believe that if this *could* happen then it *must* always happen.

Neither Darwin nor the modern neo-Darwinists, who have taken this assumption to be the core of natural selection, realize that procreation is a periodic strategy to enable individual survival and prosperity rather than a genetically programmed continuous activity. According to Darwinism, individuals in nature and, by implication, in human society are merely mindless robots when it comes to procreation. Indeed, the neo-Darwinist Richard Dawkins in his popular book *The Selfish Gene* (1976/1989) refers to individuals in both nature and society as "gene machines," "survival machines," and even "lumbering robots." The sociobiologists E. O. Wilson (1975)—the author of the "genetic leash" view—and R. Trivers (1985) also adopt the idea of genetically determined animal behavior.

Inadvertently, Darwin presents evidence in a later chapter of *Origin* that is sufficient to destroy his assumption that every organism tends to increase at a high geometric rate. When discussing circumstances favorable to natural selection, Darwin draws our attention to the role of isolation. While isolation should prevent "intercrossing" (which retards natural selection), expose all individuals to uniform conditions of change, and provide a degree of protection for newly emerging species, Darwin thought that, on balance, *the less severe competition from external sources* would lead to slower rates of "modification." For example, Darwin argues that "on a small island, the race for life will have been less severe, and there will have been less modification and less extermination" (*Origin*: 151). But how could the race for life even on a small isolated island be "less severe" if "all organisms" procreate at a geometric rate with "no exception to the rule"? The point that Darwin fails to realize is that the degree of external competition, which is dramatically reduced by the fact of isolation, is irrelevant in a world in which all organisms, "without exception," procreate at a high geometric rate. Indeed in a Malthusian world, severe competition will be generated more quickly on a small isolated island than on a large open continent.

But just in case it is (wrongly) thought that there is something atypical about small islands, a few pages later in *Origin*, Darwin, when discussing the "principle of divergence," marshals evidence *against* his procreation assumption for continents as well. To illustrate his thesis that animals with a more highly diversified structure will possess an advantage in the struggle for existence, Darwin draws our attention to the case of Australian marsupials. We are told that

> It may be doubted, for instance, whether the Australian marsupials, which are divided into groups differing but little from each other, and feebly representing . . . our carnivorous, ruminant, and rodent mammals, could successfully compete with these well-pronounced orders. In the Australian mammals we see the process of diversification in an early and incomplete stage of development. (*Origin*: 159)

The clear implication is that this is an outcome of Australia's isolation from the rest of the Old World. Yet how could isolation—the absence of external competition—explain this retarded diversification given the tendency to a geometric rate of procreation, the large size of the Australian land mass, and the very long period that marsupials have occupied Australia? Clearly it could not. There is, therefore, a self-destructive tension in *Origin* between what Darwin knew about reality and the procreation assumption that he was forced to import from political economy in order to make his farmyard theory operational in nature.

Darwin appears not to have noticed that in *Origin* he had presented more than sufficient evidence to destroy the big assumption on which the legitimacy of his entire theory depended. It will be argued that he completely misunderstood the nature of competition for scarce resources in both human and non-human societies.

4 Survival of the Fittest

"Survival of the fittest" is the term coined by Herbert Spencer (1820–1903) to refer to Darwin's concept of natural selection. Spencer had developed a general theory of evolution as early as 1852 to explain the "progress" of both the inorganic and organic worlds. This was some seven years *before* the publication of *The Origin of Species*. Darwin even sent him a letter of warm approval regarding his evolutionary thesis (Carneiro 1974: ix). What Spencer failed to do was to devise a scientific, as opposed to a metaphysical, mechanism to explain the process of evolution. Later, in his *First Principles* (1860–1862), Spencer grafted Darwin's concept of natural selection, which he preferred to call survival of the fittest, onto his preexisting general "law" of progress. Even so, in Spencer's theory, natural selection was merely the mechanism through which his earlier metaphysical "law of progress" operated (Snooks 1998a: 59–60).

The point of this story is that, in the fifth edition of *Origin*, Darwin adopted Alfred Wallace's suggestion that he use Spencer's term to describe his own evolutionary mechanism on the grounds that it avoided the damaging implication that natural selection involved a divine selector (Carneiro 1974: xix–xx). Wallace's fears, as shown in the next chapter, were justified. This change in terminology could do no more than paper over the fundamental flaw in Darwin's theory of natural selection.

Natural selection operates by organisms competing to adapt to the physical and social environment. In Darwin's mind the physical changes, largely climatic, were slow acting and had only a "slight effect," while "changes in the numerical proportions of some of the inhabitants independently of the change in climate" had the main effect. Natural selection, as mentioned earlier, acts on "profitable" variations—those "extremely slight modifications"—that provide some individuals with a slight advantage over others in the continuous "war of nature." Darwin claimed that the circumstances favorable to natural selection included "a large amount of inheritable and diversified variability," a large population, effective barriers to "intercrossing," large and diversified geographical habitats, and very long periods of time to allow for the accumulation of very slight individual variations.

It is interesting that Darwin felt the need to personify his abstract conception of natural selection, possibly in anticipation of the criticism that he did receive about his vision of a bleak world devoid of direction and hope. In *Origin* Darwin identifies natural selection with nature. An example is when he writes: "as man can produce . . . what may not nature effect" (*Origin*: 132). And on occasions he refers to natural selection or nature as "she" and "her." Nature appears to be given an active, even moral, role at times:

> It may be said that natural selection is daily and hourly scrutinising, throughout the world, every variation, even the slightest; rejecting that which is bad, preserving and adding up all that is good; silently and insensibly working, whenever and wherever opportunity offers, at the improvement of each organic being in relation to its organic and inorganic conditions of life. (*Origin*: 133)

But these are only rhetorical flourishes. In reality natural selection is a passive filter that sorts out profitable from unprofitable variations and allows the former to accumulate slowly but continuously over vast periods of time. Natural selection, therefore, is not an active principle in life. That role is played by the assumed Malthusian propensity to populate and perish. By personifying an abstract filtering device rather than the organisms involved in the "war of nature," Darwin underlined the absence of any endogenous driving force in his system. Also he ran the risk of giving the impression that natural selection operated via a divine selector.

For Darwin, natural selection was not a complete explanation of "descent with modification." Much to the concern of Alfred Wallace, the coinventor of this theory, Darwin persisted, particularly in *The Descent of Man* (1871), with the idea that some accumulated variations affecting only the male of the species were the outcome not of natural selection but of "sexual selection." While Darwin attributed only a minor role to sexual selection, his reluctance to abandon it at Wallace's suggestion makes clear the fact that he did not regard natural selection as a general theory. A general theory of selection would need to integrate both the "natural" and the "sexual" elements, which was something Darwin failed to do. And, as Wallace feared, it provided the excuse that sociologically inclined biologists needed in order to abandon the economic foundations of natural selection—the struggle for control of scarce resources. The neo-Darwinian distortion of Darwin's theory is dealt with in part II.

What does Darwin say about sexual selection in *Origin*? He claims that sexual selection

> depends, not on a struggle for existence, but on a struggle between the males for possession of the females; the result is not death to the unsuccessful competitor, but few or no offspring. Sexual selection is, therefore, less rigorous than natural selection. Generally, the most vigorous males, those which are best fitted for their places in nature, will leave most progeny. But in many cases, victory will depend not on general vigour, but on having special weapons, confined to the male sex. (*Origin*: 136)

Accordingly, the males develop unique physical attributes devoted solely to attracting the attention of the females and/or driving off their sexual competitors. And in *Descent* (1871) Darwin not only defines natural selection more narrowly, as those changes that enabled organisms to adapt more closely to their environment, but also swept everything else into the sexual selection grab bag. Interestingly, Wallace believed that these gender differences could be explained as the outcome of natural selection and that there was no need for an additional concept (Wallace 1871: 127, 156, 247, 283). Indeed, it would only weaken their theory of evolution.

Wallace was correct. Rather than being an "aid to ordinary selection" (*Origin*: 170), the subsidiary concept of sexual selection actually works to undermine Darwinian natural selection. What it suggests is that there must be a more general theory of life that can encompass the issues Darwin has identified under both forms of selection, and that this more general theory would need to reject the ideas of both natural and sexual selection. Such a general theory—the dynamic-strategy model—is presented in part III. This new theory of life shows that genetic change and procreation are just two of the four dynamic strategies that are pursued by organisms to meet their objective of survival and prosperity.

5 The Divergence of Life

It is the divergence of life that natural selection is supposed to explain. Darwin's great purpose was to show that the vast diversity of life today is the outcome not of the work of a divine creator some six thousand years ago but of the "action of natural selection through divergence of character and extinction on the descendants from a common parent." To a large degree it was the religious orthodoxy of his day that led Darwin to focus on what we now think of as the outcome of genetic change—the transmutation and succession of species—rather than on the entire dynamics of life of which genetic change is but one element. In part III I argue that it is a focus that distorts our view not only of the dynamics of life but, ironically, also of the nature and role of genetic change.

Darwin regarded the "principle" of the "divergence of character" as a matter of "high importance" to his theory. For this reason he devoted 30 percent of the "Natural Selection" chapter and the only diagram in the entire book to this issue. This famous branching-tree diagram illustrates how, through natural selection, individuals in a species diverge from their parents and give rise to new varieties and species, which in their turn generate further divergence. In this way Darwin purports to show how small differences can be turned into separate species over vast periods of time under pressure from the ever present struggle for existence.

Darwin argued that the success of individuals in this struggle depends on the degree of divergence of character that they experience. The underlying reason is that this growing degree of divergence provides individuals with greater access to the natural resources required for survival. Owing to less competition with the parent species, these modified offspring are less likely to be exterminated by them. Darwin explains:

> as a general rule, the more diversified in structure the descendants from any one species can be rendered, the more places they will be enabled to seize on, and the more their modified progeny will be increased. (*Origin*: 163)

Indeed, these modified descendants will be able to eliminate the weaker sibling varieties and even the parent species. Darwin continues:

> The modified offspring from the later and more highly improved branches in the lines of descent, will, it is probable, often take the place of, and so destroy, the earlier and less improved branches. (*Origin*: 163)

Divergence, therefore, increases the probability of a favored line of descent "succeeding in the battle of life." In Darwin's mind the emergence of new species through divergence implied the elimination of old species. As we shall see, this view led him to distort the pattern of life.

This preoccupation with the divergence of life also led Darwin to overlook the true significance of genetic change in the wider dynamics of life. The same is true of his followers down to the present time. Darwin refers to what I see as the real role of divergence only in the summary to the "Natural Selection" chapter. He does not return to it again. "Natural selection," he tells us, "also, leads to divergence of character; *for more living beings can be supported on the same area* the more they diverge in structure, habits, and constitution" (*Origin*: 170; my emphasis). As I show in *The Dynamic Society* (Snooks 1996: chs. 4, 12), genetic change plays the same role in the dynamics of life that technological change plays in the dynamics of human society: it enables individuals to gain greater access to the natural resources required to survive and prosper. At other times nongenetic/nontechnological dynamic strategies will be pursued without the need for transforming genetic/technological change. This is a central part of the new dynamic theory of life—the dynamic-strategy theory—developed in part III.

The explanation of "divergence of character," which to Darwin's mind resembled the irregular ramifications of a tree, was similar to Adam Smith's explanation in *The Wealth of Nations* (1776) of the expansion and prosperity of the economy. Classical political economy emphasized the growth process of specialization and division of labor. Perhaps if Darwin had instead read and thoroughly absorbed the economics of Karl Marx (1818–1883), which, unlike Smith's, emphasized technological change, he may have equated biological variations with the variations of productive technique and have realized that both are only part of a wider dynamic. But, of course, this would have been fatal to his natural selection hypothesis. It is doubtful that Darwin would have recovered from the blow that such a revelation would have dealt, as he had no understanding of the nature of dynamic systems. In any case, the first volume of *Capital*—the only volume published in Marx's lifetime—did not appear until 1867 and then in German which Darwin, who was sent an inscribed copy by Marx, found difficult to read. Certainly it had no influence on *The Descent of Man* that appeared some years later. There appears to have been no meaningful interchange between the Victorian creators of the two big ideas that dominated the twentieth century.

Of course, this type of game, which biographers love to play, can be taken too far. As already noted, while great thinkers such as Darwin may use the ideas of others that are doing the rounds in contemporary society, they are certainly not driven by them. Darwin's theory of natural selection was based, as shown in the next chapter, on his understanding of the farmyard model—the model of "variation under domestication." He only borrowed ideas from elsewhere to resolve problems experienced in projecting the farmyard onto the natural world.

Conclusions

With his brilliantly simple model of natural selection, Charles Darwin provided the intelligent reader with a persuasive, logical and, apparently, realistic reason for believing in evolution. The impact of simple ideas launched at the right time can be momentous. They can also be extremely tenacious, as demonstrated by the survival of Darwinian natural selection in scientific and popular circles for one-and-a-half centuries despite a series of, admittedly minor, challenges.

Yet in describing this evolutionary model we have discovered a number of flaws in the theory that should have alerted its author and his followers to its severe limitations. But, as is well attested, even deficient theories will continue to command widespread support if they have no real competitors. What is it, one is led to ask, about the biology discipline that has prevented the emergence of any serious rival to Darwinism in all this time? Just imagine how backward physics (and modern society) would be if it were dominated today by ideas that had emerged in the mid-Victorian period. While this question will be taken up in part II, we first need to explore the method employed by Darwin to construct and launch his big idea.

Chapter 3

Flaws in the Foundations:
Darwin's Method of Theory-Building
and Persuasion

With the theory of natural selection, Darwin was able to persuade the world that he had fashioned the key to unlock the secrets of life on Earth. In this chapter we will consider two closely related questions: How did Darwin arrive at this momentous idea and why was this idea so successful? These issues are closely related because they both depend on the method of analogy he employed in *The Origin of Species*. But both the intellectual breakthrough and its success came at a cost: the theory of natural selection has distorted our view of the dynamics of life. There are major flaws in the foundations of Darwinism.

Darwin, as we have seen, begins *Origin* with a homely discussion of "variation under domestication" with which most of his readers—gentlemen farmers, country parsons, country sportsmen, and bird and plant fanciers—would have been familiar. In subsequent chapters he skillfully extends the farmyard analogy to explain "variation under nature" and, finally, "natural selection." With this technique, Darwin takes his readers through a well-known process close to home and then leads them to imagine parallels between the farmyard and nature. Simple but effective.

Yet this was not just a tactic of persuasion. It was the method Darwin had employed two decades earlier to make sense of the vast amount of evidence he had collected on the voyage of the *Beagle*—a body of material that was threatening to overwhelm him. For a naturalist, the method of analogy was an intellectual shortcut—like slashing the Gordian knot—that appeared to pay high dividends both scientifically and popularly. But there is a problem. Darwin's analogy, while appearing appropriate at first sight, is totally misleading. Genetic change in the natural world is nothing like artificial breeding in the farmyard. When we subject Darwin's farmyard analogy to close scrutiny, it crumbles to dust.

Our examination of Darwin's dual method of theory-building and persuasion involves three main steps: the fieldwork undertaken by Darwin on the five-year voyage of the *Beagle* that convinced him of the truth of evolution; the employment of the farmyard analogy to explain evolution; and the theory-testing, primarily involving the results of farmyard breeding experiments that occupied Darwin for the rest of his life. Darwin's method, therefore, involved a high degree of circularity: natural selection emerged from the farmyard analogy and it was verified using evidence from farmyard experiments. Little wonder he was unable to recognize its limitations.

Voyage of the *Beagle*

Darwin's five-year voyage in the *Beagle*, 1831 to 1836—which took him down the east coast of South America, around Cape Horn and back up the west coast to the Galapagos, across the Pacific to New Zealand and southern Australia, north-west to the Cocos Islands, west to Mauritius and Capetown, northwest to St. Helena and Ascension Island, west to South America, and then north to England—changed his life completely. The traditional view is that in the process he was transformed from an unfocused and drifting member of the English gentry to a man driven to explain the "mystery of mysteries," the transformation of life on Earth. But for that fortuitous voyage, we are told, Darwin would most likely have spent his days as a country parson, collecting specimens randomly in his rambles across the countryside. More recently it has been persuasively argued that by the time Darwin boarded the *Beagle* he "was already a well-prepared young scientist, and not aimless at all" owing to his enthusiastic and dedicated informal (since the age of ten years) and formal (with Dr. Robert Grant at Edinburgh and Professor John Henslow at Cambridge) nature studies (Howe 2001: 45–57). Whatever view we may take of Darwin's early training as a naturalist, without the voyage of the *Beagle* it is highly likely that the theory of natural selection would have fallen to the lot of Alfred Russel Wallace.

The voyage of the *Beagle* provided a unique environment for someone of Darwin's temperament. It exposed him to abundant evidence of the transmutation of life and led to an isolation that forced him to think about little else for five years. Darwin's isolation was, for much of the voyage, complete. Not only was he destined to share a cabin and a table with a captain—Robert Fitzroy—who was difficult to communicate with, Darwin was also cut off from shipboard activities by being continuously and violently seasick. He could find relief only by retiring to his hammock, where he was left alone with thoughts of what he had observed and collected. And on land he was preoccupied with exploration of the interior of the land they were visiting together with the time-consuming task of collecting fossils, plants, and animals. It was a routine broken rarely by discussions at dinner with his foreign and colonial hosts.

In other words, Darwin spent the first half of the 1830s—during his formative early to late twenties—in almost continuous contemplation, both mental and written (some 2,500 manuscript pages), about the way the world and life had changed over vast periods of time. The *Beagle* became, in effect, an incubator of evolutionary ideas. In a curiously parallel way this reflected the role that Darwin claimed for isolation in assisting the emergence of new species.

To what did Darwin's thoughts turn during this voyage? As this has been dealt with exhaustively by his various biographers, I will offer just a short commentary to establish the main features (Desmond and Moore 1991: 101–82). Essentially Darwin encountered a fascinating sample of evidence about the transformation of both the physical world and life over long periods of time, a demonstration of the powerful forces (earthquakes and volcanoes) acting on the "conditions of life," and a sample of the abundant variety of life existing in the world. By confronting this bewildering variety of evidence, and being forced to think endlessly about it, Darwin began to fashion a new perspective on the "mystery of mysteries."

There appear to be four main themes that occupied Darwin's mind during this voyage of destiny. The first concerns the question of whether the physical world was created by single catastrophic events as the traditionalists insisted, or whether it was undergoing continual gradual change as Charles Lyell had recently argued in his *Principles of Geology* (1830–1833). Familiar with Lyell's work, Darwin discovered that the evidence he was collecting supported the great man's argument. At firsthand, Darwin witnessed the impact of the forces that are constantly changing the face of the Earth—earthquakes and volcanoes—which were devastating for human settlement (he was astounded at the collapse of the cathedral at Concepción—Beagle *Diary*, 5 March 1835) but only changed the physical environment in marginal ways. Yet, he noted, while the physical forces of life operated through small steps, the accumulation of those small steps over very long periods of time brought about remarkable changes in the environment. These remarkable changes were reflected in his discovery of fossilized trees (formerly at sea level) in the Andes at 7,000 feet, of sea shells on St. Helena at 2,000 feet, and of coral atolls that, he theorized, had been built up gradually as former mountains sank slowly into the sea.

The second theme concerns whether animal and plant life was also transformed slowly over time by the accumulation of very small changes. Curiously Lyell, who had revolutionized thinking about the physical world, was a conservative when it came to the question of the transformation of life. He argued, in a second volume of the *Principles* read by Darwin on the *Beagle*, that each species of animal and plant life was entirely dependent on its geographical "center of creation," and that any change in the physical conditions of life would exterminate rather than transform it. But the evidence that Darwin was collecting caused him to question and then to reject Lyell's views on life. In the first place, Darwin's observation of primitive life forms on coral atolls suggested that both plant and animal life had a common origin. And in the second place, he became convinced that new varieties emerged from parent species. In particular this appeared to be the case for bird and animal life on the Galapagos Islands that straddled the equator off the coast of Ecuador. Life, rather than being fixed as Lyell claimed, appeared to possess the ability to adapt to changing conditions. Darwin thought he could even detect variations in species where the physical landscape remained unchanged.

The third theme concerns the causes of the extinction of species whose fossilized bones were uncovered by Darwin on his treks into the hinterland. Increasingly he became dissatisfied with the traditional argument that these species had been exterminated by catastrophic events such as the biblical flood of forty days, earthquakes, and volcanic eruptions. His accumulating evidence seemed to suggest that these catastrophes were not sufficient to destroy life throughout large regions of the Earth. He began to believe that, over very long periods of time, new species replaced earlier ones in a gradual and continuous way. Darwin's hypothesis appeared confirmed when his fossil collection, examined by experts back in England, turned out to be closely related to living animals in the same regions of South America.

Finally, Darwin's observations convinced him that life involved an intense struggle for existence. He thought he was witnessing this in the jungles of South America and even in the tribes of indigenous peoples of South America, New Zealand, Australia, and South Africa.

While these various themes emerged during the voyage, they, as yet, did not form the basis of a grand theory. There were signs, however, that this might not be long in coming. Darwin's biographers tell us that he had, in his journals, begun to show an interest in generalizing his ideas from small systems to large systems: "He had started a lifelong trend, extrapolating from small origins to big outcomes, from microscopic corals to huge reefs, from crustal twitches to the Andes" (Desmond and Moore 1991: 190). This "tendency" led, within a few years of arrival back in England, to his projection of the farmyard onto nature.

Darwin's Farmyard Analogy

How was Darwin to make sense of all he had seen and collected on the voyage of the *Beagle*? Where could he find answers to the "mystery of mysteries"? Various authors have suggested that he found answers in the main intellectual fashions of the day: that the "struggle for existence," the "survival of the fittest," and the "divergence of character" were derived from the political economy of Adam Smith and Malthus. As if genius is merely the ultimate expression of the intellectual concerns of the day. To the contrary, I will suggest that Darwin began closer to home and only borrowed other ideas when it helped either to resolve a problem arising from his analogical method or to persuade others to accept his theory of evolution.

The first six months after his return from abroad were, quite naturally, filled with the excitement of seeing family and friends, discussing his discoveries with new scientific colleagues, and finding good homes for his valuable collections. While Darwin's main intention was to catalogue and describe his discoveries and collections, he also began to ask larger questions about the nature of life. From mid-1837 he began jotting down his thoughts on these larger issues in the first of his "transmutation" notebooks (the "B" notebook).[1] From the beginning it was clear that he had accepted the idea of evolution, or transmutation as it was then called, and was attempting to discover a scientific explanation. Initially his thoughts were random and quite wild—ranging from spontaneous generation to perpetual becoming—and throughout the second half of 1837 he made little progress.

By the time Darwin began his second transmutation notebook (the "C" notebook) in February 1838, his thoughts had become more systematic and were focused on the causes of heritable variation. But how was he to resolve this "mystery of mysteries"? To do so Darwin finally turned from the exotic jungles of South America to the familiar Victorian world of landowners and plant and animal breeders. In the end this was to become the experimental "laboratory" from which his theory of natural selection emerged. And it was this laboratory that enabled him to marginalize, but not eliminate, the Lamarckian ideas that he had adopted soon after the return of the *Beagle*.

Darwin was familiar with the breeding of gundogs, game birds, and horses—indeed, before the voyage of the *Beagle* it seemed that this would be an important part of his life—and had an acquaintance with the breeding activities of gentlemen farmers. It must have seemed a good starting point for pursuing the larger issues that had begun to preoccupy him. Over the following four years Darwin gave considerable attention to the way breeders attempted to achieve their objectives by systematically selecting "useful" traits and eliminating the rest. In addition to developing the habit of questioning farmers, breeders, and nurserymen—which continued throughout his life—Darwin began undertaking his own plant and animal experiments, which he wrote up in his "Questions and Experiments" notebook begun in early 1839. He also began regularly reading breeders' manuals on pigs and poultry, attending agricultural shows, and participating in the formal and informal activities of a number of breeders' clubs. There can be no doubt that these questions, experiments, and activities concerning the methods and outcomes of domestic breeding became the foundation for Darwin's theory of natural selection.[2]

When in May 1842 Darwin came to write the first version—a 35-page handwritten sketch—of what was to become *The Origin of Species* (1859), it is clear that his theory of natural selection owed its origin to the farmyard analogy. As in the published version, he began his account with a discussion of domestic breeding. The same was true of the second version—a 189-page essay—written in the spring of 1844. It is significant that these early versions, which employ "variation under domestication" as the analogy for "variation under nature," were written at a time when Darwin doubted that he would ever dare publish his new theory, for fear that he would be ostracized by his social class and by the conservative intellectual establishment that had only recently embraced him. As these groups were already under attack from the rising industrial strategists (Snooks 1997: ch. 10) they were particularly sensitive to any threats from within their ranks. And Darwin was no rocker of boats: they quite literally made him sick. My point is that, at this stage, Darwin was not employing the farmyard analogy for persuasive purposes only as an analytical device.

Even once the core of the theory had been written down, Darwin did not abandon his interest in domestic breeding. Time only served to intensify it. The imaginative leap from the farmyard to nature was relatively easy: the difficult task was to establish his case beyond reasonable doubt. To do this he decided to investigate domestic breeding deeply and systematically. From the early 1850s, therefore, Darwin focused his attention on the selective breeding of pigeons and ducks, which he "skeletonized" and subjected to detailed measurement and scrutiny in order to document the aggregation of slight internal and external variations.

Before the final version of *Origin* was written in 1858 to 1859, therefore, the farmyard had become Darwin's evolutionary laboratory. Even after publication in 1859 he continued these experiments both to counter criticism and to write *The Variation of Animals and Plants Under Domestication*, published in 1868, that had been promised in *Origin* as the source of the detailed evidence that supported his theory of natural selection. That it was one of his largest books—two volumes of more than four hundred pages each—shows the importance of this material for his thinking about evolution. It was meant to justify his farmyard analogy and to empirically verify his selection thesis, but in reality it merely demonstrated the circularity of his method.

The *Domestication* book focused on the great variability of domestic species, the problems of interbreeding, the rejection of the popular idea of a "divine designer" as the cause of variation, together with a speculative theory about the transmission of variations (involving "gemmules") that he christened "pangenesis." Interestingly this large, detailed book sold rapidly, substantiating my argument that the farmyard analogy, initially employed as an analytical device, was also an important means of persuasion. It is doubtful that Wallace's more "scientific" approach—although he too refers to domestic breeding—would have attracted as much initial attention.

Although Darwin did not formally discuss his method in *Origin*, he scattered comments throughout the text about the role and significance of his domestication analogy. At his most revealing, Darwin, in the "Natural Selection" chapter, writes:

> As has always been my practice, let us seek light on this head from our domestic productions. We shall here find something analogous. . . . But how, it may be asked, can any analogous principle apply in nature? (*Origin*: 155–56)

Here Darwin asks the crucial question, but his answer, an expression of faith, is disappointing. He tells us: "I believe it can and does apply most efficiently." By now he is convinced that the farmyard is the appropriate model of nature.

Also, scattered throughout *Origin*, usually before or after he has appealed to evidence from breeding experiments, Darwin inserts comments of the following kind: "There is no obvious reason why the principles which have acted so efficiently under domestication should not have acted under nature"; or "we might have expected that organic beings would have varied under nature, in the same way as they generally have varied under the changed conditions of domestication"; or "can the principle of selection, which we have seen is so potent in the hands of man, apply in nature? I think we shall see that it can act most effectually"; or "as man can produce and certainly has produced a great result by his methodical and unconscious means of selection, what may not nature effect?"

(*Origin*: 441; 442; 130; 132). While further examples of this type could be quoted, the point has been made: the farmyard analogy is the centerpiece of Darwin's theory of natural selection.

Why labor the point? It is well-known that Darwin employed evidence from domestication experiments. My reason is that the farmyard analogy is fatally flawed. Even Darwin and his closest supporters feared that this might be the case. It was a fear they suppressed. There are two illustrations of this anxiety. The first concerns a response made by Darwin to Wallace's argument that the use of Spencer's term "survival of the fittest" in place of "natural selection" would avoid unnecessary criticism about a divine selector. While accepting the force of the argument, Darwin said, quite revealingly, that "Survival of the Fittest" lost the analogy between nature's selection and the fanciers' (Desmond and Moore 1991: 535). For Darwin this substitution of terms was the thin end of the wedge, because it severed the essential connection between his speculative theory of evolution and the known system and evidence of domestic breeding.

The second comment, this time in *Origin*, is a confession by Darwin about the dangers inherent in the method of analogy. Somewhat unguardedly, Darwin writes:

> Analogy would lead me one step further, namely, to the belief that all animals and plants have descended from some one prototype. *But analogy can be a deceitful guide.* (*Origin*: 455; my emphasis)

This is a truly ironical statement. On the one hand Darwin was aware that analogy can be a "deceitful guide," but on the other his grand theory of evolution owed its existence to this suspect method.

That this was not an isolated example can be seen from the personal correspondence between Darwin and Thomas Huxley ("Darwin's bulldog") after the latter had read the proofs of *Origin*. Huxley was very supportive but he had one reservation. He thought that the domestication analogy was incomplete in at least one important respect. At this suggestion Darwin retreated in panic, as the following extract shows:

> You speak of finding a flaw in my hypothesis . . . and this shows you do not understand its nature. It is a mere rag of an hypothesis with as many flaw and holes as sound parts. . . . [But] I can carry in it my fruit to market for a short distance over a gentle road; not I fear that you will give the poor rag such a devil of a shake that it will fall all to atoms; and a poor rag is better than nothing to carry one's fruit to market in. (Quoted in Desmond and Moore 1991: 475)

The road since then has been long and hard, and Darwin's "poor rag" less than robust.

How does the domestic-breeding analogy in *Origin* work? To the best of my knowledge this has not been examined, probably because the Darwinists have been concerned that if they shake Darwin's "poor rag" it will indeed "fall all to atoms." They have been supported in this resolve by the dominant hypothetico-deductive method championed so effectively by Karl Popper (1902–1994) and his followers during the second half of the twentieth century.

In essence, the hypothetico-deductive advocates argue that it does not matter how scientists arrive at their hypotheses because if incorrect they will be falsified by real-world evidence and be rejected by the science community. Popper (1965: 192) asserts that hypotheses or laws are *"free* creations of our own minds, the result of almost poetic intuition"; the philosopher C. G. Hempel (1966: 15; emphasis in original) claims, in a similar way, that "scientific hypotheses and theories are not *derived* from observed factors" but "constitute guesses at the connections that might obtain between the phenomena under study"; and the physicist R. Feynman (1967: 156) when discussing the origin of laws asserts: "First we guess it. Then we compute the consequences of the guess to see what would be implied if this law that we guessed is right." According to Popper (1971: 30–31), this part of the generation of scientific theory belongs not to the "logic of knowledge," which is amenable to rational proof, but to the "psychology of knowledge," which can be conveniently ignored because it is essentially irrational. Herein lies the "scientific" *excuse* for not scrutinizing the way Darwin arrived at his natural selection hypothesis.

But this is merely sleight of hand. For the hypothetico-deductive approach to work, two critical conditions must be met. The first condition is that the process of falsification must be employed systematically and it must be highly effective in weeding out false hypotheses. In other words, every time scientists guess wrongly, every time they are led astray by their "poetic intuition," their error must be exposed and their hypotheses must be amended or abandoned. The second, and even more important, condition is that a scientist's guess must not only withstand falsification but also be sufficiently general to embrace the entire causal system in reality. If these two critical conditions are not met, the resulting scientific theory and laws will inhabit the realm of fantasy rather than science.

The problem is that, in reality, scientific theories, particularly in the fields of the biological and social sciences, often fail to meet these conditions (Snooks 1998a: ch. 2). The usual response to unaccommodating evidence is to protect the core of the disputed theory by surrounding it with ad hoc supporting arguments. What I call flying buttresses. This response, however, is a clear admission that the original theory is unable to explain reality. If the supporting arguments, or flying buttresses, are removed, the entire structure will fall apart like a wounded Gothic cathedral. Old theories are only replaced when new and more intellectually persuasive, but equally unverified, theories come along. This is particularly the case with Darwin's theory of natural selection, owing to the greater difficulty of deducing consequences that can be subjected to falsification. ·

Further, the "poetic intuition" method invariably leads to the construction of theories that are insufficiently general, inclusive, or dynamic to handle the targeted issue in reality. An interesting contemporary example of this is the reaction of scientists to the new information about the outer planets in our solar system that were sent back to Earth by *Voyager* after its launch in 1977. When data from each of the outer planets and their moons were received by the mission scientists, they invariably exclaimed that they "hadn't anticipated anything like this!" Despite well-established physical laws and sophisticated calculations, their theories distorted our perception of the reality of the outer planets. Their theories, in other words, were insufficiently general to encompass the reality of this outer zone. It is for this reason that fact is usually stranger than fiction. This is true also of all other areas of scientific theory-building, including Darwin's theory of evolution, which distorts biological reality by focusing on only one element in the dynamics of life.

Because these critical conditions are rarely met, the *way* in which a scientific theory is constructed is absolutely vital to its success in revealing reality. It is all a matter of probabilities. Some methods of theory construction have a much higher probability of being realistic than others. As I argue in *The Laws of History* (Snooks 1998a), the deductive method has a much lower probability of providing a basis for realistic theory-building than the inductive, or historical, method. While the "problem of induction"—the absence of mechanical rules for generalizing from empirical data—must be taken into account, it is not as debilitating as the **problem of deduction**—the absence of rules for ensuring that all relevant explanatory variables have been recognized. The same argument can be applied to the analogical method: there are no rules to ensure that the analogy is completely relevant to the larger system under study. And the problem is compounded if, as was Darwin's practice, a systematic attempt is made to verify (rather than to falsify) the resulting theory using data from the analogical system. Clearly Darwin's methodological method of theory-building and theory-verification was circular.

The operation of Darwin's farmyard analogy can best be understood when set out in tabular form (Table 3.1), comparing the three main steps in his thinking about both artificial and natural selection. This shows quite clearly how the farmyard analogy breaks down.

Step 1—The Raw Material for Modification

This step in the analogy is quite straightforward. Domestic breeders must work with the variations in the species of plants and animals provided by nature, although through their artificial breeding programs they are able to influence the nature of these variations. While Darwin concedes that the "laws of variation" are "quite unknown, or dimly seen" (*Origin*: 75), he believes that in the farmyard "some slight amount of change may . . . be attributed to the direct action of the conditions of life—as, in some cases, increased size from amount of food, color from particular kinds of food and from light, and perhaps the thickness of fur from climate" (*Origin*: 74). Also he believed that changes in "habit" enforced by man would impose changes on domestic animals and plants. In a similar way, Darwin argued, variation in nature might arise from the slowly changing conditions of life, from changing habits, and from changing use/disuse of particular organs. The Lamarckian elements in this argument are obvious.

Table 3.1   Darwin's farmyard analogy

	Logical steps	Artificial selection	Natural selection
1	Raw material for modification.	Variation in nature, forced by domestic breeding.	Variation in nature arising from: • changing conditions of life • habit • use/disuse.
2	Condition required to activate selection.	Forced breeding conditions set by man.	Maximization of reproduction by all organisms (Malthus).
3	Filter enabling accumulation of "profitable" variations.	Man selects "useful" variations, culling the rest (man's methodological selection).	Struggle for survival leads to survival of the fittest and to the accumulation of "profitable" variations through reproduction.

Based on first edition of *Origin*.

Having conceded that changing external conditions under domestic breeding can have an effect, Darwin draws attention to differences that occur when these conditions are exactly the same. He tells us:

> Seedlings from the same fruit, and the young of the same litter, sometimes differ considerably from each other, though both the young and the parents, as Müller has remarked, have apparently been exposed to exactly the same conditions of life; and this shows how unimportant the direct effects of the conditions of life are in comparison with the laws of reproduction and of growth, and of inheritance; for had the action of the conditions been direct, if any of the young had varied, all would probably have varied in the same manner. (*Origin*: 73)

The lack of understanding of the "laws of variation" continued to concern Darwin and led to his unsuccessful attempts to explain them by his "pangenesis" theory in *Domestication* (1868). As we know, these laws continued to evade biologists until the rediscovery of Mendel's (1866) genetic experiments at the end of the nineteenth century and the discovery of DNA in the mid-twentieth century. The important point to realize here is that Darwin's views concerning variation in nature arose from his farmyard analogy.

The farmyard analogy was also the source of Darwin's concept of "sexual selection," which was to cause him so much anxiety. Later it provided the emerging discipline of genetics with the means to hijack the mantle of Darwinism. In the domestication chapter of *Origin*, Darwin writes:

The laws governing inheritance are quite unknown. . . . It is a fact of some little importance to us, that peculiarities appearing in the males of our domestic breeds are often transmitted either exclusively, or in a much greater degree, to males alone. (*Origin*: 76)

By the time we reach the chapter "Natural Selection," this concept has been given a separate, if minor, existence from natural selection. We are told that it is "less rigorous than natural selection" and that it "will give its aid to ordinary selection." This, as suggested in chapter 2, is evidence that natural selection cannot be regarded as a general theory. There are clearly matters outside its influence, and sexual selection, according to Darwin, is one of them. In part III I will show how genetic change and procreation—but not natural selection or sexual selection, which must be rejected—fit into a more general and dynamic model of life.

We have seen how Wallace tried to warn Darwin of the trouble this concept would bring. He argued that the physical attributes that Darwin thought were the outcome of sexual selection could be explained as part of the struggle for existence (Wallace 1871: 127, 156, 247, 283). In this respect Wallace was more of a Darwinian than Darwin himself. The interesting questions are: Why did Darwin cling so tenaciously to a second, if minor, mechanism of biological change? Why didn't he include both elements in a more general dynamic model? Or alternatively, why not follow Wallace's urging and abandon the separate concept of sexual selection? The answer to all these questions is the same—Darwin was a prisoner of his own analogy. This is discussed further in the next chapter.

Step 2—The Condition Required to Activate Selection

In the farmyard, as in nature, there is, we are told, no systematic direction in biological variation (mutation). Some variations are "useful" to the breeder and others are not, just as in nature some variations are "profitable" to the organism and some are not.[3] Where there is no pressure, either artificial or natural, to force the systematic selection of some variations and rejection of the rest, according to Darwin there will be no accumulation of variations, no directional biological change. What is it, Darwin asked in those early months after returning from the voyage of the *Beagle*, that generates the condition required to bring selection into play?

Quite clearly in the case of the farmyard the selective force is the desire of the breeder to produce certain "useful" or "desirable" physical and instinctual characteristics. This desire leads to carefully controlled breeding conditions. Under domestication, then, continuous small variations are accumulated by the breeder not only by culling but by continuous forced breeding. But what is it, Darwin asked, that activated the systematic accumulation of variations in nature? It was at this stage that Darwin's farmyard analogy ran into difficulties. The most common response of the day by those who had been led through the domestication exercise was that it is man who selects in the farmyard and God who selects in nature. Indeed, following the publication of *Origin*, many Christian readers who were prepared to accept the idea of evolution substituted the "divine designer" for natural selection. Darwin was forced to respond to this theory—a heresy encouraged by the nature of his analogy—in *Domestication* (1868).

Darwin needed to rescue his analogy from this obvious Victorian response. His solution had its origins in his observations on the voyage of the *Beagle* and in his early reflections after completing the voyage—the struggle of organisms to survive. But it was only after reading Malthus's *Essay on the Principle of Population* in September 1838 that the penny finally dropped. The solution was simple and persuasive. It must be the "tendency" for organisms to multiply geometrically that created the struggle for existence—in which those with a slight physical or instinctual advantage survived while the rest perished—that generated the condition bringing "selection under nature" into play. Hence, Darwin adopted the "doctrine of Malthus" not because it was fashionable to do so but because it enabled him to employ a widely accepted argument to prevent his farmyard analogy breaking down.

What was more, Darwin's knowledge of domestic breeding suggested that, as variations between one generation and the next were extremely slight, with only the most experienced breeders able to recognize them, it would require severe competition operating continuously over vast periods of time to enable life on Earth to progress from a single-celled organism to the bounty that we know today. In other words, Darwin's farmyard model requires not only that all organisms be able to breed at a geometric rate but that they actually do so at all times and in all places. Otherwise there will be periods when biological change is directionless, an outcome that Darwin realized would undermine his thesis. The bottom line is that the critically important concept of geometric population increase had to be imported from Victorian political economy to shore up his farmyard analogy. It was an assumption that he, like Malthus, never attempted to test, because it would have been falsified by real-world evidence.

Step 3—The Biological Filter

The central question that Darwin needed to address was: What is the mechanism by which "useful" or "profitable" variations are allowed or forced to accumulate over very long periods of time? Turning as he always did to the farmyard, Darwin demonstrated that the breeder selects those variations that are useful to his economic life or that tickle his fancy, while ruthlessly culling the rest. He tells us:

> One of the most remarkable features in our domesticated races is that we see in the adaptation, not indeed to the animal's or plant's own good, but to man's use or fancy. . . . The key is man's power of accumulative selection: nature gives successive variations; man adds them up in certain directions useful to him. In this sense he may be said to make for himself useful breeds. (*Origin*: 89–90)

He emphasizes that this "principle of selection" is not "hypothetical." Of course he was justified in claiming that leading breeders had, within a single lifetime, "modified to a large extent" some breeds of cattle and sheep. In their hands, domestic species were "quite plastic," and their transmutation could be accelerated beyond what nature could effect in a man's lifetime.

Darwin needed to persuade his readers that what was true in the farmyard was true in nature. He did so by taking a number of small and seemingly logical steps. In an effort to shift the farmyard analogy closer to nature, he suggests that the process of selection under man does not need to be quite as deliberate and "methodical" as in Victorian Britain. What he has in mind is "a kind of selection, which may be called Unconscious," by which less scientific men do not consciously attempt to permanently alter a breed but merely to retain the best newborn animals. In this way directional biological change will still be achieved, but only in a "slow and insensible" way. Even in "barbarous" lands, we are told, "savages" who keep animals will improve their breeds not through any forethought but merely by selecting their best animals for preservation during severe winters, famines, and "other accidents." This generates extremely slow accumulation of profitable variation, but directional variation nonetheless. From here it is but a small step to consider the process of "unconscious" selection in nature itself. Darwin was a master of persuasion. The irony is that he would have been an excellent clergyman—a salesman for God.

But before taking this last small step, Darwin discusses the conditions affecting "man's power of selection" and, by analogy, of nature's also. By closely investigating the process of domestic breeding, Darwin concludes that six factors are involved: the degree of variability in a species for selection to work on; the size of the population (or number of breeding experiments); how useful is the species to man; the constancy of the selective procedure; the isolation of the superior variants (that is, the prevention of interbreeding); and the ruthless elimination of inferior variants. With the analogy complete—or so he hoped despite Huxley's private concerns—Darwin was ready to lead his readers gently through the farmyard gate and into the world of nature beyond.

Just as man, either consciously or unconsciously, selects and accumulates useful variants in the above manner, so in nature, Darwin maintains, the constant struggle for existence, driven by populations that are always out of control, leads to the survival of those individuals with even a slight advantage and to the extinction of the rest. And those that survive are able to pass on their advantages to their offspring. While this leads to the extremely slow accumulation of variations required to survive—much slower than conscious or even unconscious selection in the farmyard—over vast periods of time, it can and does lead, Darwin insists, to the transformation of species. While *Origin* is a triumph of persuasion, there are fundamental flaws in its analytical foundations.

It is time to test the strength of Darwin's "poor rag" to see where the "flaw and holes" might be. As Darwin was well aware, analogy can be a "deceitful guide." While analogy is a very crude approach to theory-building, it has the virtue of simplicity. If the reality one is attempting to explain appears too complex to understand, or the evidence is too fragmentary, it is possible to take a shortcut by first identifying a simple, well-known system that hopefully has many characteristics in common with the more complex and distant reality that requires explanation, and then employing it to construct one's theory. Invariably this method breaks down because the analogy is incomplete or inappropriate. The end result is that it leads to the distortion of reality. The analogical method is only really appropriate if the two mechanisms being compared are both part of a more general dynamic process. This was not the case with the farmyard.

The argument here is that Darwin's farmyard analogy was entirely inappropriate to his great task. There is a critical difference between the process of domestication and the process of life because there is no force in nature that corresponds to the role of the breeder in the farmyard. As Darwin's contemporaries sensed, there is a fundamental irony here. The farmyard analogy requires a role for a godlike interventionist to make the system work, while at the same time claiming to eliminate the hand of God.

In Darwin's system the divine selector could only be hidden from view by making the very strong and critical assumption that all individuals in all places and at all times tend to procreate at a geometric rate. "There are," Darwin insists, "no exceptions." This strong assumption was essential if organisms were to be continually involved in the struggle for existence that was necessary to bring the godless mechanism of natural selection into play.

As we now know, there are numerous examples of species, human and nonhuman, that do not procreate continuously at a geometric rate. In human society even the least technologically sophisticated—the Aboriginal Australians, the Kalahari Bushmen, and the Alaskan Inuit—pursued long-run policies of population control. And in more technologically sophisticated societies, such as Britain in the seventeenth to the nineteenth centuries (overlapping the time of Charles Darwin), population change can be shown econometrically to be a function of the rise and fall of that country's dynamic strategies of commerce and technological change (Snooks 1999: 239–50). The same is true of nonhuman society at even the most basic level. For viruses it is known that rapid procreation is turned on and off by designated genes in response to environmental conditions. And there is widespread evidence that plants and animals are capable of self-regulation where their population densities are concerned.[4] Far from there being "no exceptions" to Darwin's rule, the exceptions are the rule. In part III I argue that procreation (and migration) is a dynamic strategy—the family-multiplication strategy—that plant and animal organizations pursue as an *alternative* to genetic change.

Hence the critical population assumption made by Darwin to complete his farmyard analogy is factually wrong. And when that prop is removed the whole structure falls to the ground. Changing the metaphor, Darwin's "poor rag" is just not up to the great task that he set himself. In reality he had no hope of carrying his "fruit" in it to market. Yet he employed his farmyard analogy so persuasively that no one appears to have noticed. Now it is no longer in their interest to avert their gaze.

At a higher level of complexity, even if Darwin's population assumption had been correct, his natural selection theory would still not have been appropriate. The fundamental problem with Darwin's farmyard analogy is that it does not acknowledge the dynamic reality of life. Rather than dealing with the fluctuating fortunes of life, it focuses on only one element in that process—the transmutation of species. By doing so this type of theory not only fails to explain the dynamics of life, but it even misinterprets the real role and influence of systematic genetic change.

Life is a complex dynamic process that is greater than just genetic change. It includes the changing quantum of life—the global biomass—as well as its changing forms. The role of genetic change in this wider context is similar—if I dare to employ an analogy—to the role of technological change in the expansion and growth of an economy. To understand this dynamic process Darwin should have employed the more difficult but less hazardous (or less "deceitful") method of induction rather than the simplistic but distorting method of analogy. This would have involved focusing directly on the evidence of life, recognizing that it was the outcome of a broader dynamic process than a change of biological forms, and constructing an endogenously driven dynamic model to explain the historical patterns—a model in which variation (genetic change) plays a significant but widely fluctuating role over time. As Darwin left it, the theory of natural selection possesses no realistic driving force and no dynamic mechanism by which such a driving force could be imparted to the larger process of biological growth. Until this is done we can have no realistic explanation of the emergence and fluctuating fortunes of life, of its changing biological forms, of the laws of life, or of what the future holds. Darwinism can tell us nothing of these things.

Darwin's Use of Evidence

By the 1840s, the nature of Darwin's theory of evolution had been firmly established. By then he was totally committed to the truth of his farmyard analogy. The task of the rest of his life was to collect as much information as possible to shore up his concept of natural selection. It was a quest not to test the theory and possibly to falsify it, but to justify it to himself and the world.

The way he went about this task is revealing. On introducing a new aspect of his theory in *Origin*, Darwin almost always appeals first to the evidence of domestic breeding and only thereafter to evidence, usually fragmentary, from the natural world. This reflects both the central role of the farmyard analogy and the circularity of his method. In effect the *same* type of facts that led Darwin to adopt the theory of natural selection are employed to support it.

After the publication of *Origin*, Darwin continued to accumulate a vast amount of information regarding both domestic and wild life from around the world. Much of this evidence was published in *Domestication* (1868)—and in the revised edition of that work seven years later—which was the forthcoming work referred to in *Origin* that would comprehensively verify his natural selection theory. It is significant that the bulk of the "supporting" evidence for verification came not from nature but from the farmyard. The problem is that this evidence is totally irrelevant if the analogy breaks down. And, as we have seen, it does.

While Darwin complained, with considerable justification, about deficiencies in the fossil record, he was really not the man to employ it as a basis of theory-building. Certainly he was no inductivist. Darwin's mind did not appear to work that way. As all his work demonstrates, he was addicted to the analogical method. Even in Darwin's time the fossil evidence challenged some of his most cherished assumptions regarding the rise and fall of species and whether the accumulation of variations occurred continuously (Falconer assisted by Walker 1859).

In *Origin*, instead of searching for patterns in the fossil record, Darwin merely attempted to explain them away. The reason being that this evidence of the real world beyond the farmyard gate constantly posed a threat to his theory. For example, Darwin ignored the contradiction between existing evidence of the collapse of dominant life forms, such as dinosaurs, and the subsequent emergence of formerly minor forms such as mammals. This threatened his "principle of divergence" which required a relatively smooth transition between species, with the emerging species gaining an advantage over and then eliminating the earlier species. Darwin's problem was that he was unable to explain the internal dynamics of the rise and fall of species. If a species fell, then it was because it could not handle competition for scarce resources generated by new species, rather than because the old species had exhausted its genetic potential (as I will argue in part III).

Darwin also glosses over the fossil evidence concerning the migration of species around the world. He asserts that it is a response to natural selection, as "selected" species eliminate neighboring species and move in to take over their territory. Rather than providing the most appropriate explanation of the dispersal of species, Darwin attempts to persuade us that it is merely further evidence of the operation of his theory of evolution. In part III I present a more general dynamic theory in which procreation and migration—which I call the **family-multiplication strategy**—is an alternative strategy to genetic change.

Finally, Darwin treats those "long and blank intervals" in the fossil record rather cavalierly. For example, in *Origin* he writes:

> During these long and blank intervals I *suppose* that the inhabitants of each region underwent a considerable amount of modification and extinction, and that there was migration to other parts of the world. (*Origin*: 329; my emphasis)

This "supposition" was necessary to justify his theory of natural selection. Had there been no genetic change during these long and blank periods, his theory would face, as it does today, a serious challenge. Even in Darwin's time the fossil evidence suggested that once a species appeared in the fossil record it changed little until its final disappearance. Most reviewers of *Origin* when it first appeared drew attention to this difficulty, as can be seen in the reviews collected by D. L. Hull (1973). My point is that, had Darwin been an inductivist rather than an analogist, and had he taken the fossil evidence seriously, he would have either developed an entirely different theory of life or abandoned the whole exercise.

Conclusions

In contrast to the huge public relations exercise conducted by the neo-Darwinists about the rigorous scientific method employed by Darwin in developing his theory of evolution, this investigation shows that his great achievement was one of persuasion rather than theory-building. Darwin adopted a very crude method to explain evolution—the farmyard analogy—which even he admitted had the potential to be a "deceitful guide." Why then did he employ the analogical method? The answer is quite simple. It was the only way that he could make sense of what he had seen and collected on the voyage of the *Beagle*. The magnitude of this evidence threatened to overwhelm him because he did not have a comparative advantage in detecting patterns in the historical record and in developing a general dynamic model to explain them. Darwin merely gathered selective evidence from nature in support of his farmyard theory. He cut the Gordian knot by employing an analogy to an activity with which he and his readers were familiar—domestic breeding. Unfortunately this analogy is completely inappropriate to an understanding of the dynamics of life and of the role of genetic change in this process. Yet, while the analogy was fatally flawed, it was a masterstroke of persuasion. As a result, the scientific world has for one-and-a-half centuries embraced a theory of life that is totally unsupportable.

Chapter 4

Stress in the Structure:
Darwin's *Descent of Man*

While *The Origin of Species* (1859) is concerned with the construction of a theory to explain the transmutation of biological forms, *The Descent of Man* (1871) attempts to apply that theory to the emergence of mankind and to the progress of human society. *Descent* is a major test for Darwin's theory of evolution. It is difficult enough to construct a plausible theory of life, but much more so to apply that theory to the life form with which we are most familiar—our own. Nevertheless, it is essential that any scientific theory of life be able to explain the fluctuating fortunes not only of biological forms in general but of the most successful of these in particular. While this was a challenge that Darwin could not ignore, the attempt to face it led to serious stress fractures spreading rapidly throughout his evolutionary structure.

The Unbalanced Structure of *Descent*

The problems that Darwin experienced in applying his theory to the "evolution" of mankind and society are reflected in the structure of *Descent*. It is curious that in a work devoted to explaining how natural selection has shaped our own species, only one-third of its pages are allocated to this subject. And of these a mere eight pages are concerned with "Natural Selection as affecting Civilised Nations." The remaining two-thirds of the book focus, curiously, on that "less robust . . . aid to ordinary selection" which Darwin called "sexual selection." Even more curious, the major part (83 percent) of the material on sexual selection is devoted to nonhuman species. Clearly Darwin had little to contribute to the understanding of our own species and even less to that of human civilization.[1]

This unbalanced structure, as will become clear, arose from the failure of Darwin's theory of natural selection to throw any effective light on the nature and progress of human society. And this was not through a lack of interest on his part. We know from his notebooks that he wanted to transform our understanding of human morals, religion, and politics by treating them like any other branch of natural history (Desmond and Moore 1991: 556–57). Just as sociobiologists have attempted, equally unsuccessfully, in the last few decades. But even Darwin's new emphasis on sexual selection was unable to do the job.

Part I of *Descent*—the first third—is concerned with "The Descent or Origin of Man." It focuses on anatomical evidence for the descent of man, the manner of our evolution, differences in intelligence between man and earlier species, the emergence of intellectual and moral faculties in civilized society, and differences between the "races of man" in terms of physical and intellectual characteristics. There is nothing here, however, about how natural selection might have shaped the sociopolitical institutions of nations and civilizations in either the Old or New Worlds. Part II on "sexual selection" is really the book's main focus. It is concerned with the principles of sexual selection and with the manner in which these have influenced the development of "secondary sexual" (or nonreproductive) characteristics—such as size, color, song, "weapons" of offense and defense, etc.—of males in the categories of insects, fish, reptiles, birds, and mammals. Only in part III on "Sexual Selection in Relation to Man," which is dealt with in just forty pages, does Darwin finally return to an examination of differences in secondary sexual characteristics between men and women. In this chapter I focus on how and why natural selection was unable to explain the fluctuating fortunes of human civilization.

Stress Fractures in Darwin's Theory

When Darwin applied natural selection to humanity, the limitations of his farmyard theory (discussed in chapter 3) became obvious to those who were willing to see. In *Descent* he avoids confronting these limitations by skating quickly over the society of man and focusing attention on sexual selection, largely in relation to nonhuman species. The theoretical limitations of natural selection can, of course, be more easily disguised in species about which detailed knowledge is less readily available. When applied to human society, the natural selection theory breaks down because of the totally unrealistic population assumption. This was predicted in our discussion (chapter 3) of the limitations of Darwin's farmyard model.

The Growth of Human Population

Darwin experienced difficulties in applying his theory to human society because the evidence was less accommodating. And it was less accommodating, not because human society is "unique" or "artificial" as modern neo-Darwinists claim, but because the evidence is more abundant. This abundance exposes the flaws in Darwin's theory. When discussing other species, or even our own in prehistoric times, it is a simple matter to disguise inadequacies in the theory by making convenient assumptions about the limited evidence. While Darwin complained, quite genuinely, about the inadequate fossil records, this situation actually made his task of persuasion about other species in the deep past much easier. But when historical data became readily available, Darwin's claim that the population of all human societies normally increased at a geometric rate could not be sustained.

When discussing population growth in the nineteenth century, Darwin employed a number of transparently invalid methods to support his natural selection theory. Five are discussed here. First, he focused on the theoretical rate at which human population *might* increase under favorable conditions, rather than on the *actual* rates at which they have increased in different societies at different times. Despite the existence of census records in Britain, other rich European countries, and some colonies, Darwin ignored these data. Second, when he does draw our attention, in a nonquantitative way, to actual rates of population growth, it is to a country like the United States that was growing far more rapidly than any other at the time because, unknown to Darwin, it was pursuing the dynamic strategy of family multiplication in order to bring the vast natural resources of North America into the production process (Snooks 1997: ch. 11).

Third, in the face of the obvious fact that actual population growth in Western Europe was considerably slower than the geometric rate required by natural selection, Darwin fudges the issue to make it appear that this was the outcome of Malthusian checks (rather than deliberate dynamic strategies), such as the "difficulty of gaining subsistence, and of living in comfort," restraint of marriage, higher infant death rates, disease, emigration, and war (*Descent*: 428). Also he leaves the reader with the impression that, despite this conscious restraint, population growth in the "civilized world" was sufficient to generate the struggle for existence necessary to bring natural selection into play.

But this impression is false. Population growth in Western Europe during the nineteenth century—as everywhere else—was the outcome of the requirements of their dynamic strategies. It was, in other words, a response to **strategic demand** generated by the unfolding dynamic strategy of technological change (Snooks 1999: 239–50). Also Darwin fails to remind his readers that in *Origin* he claimed that the struggle for existence in nonhuman species arose because they were incapable of adopting artificial restraints! It appears to have been enough for Darwin that the kingdoms of Europe had been at each other's throats—certainly a serious *political* struggle—for almost a thousand years. The problem for Darwin, as discussed in part III, is that this was an outcome not of automatic population increase but of the pursuit of the dynamic strategy of conquest.

All three points are reflected in the following extract from *Descent*:

> *Rate of Increase.*—Civilised populations have been known under favourable conditions, as in the United States, to double their numbers in twenty-five years; and, according to a calculation by Euler, this *might* occur in a little over twelve years. At the former rate, the present population of the United States (thirty millions), would in six hundred and fifty-seven years cover the whole terraqueous globe so thickly, that four men would have to stand on each square yard of surface. The primary or fundamental check to the continued increase of

man is the difficulty of gaining subsistence, and of living in comfort. We *may infer* that this is the case from what we see, for instance, in the United States, where subsistence is easy, and there is plenty of room. If such means were suddenly doubled in Great Britain, our number would be quickly doubled. With civilised nations this primary check acts chiefly by restraining marriages ... greater death rate of infants ... severe epidemics and wars ... [and] emigration. (*Descent*: 428; my emphasis)

It did not occur to him that population growth is a response to the needs of a society's dynamic strategy rather than the automatic exogenous force that brings natural selection into play.

Fourth, when admitting that "barbarous" races grew even less rapidly than "civilized" races, this is passed off, with reference to Malthus, as the outcome of a lesser "reproductive power." To "explain" this difference in reproductive power, Darwin employs his farmyard analogy to suggest that "civilized" races are like domestic species, and "barbarous" races are like wild species. The former are more productive because of the ideal conditions under which they live. What then would Darwin have thought on finding that a century later it was the poor ("barbarous") rather than the rich ("civilized") countries that experienced high rates of population growth? Once again the farmyard model breaks down.

Despite Darwin's assurances to the contrary, it is transparently clear that no viable human society has ever experienced population growth at a Malthusian rate, and that few have experienced even high rates for more than short periods of time (Snooks 1996: 380–88). Darwin argues erroneously that

There is great reason to suspect, as Malthus has remarked, that the reproductive power is actually less in barbarous, than in civilised races. . . . This may be partly accounted for, as it is believed, by the women suckling their infants during a long time; but it is highly probable that savages, who often suffer much hardships, and who do not obtain so much nutritious food as civilised men, would be actually less prolific. I have shewn in a former work [*Domestication* 1868], that all our domesticated quadrupeds and birds, and all our cultivated plants, are more fertile than the corresponding species in a state of nature. . . . We might, therefore, expect that civilised men, who in one sense are highly domesticated, would be more prolific than wild men. It is also probable that the increased fertility of civilised nations would become, as with our domestic animals, an inherited character. (*Descent*: 428–29)

Here is a clear-cut prediction—something that Darwin was usually careful not to make—and it is wrong. Not wrong in the trivial sense of timing or degree of precision, but wrong in the fundamental sense of creating a false reality. It is the type of error that would cause a serious hypothetico-deductivist to reject the underlying theory completely. Not only was Darwin wrong about future (that is, twentieth-century) fertility rates in rich and poor countries but, more importantly, he was wrong in claiming that fertility is driven by (in modern terminology) genetics. In reality, as we shall see in part III, fertility is driven by strategic demand that changes as a society's dynamic strategy unfolds.

Finally, taking advantage of the total lack of evidence, Darwin *assumes* that in the era before the emergence of man, our ancestors propagated rapidly, checked only by starvation and disease. This, he claimed, would have brought natural selection into play, thereby shaping the physical and intellectual characteristics of what was to become Homo sapiens. Darwin explains:

> If we look back to an extremely remote epoch, before man had arrived at the dignity of manhood, he would have been guided more by instinct and less by reason than are the lowest savages at the present time. Our early semi-human progenitors would not have practised infanticide or polyandry [as do Darwin's modern "savages"]; for the instincts of the lower animals are never so perverted as to lead them regularly to destroy their own offspring, or to be quite devoid of jealousy. There would have been no prudential restraint from marriage, and the sexes would have freely united at an early age. Hence the progenitors of man would have tended to increase rapidly; but checks of some kind, either periodical or constant, must have kept down their numbers, even more severely than with existing savages. (*Descent*: 430)

These checks were, according to Malthus who Darwin accepted as the authority on human populations, starvation and disease. Neither Malthus nor Darwin realized that population was an instrument employed by individual organisms in the strategic pursuit.

Man's Struggle for Existence

Evidence available to Darwin on population growth makes it quite clear that the maximum theoretical rate was not even approached in the longer term in human society. Although he does not make this explicit, it is obvious that he recognizes the problem. To acknowledge this matter openly would have meant abandoning the theory of natural selection. When discussing natural selection in relation to mankind in *Descent*, Darwin does not insist on the constancy of the struggle for existence as he did in *Origin*. In fact he suggests that man and his progenitors have been exposed to the struggle for existence only "occasionally."

Possibly Darwin thought he would be charged with historical inaccuracy in *Descent* if he insisted that "there is no exception" to the Malthusian rate of population increase and, hence, that the struggle for existence is a continuous and intensive process in human society. But in his theoretical analysis, even after this time, Darwin continued to propagate the myth of rapid and continuous population growth. In the much revised sixth edition of *Origin*, published after *Descent* in 1872, he continued to insist that all populations increased geometrically (by which he meant at or near the maximum rate) and that "there is no exception to the rule that *every* organic being naturally increases at so high a rate" (*Origin*, 6th ed.: 53; my emphasis). Darwin never attempted to explain the inconsistency between *Origin* and *Descent*, because it would have meant acknowledging the fallacy of his farmyard theory.

In *Descent*, however, Darwin does make a fatal admission. He tells us that the early progenitors of man must also have tended, like all other animals, to have increased beyond their means of subsistence; they must, therefore, *occasionally* have been exposed to a struggle for existence, and consequently to the rigid law of natural selection. Beneficial variations of all kinds will thus, either *occasionally* or habitually, have been preserved and injurious ones eliminated. (*Descent*: 431; my emphasis)

This is a fatal admission for at least two reasons. First, in the case of mankind and our early progenitors, it releases the iron assumption of continuous and rapid population increase that is essential for bringing natural selection into play. This opens the way for alternative theories not only to explain the fluctuating fortunes of mankind during those periods at least when natural selection is not operating (such as the "punctuated equilibria" theory of the paleontologists), but also to displace natural selection entirely (such as the dynamic-strategy theory presented in part III below). Second, if the struggle for existence operates only "occasionally" in the case of mankind, why not also in the case of all other species? This question is particularly pertinent owing to Darwin's persistent attempt in *Descent* to show how closely man is related to the other mammals and of the other mammals to the "lower" species.

Natural Selection and Human Civilization

In *Descent* Darwin regards it as self-evident that natural selection has been responsible for the development of those physical characteristics that enabled early man and his progenitors to adapt to their environment. Darwin is forced to treat this matter as self-evident because he is unable to marshal any direct evidence. He is forced to assume that all biological modifications—such as the size of man's brain, the development of an effective hand, the ability to stand erect while using the hands, etc.—that enabled our species to adapt more closely to the environment are due to the operation of natural selection. It becomes something of a Victorian parlor game to chant endlessly that this and that variation is due to natural selection. There is no evidence, no proof, only ritualistic assertion. In particular he fails to explain how natural selection could fulfil this role if the condition required to bring it into play—the struggle for existence—operates only "occasionally."

By the time Darwin had finished writing *Descent*, the extent of his claim for natural selection as the mechanism of evolution had been reduced in a further important respect. While he still regarded it as the most "rigorous" form of selection, from the fifth edition (1869) of *Origin* he defined it exclusively in terms of those modifications that enable an organism to better adapt to its environment. By better adaptation he meant gaining better access to natural resources, particularly nutrients.[2] In this respect Darwin confesses:

I now admit, after reading the essay by Nägeli on plants, and the remarks by various authors with respect to animals, more especially those recently made by

Professor Broca, that in the early editions of my 'Origin of Species' I perhaps attributed too much to the action of natural selection or the survival of the fittest. I have altered the fifth edition of the 'Origin' so as to confine my remarks to adaptive changes of structure. . . . I did not formerly consider sufficiently the existence of structures, which, as far as we can at present judge, are neither beneficial nor injurious; and this I believe to be one of the greatest oversights yet detected in my work. (*Descent*: 441–42)

It is an oversight, he explains, resulting from an enthusiasm to show that the Earth's species had not been separately created.

To cope with these newly acknowledged, nonadaptive variations, Darwin flagged a greater role for sexual selection, which in *Origin* was little more than an afterthought, taking up less than two pages of text. In *Descent*, as we have seen, it was to take up two-thirds of the book. When Darwin came to summarize the chapter entitled "On the Manner of Development of Man from Some Lower Form," he wrote:

As all animals tend to multiply beyond their means of subsistence, so it must have been for the progenitors of man; and this would inevitably lead to a struggle for existence and to natural selection. The latter process would be greatly aided by the inherited effects of the increased use of parts, and these two processes would incessantly react on each other. It appears, also, as we shall hereafter see, that various *unimportant* characters have been acquired by man through sexual selection. An unexplained residuum of change must be left to the assumed uniform action of those unknown agencies, which occasionally induce strongly marked and abrupt deviations of structure in our domestic productions. (*Descent*: 443; my emphasis)

In *Descent*, therefore, Darwin was more willing to acknowledge that the evolutionary mechanism of "selection" extended beyond natural selection. This was despite natural selection being the most "rigorous" part of selection responsible for all "important" modifications. Yet he was unable to develop a general theory of selection that included both the "natural" and the "sexual" elements. The reason is that there is no general theory of selection that can be developed from Darwin's farmyard analogy. My concept of **strategic selection** introduced in chapters 10 and 12 has a very different source in the dynamic-strategy theory.

While Darwin's analysis of the "evolving" physical characteristics of mankind had to negotiate unexpected difficulties, his discussion of human "mental powers," "moral faculties," and "civilization" ran into even greater problems. So great that they caused a breakdown in his theory. In a large book of 529 pages (in the Random House Modern Library edition) on the descent of man, only eight pages, or 1.5 percent, were devoted to human society. This was a matter not just of preference but of necessity. As natural selection is not a dynamic theory—it has no endogenous driving force or dynamic mechanism—it is incapable of explaining the fluctuating fortunes of human society or, indeed, of life itself. It is just that the deficiencies of Darwin's model are more apparent in the case of human society, because we know so much more about it.

Darwin moves uneasily and unconvincingly from the "lower animals" to mankind. While he is at pains to emphasize the continuity between the "lower" and "higher" animals, Darwin claims that mankind is able to reduce the burden of the struggle for existence through the use of greater intellectual abilities. Yet at the same time he assures us, without providing any evidence, that human civilization owes its development to natural selection. To avoid facing this inconsistency, Darwin treats human society in a very superficial and incomplete way.

Rather than examining the recurrent patterns in human history and developing a general dynamic model to explain them, Darwin merely focuses on social customs, which he asserts "evolved" under the abstract hand of natural selection. This was, of course, his usual practice for other species as well: rather than examining patterns in the fossil evidence he merely focuses on farmyard experiments. But these experiments could not be employed as evidence for his ideas on human society. He was at least prepared to admit that he was unable to explain why some tribes were successful and others were not: "The problem . . . of the first advance of savages towards civilization is at present much too difficult to be solved" (*Descent*: 501). More difficult than explaining the origin of species? Not more difficult conceptually, just more difficult empirically, because the greater evidence available concerning human society exposes the flaws in Darwin's theory of evolution.

What does Darwin have to say about the dynamics of human society? We are told:

> Obscure as is the problem of the advance of civilisation, we can at least see that a nation which produced during a lengthened period the greatest number of highly intellectual, energetic, brave, patriotic, and benevolent men, would generally prevail over less favoured nations. (*Descent*: 508)

Darwin believed, but did not demonstrate, that "natural selection, aided by inherited habit" was responsible for the development of these "social virtues" that gave nations a comparative advantage in the struggle for regional and global dominance (*Descent*: 498). Quite obviously this tells us little about the sources or process of social dynamics.

A further critical test for any dynamic model is what it can tell us about likely future change. We need to ask, therefore, what use can be made of Darwin's theory of evolution to predict the future of individual nations, of human society as a whole, and of the human race. Unfortunately we have to conclude that it is of no use at all. As Darwin admitted about the least difficult of these future issues:

> Who can positively say why the Spanish nation, so dominant at one time, has been distanced in the race. The awakening of the nations of Europe from the dark ages is a still more perplexing problem. (*Descent*: 507)

Any realistic dynamic model—such as my dynamic-strategy theory (Snooks 1996: ch. 13; 1997: chs. 12 and 13)—should be able to easily explain these matters. The fact that Darwin's theory of evolution cannot be used for this purpose—nor to predict the future of human society or of the human species, as has been done in chapter 16 of this work—is a major defect. This is why recent attempts by sociologists and institutional economists to employ evolutionary theory to analyze the dynamics of human society have also failed. They could have saved themselves the trouble had they carefully examined Darwin's own failure in *Descent of Man*.

How did Darwin develop his very limited argument about the dynamics of human society? He claimed that a tribe will only survive the struggle for existence if it is able to develop, under the operation of natural selection, the necessary "social virtues." By social virtues, Darwin meant cooperation between individuals and groups. He argues that

> When two tribes of primeval man, living in the same country, came into competition, if (other circumstances being equal) the one tribe included a great number of courageous, sympathetic and faithful members, who were always ready to warn each other of danger, to aid and defend each other, this tribe would succeed better and conquer the other. Let it be borne in mind how all-important in the never-ceasing wars of savages, fidelity and courage must be. . . . Selfish and contentious people will not cohere, and without coherence nothing can be effected. A tribe rich in the above qualities would spread and be victorious over other tribes: but in the course of time it would, judging from all past history, be in its turn overcome by some other tribe more highly endowed. Thus the social and moral qualities would tend slowly to advance and be diffused throughout the world. (*Descent*: 498)

While Darwin appeals generally, and hopefully, to the witness of "history," he makes no attempt to marshal its evidence about the fluctuating fortunes of human society. That would have destroyed his faith in the theory of natural selection. Instead he merely speculates rather weakly about the role of cooperation in leading to the success of some tribes over others.

In effect, Darwin employs an evolutionary institutional argument of the kind that in *The Ephemeral Civilization: Exploding the Myth of Social Evolution* (Snooks 1997) I have shown is incapable of explaining either institutional (laws and customs) or organizational (social grouping) development, let alone fundamental dynamic change. As demonstrated in that work, cooperation is the outcome not of evolutionary processes but of the confidence that a society has in its prevailing dynamic strategy. I call this **strategic confidence**. Once a society's dynamic strategy has been exhausted and strategic confidence collapses, cooperation instantly evaporates, and only reappears when a new and viable dynamic strategy emerges. This is discussed in greater detail in chapter 14 below.

Darwin's "social virtues," therefore, are not an accumulating stock of social and moral characteristics but rather an ephemeral flow of strategic requirements. Hence, human institutions and organizations do not evolve according to some Darwinian principle; they respond directly to changes in strategic demand as the dominant dynamic strategy unfolds. As I demonstrate in *Ephemeral Civilization*, sociopolitical institutions—such as political democracy—can and do reverse back on themselves whenever the long-run sequence of dynamic strategies—such as conquest → commerce → conquest—reverses itself, as it often did in the premodern world (ancient Greece and medieval Venice) and even in the modern world (Germany and Japan). Reversals of this nature—which would be equivalent to a species reversal—are not possible in evolutionary theory because backward steps are "unprofitable" and as such would be eliminated. Alfred Wallace (1871: 36) was quick to realize this.

Interestingly, Darwin raised difficulties concerning his own explanation of the progress of social and moral qualities. While he was convinced that tribes possessing superior qualities would triumph and that these qualities would be carried forward, he was less sure about "how within the limits of the same tribe did a large number of members first become endowed with these social and moral qualities, and how was the standard of excellence raised?" (*Descent*: 498). What he was unsure of, in other words, was how natural selection enabled altruism to overwhelm selfishness in any tribe. He was acutely aware his theory suggested that those individuals willing to sacrifice themselves for the common good would not leave as many offspring as those determined to survive at the expense of others. Of course, this undermines his argument about the accumulation of social virtues.

Darwin's attempt to resolve this critical difficulty is not at all persuasive. He argues that

> as the reasoning powers and foresight of the members became improved, each man would soon learn that if he aided his fellow-men, he would commonly receive aid in return. From this low motive [lower than survival of the fittest?] he might acquire the habit of aiding his fellows. . . . Habits, moreover, followed during many generations probably tend to be inherited. (*Descent*: 499)

Darwin further thought that this tendency would be reinforced by the much more powerful stimulus of "the praise and blame of our fellow men" to which we respond through instinct "no doubt . . . originally acquired, like all other social instincts, through natural selection." This rather weak argument, which is strongly Lamarckian, has not even convinced the neo-Darwinists who, since the mid-1960s, have adopted the genetically based kinship-selection argument for the emergence of "altruism." But, as shown in part II, this modern argument is even less convincing than Darwin's. All Darwinians have difficulty in reconciling competition, which is supposed to drive evolution, and cooperation, which holds societies together. No such difficulty characterizes the dynamic-strategy theory presented in part III.

When turning from early man to modern civilization, Darwin merely focuses on the extent to which contemporary social customs strengthened or weakened the progress of mankind physically and intellectually. In general he concludes that civilization, by allowing physically and intellectually weak individuals to survive, was working against the operation of natural selection (another defeat for his farmyard theory). He explains with unfortunate clarity:

With savages, the weak in body or mind are soon eliminated; and those that survive commonly exhibit a vigorous state of health. We civilised men, on the other hand, do our utmost to check the process of elimination; we build asylums for the imbecile, the maimed, and the sick; we institute poor-laws; and our medical men exert their utmost skill to save the life of everyone to the last moment. There is reason to believe that vaccination has preserved thousands, who from a weak constitution would formerly have succumbed to small-pox. Thus the weak members of civilised societies propagate their kind. No one who has attended to the breeding of domestic animals will doubt that this must be highly injurious to the race of man. (*Descent*: 501)

This is the type of notorious conclusion to which a simple farmyard analogy can bring us. The policy implications of such a model are simply unthinkable. Not that Darwin attempted to draw explicit policy conclusions from his model. Indeed he expressed the view that perpetuation of the weak was a cost that human society had to bear stoically or suffer a "deterioration in the noblest part of our nature." But the policy implications are very clear. And less humanitarian and more ruthless men, such as Adolf Hitler and his henchmen, have used Darwin's theory as the basis for and justification of their evil policies of eliminating the weak in mind and body. This is precisely why simplistic theories from the natural sciences should never be applied in a cavalier way to human society.

Owing to the totally unacceptable social policy conclusions that arise from Darwin's theory of natural selection, most thinkers have opted either to limit the application of Darwinism to biology or, like Charles Darwin himself, to employ it to "explain" social development while ignoring its social policy implications. This has led to the castigation of the so-called "social Darwinists" who have dared to draw social policy conclusions. But to condemn "social Darwinism" while accepting "biological Darwinism" is totally inconsistent. It is also unnecessary because Darwinism is relevant to neither area. It is a sign, however, of the stresses in the structure of Darwinism.

A Preoccupation with Sex

Why did Darwin devote two-thirds of the *Descent of Man* to sexual selection? This question takes on added significance when it is recalled that he regarded it as a "less rigorous" mechanism, as merely an "aid to ordinary selection," and as responsible only for "unimportant" differences between both the sexes and the races of mankind. While it might be claimed that Darwin was merely treading a scientifically objective path, I will argue that his preoccupation, criticized by Wallace, was an attempt to buttress the core concept of natural selection, which was showing increasing signs of stress.

From the fifth edition of *Origin*, as we have seen, Darwin's efforts to define natural selection more precisely left a potentially large range of biological variations unaccounted for. Darwin appears to have thought that this would undermine his concept of natural selection—as indeed it does—as the central mechanism behind "descent with modification." To prevent this he attempted to provide greater justification for sexual selection than given to it (just two pages) in the first edition of *Origin*. He was determined to show in *Descent*, at considerable length, that natural selection and sexual selection were partners, even if highly unequal, in the more general category of evolutionary selection. Whatever the senior partner was unable to explain could be left to sexual selection, which after all was the "aid to ordinary selection."

Darwin did not, however, attempt to develop a general theory of evolutionary selection. It was neither scientifically possible, because Darwinian selection is a farmyard rather than a natural process, nor was it his forte. He merely proposed a simple hierarchy, with natural selection as the dominant mechanism (hedged around with changing external conditions, habit, and "growth correlation"), and with sexual selection in reserve to sweep up all the remaining loose ends. And there were plenty of those.

Alfred Wallace (1871) employed an entirely different approach in support of natural selection. He made it very clear to Darwin that he thought this excessive preoccupation with sexual selection would merely undermine the dominant evolutionary mechanism. Wallace argued that most, if not all, of the variations that Darwin took to be the result of competition between the males for possession of the females could be interpreted as part of the struggle for existence. Darwin, however, was unmovable. If other naturalists were able to identify variations in the natural world that appeared to have little to do with the struggle for existence, the entire concept of natural selection would be at risk. For Darwin, sexual selection was the evolutionist's insurance policy.

The physical and instinctual characteristics that Darwin was convinced needed explanation by sexual selection are the "secondary sexual" (or nonreproductive) characteristics between males and females of the same species that cannot be explained by "different habits of life" (*Descent*: 568). These include the "greater size, strength, and pugnacity of the male, his weapons of offence or means of defence against rivals, his gaudy colouring and various ornaments, his power of song, and other such characters" (*Descent*: 567). By "habits of life" Darwin meant the manner in which individuals in a species achieved sustenance and shelter. This is similar to the role played in modern human society by "methods of production" in agriculture and industry: a change in "habits of life" is achieved through genetic change, whereas a change in the "methods of production" is achieved through technological change.

Darwin had convinced himself that in those species where the "habits of life" of the sexes were identical, any uniquely male characteristics could only be useful in gaining access to the females. He argued that as the females are able to survive and procreate without the secondary sexual characteristics possessed by the males of the species, these characteristics cannot be necessary in the "battle for life," only in the battle to "possess" the females. It is the battle for the females that, through a highly specialized type of selection, shapes the nonreproductive characteristics of the males. His argument is best summarized in the following passages from *Descent*:

When the two sexes follow exactly the same habits of life, and the male has the sensory or locomotive organs more highly developed than those of the female . . . they serve only to give one male an advantage over another, for with sufficient time, the less well-endowed males would succeed in pairing with the females; and *judging from the structure of the female*, they would be in all other respects equally well adapted for their ordinary habits of life. Since in such cases the males have acquired their present structure, not from being better fitted to survive in the struggle for existence, but from having gained an advantage over other males, and from having transmitted this advantage to their male offspring alone, sexual selection must have come into action. It was the importance of this distinction which led me to designate this form of selection as Sexual Selection. (*Descent*: 569; my emphasis)

Darwin continues:

It is clear that these [secondary sexual] characters are the result of sexual and not of ordinary selection, since unarmed, unornamented, or unattractive males would succeed equally well in the battle for life and in leaving a numerous progeny, but for the presence of better endowed males. *We may infer that this would be the case, because the females, which are unarmed and unornamented, are able to survive and procreate their kind.* (*Descent*: 570; my emphasis)

This argument about sexual selection is fatally flawed. The most obvious problem is that Darwin does not understand the concept of gender specialization and division of labor according to comparative advantage. The real reason that females are able to survive and procreate despite the absence of weapons of offense and defense that are possessed by the males of the same species is that they cooperate with the males in the joint enterprise of survival and prosperity by trading companionship and sex for food, shelter, and protection (Snooks 1994: ch. 3). This division of labor and cooperation is based on the sexes specializing in what they do best. In this way the joint probability of survival and prosperity is greater than the sum of the individual probabilities without cooperation between the sexes. Accordingly, Darwin's argument that the survival of the females proves that these weapons are not necessary in the struggle for existence but only in the race for sex is invalid. Darwin has missed the fundamental point that the struggle to survive and prosper is a joint operation in which only the most aggressive gender needs to possess these weapons, or indeed the various methods of attracting the attention of the females.

The other less obvious but more important problem in Darwin's sexual selection argument—indeed, in his more central natural selection argument as well—is the farmyard assumption that the struggle for both existence and sex is determined solely by possessing a physical or instinctual advantage over others. Or in more modern terms that the struggle in life is won by those "favored" by genetic advantage. Because of this farmyard assumption, Darwin overlooked the possibility that physical/instinctual characteristics possessed by the males of a species are employed primarily not in the race for sex but in the struggle to survive and prosper by pursuing the *nongenetic* dynamic strategies of family multiplication, conquest, and commerce.

Herein lies the core of my disagreement with Darwin. An individual's role in life is a function of its contribution to the **strategic pursuit**, which in turn depends on its physical/instinctual/intellectual characteristics. Individuals specialize according to comparative advantage and cooperate in their society's strategic pursuit in order to maximize the probability of their survival and prosperity. This specialization takes place along both gender and nongender lines. In turn the strategic pursuit, rather than natural or sexual selection, leads to the enhancement of biological comparative advantage through subsequent genetic change. It all depends on the dynamic strategy they are pursuing. These issues are discussed in detail in part III.

The flaws in sexual selection can be seen most clearly when Darwin attempts to apply it to our own species, albeit briefly. He begins by outlining the different physical and intellectual characteristics of a nonreproductive nature between men and women. While his survey of physical differences—"man on average is considerably taller, heavier, and stronger than woman"—is not in dispute, the same cannot be said for his depiction of the moral and intellectual differences between the sexes. Darwin characterizes women as having less courage and lower intellectual powers than men. For example he tells us:

> The chief distinction in the intellectual powers of the two sexes is shown by man's attaining to a higher eminence, in whatever he takes up, than can woman—whether requiring deep thought, reason, or imagination, or merely the use of the senses and hands. . . . We may also infer . . . the average in mental power in man must be above that of woman. (*Descent*: 873)

We are also told that "woman seems to differ from man in mental disposition, chiefly in her greater tenderness and less selfishness," and that "man has ultimately become superior to woman" (*Descent*: 873, 874). Needless to say, we are unable to endorse Darwin's observations on these alleged moral and intellectual differences, which reflect his complete lack of appreciation of the role of gender specialization and division of labor under different dynamic strategies or different stages in the same dynamic strategy. As the technological strategy, which was introduced with the Industrial Revolution, has unfolded, women have been drawn from the household into the market sector (Snooks 1994: ch. 4). Women now stand alongside men in the marketplace.

As we have come to expect, Darwin attempts to explain these alleged gender differences—together with superficial differences (body and facial hair, face and head shape, and skin color) between the races of mankind—in terms of the natural/sexual selection hypothesis. For good measure he also throws in the additional Lamarckian argument about changes in habit eventually becoming inherited. In the brief section on sexual selection in our own species, for example, he writes:

> We may conclude that the greater size, strength, courage, pugnacity, and energy of man, in comparison with woman, were acquired during primeval times and have subsequently been augmented, chiefly through the contests of rival

males for the possession of the females. The greater intellectual vigour and power of invention in man is probably due to natural selection, combined with the inherited effects of habit, for the most able men will have succeeded best in defending and providing for themselves and for their wives and offspring. (*Descent*: 907)

And of different racial characteristics he tells us:

We have thus far been baffled in all our attempts to account for the difference between the races of man; but there remains one important agency, namely Sexual Selection, which appears to have acted powerfully on man, as on many other animals. . . . It can further be shewn that the differences between the races of man, as in colour, hairiness, form of features, &c., are of a kind which might have been expected to come under the influence of sexual selection. (*Descent*: 556)

Darwin stresses, however, that his views "on the part which sexual selection has played in the history of man, want scientific precision" (*Descent*: 908) and that, in the absence of the intensive study of mankind, they can only be an extrapolation based on his more detailed work undertaken in the farmyard and buttressed by the more fragmentary observations of naturalists. One can but agree with him.

Conclusions

There can be no doubt that Darwin's farmyard theory of natural selection is unable to explain the dynamics of human society. This is a major stumbling block, because it is the most important test that any theory of life must face. The reason is that we know more about our own species than any other. A theory of life that is unable to explain and predict the changing fortunes of the most sophisticated and dominant species on Earth is unlikely to be able to explain the dynamics of other species. A careful examination of human society makes it clear that the central assumption of Darwin's theory of evolution—the doctrine of Malthus—has no relevance to our own species. Accordingly his farmyard model breaks down completely, because this assumption is critical to the creation of the continuous struggle for existence that is required to bring natural selection into play.

Darwin appears to sense this problem in *Descent*, where, without explanation, he retreats to statements about the "occasional" operation of mankind's struggle for existence. But all the while he maintains the official Darwinian line in later editions of *Origin*. Darwin is, however, unable to disguise the breakdown in his evolutionary theory of mankind. And if it is not relevant for our own species, for which we have more extensive and reliable data, why should we expect it to be applicable to other species? Clearly we should not. This realization makes nonsense of the lipservice that generations of biologists and paleontologists have paid to the theory of natural selection as an explanation of changing life forms.

There is a problem with the evidence as well as the theory. In defining natural selection with greater precision in *Descent*, Darwin was left with many additional unexplained physical and instinctual characteristics in both nature and mankind. To deal with these, he devotes most of *Descent* to a discussion of the formerly minor concept of sexual selection.

In its expanded form, the concept of sexual selection is really just a convenient grab bag into which Darwin is able to sweep all those biological modifications that cannot be accounted for by natural selection. Sexual selection, therefore, is merely a buttress to prevent the structure of natural selection, by now crazed with stress fractures, from crashing to the ground. Wallace, the joint inventor of natural selection, clearly recognized this problem and warned Darwin that, rather than supporting natural selection with this technique, he would probably end up by leveling it. Although in the short run Wallace was wrong—for reasons discussed in part II—in the longer term he was correct—for reasons discussed in part III.

Darwin's buttressed evolutionary structure shows us that there are major forces other than speciation involved in the dynamics of life. What we need is a general dynamic theory that can integrate them all. To achieve this, the Darwinian theory of evolution must be abandoned. Darwin's "poor rag" is unable to do the job that we continue to ask of it.

Darwin begins his conclusion to *Descent* with a caution: "Many of the views which have been advanced are highly speculative, and some no doubt will prove erroneous." Yet he believes that false views, unlike false facts, are unlikely to do any harm. He assures us that

> False facts are highly injurious to the progress of science, for they often endure long; but false views, if supported by some evidence, do little harm, for everyone takes a salutary pleasure in proving their falseness: and when this is done, one path towards error is closed and the road to truth is often at the same time opened.

History has proved Darwin wrong in this forecast. Despite its fundamental flaws his theory of evolution has triumphed over reality for the past 150 years. It is a prime example of the hypothetico-deductive fallacy about theory-building and falsification. To show how the myth of natural selection has been perpetuated throughout this time we turn now to the followers of Darwin.

Part II

The Collapsing Cathedral of Darwinism

The new order . . . had plans to build a great cathedral where the chapel stood, and to take the word of their master to the people. They intended to throw this new structure like a mantle over the simple chapel. . . . With the great cathedral approaching completion, even members of the new order began to feel uneasy. Not with the cathedral's underlying philosophy but with the widening cracks that had appeared in some of the load-bearing walls. . . . Just when the purists had put the finishing touches to the great cathedral something completely unexpected happened . . . there was a sudden reverberating rumble of rupturing masonry . . . the cathedral walls collapsed outwards and the roof came crashing down. One of the observers, a white-bearded visitor from a foreign land, exclaimed that it was "the grandest pile of ruins I ever saw."

Prologue

Chapter 5

The Buttress-Builders:
Modern Paleontology

The buttress-builders in the real world are the paleontologists, who are aware of both the serious flaws in the foundations of Darwinism and the unrealistic ambitions of the cathedral-reconstructionists, or neo-Darwinists. But instead of razing the old structure to the ground and beginning again, they have attempted to shore it up with a series of flying buttresses. Unfortunately these products of the quarry have failed to either disguise or support a theory that escaped into nature from the farmyard.

The paleontologists, led by Niles Eldredge and Stephen Gould, have developed a host of ad hoc arguments to protect and extend the central Darwinian concept of natural selection. They have made a heroic attempt to prevent the fractures in the structure of Darwinism from widening. Yet, ironically, it is the buttress-builders who have drawn attention to these flaws. In the first place the building of buttresses has suggested to onlookers that there is something wrong with the original structure. And secondly, these paleontologists have uncovered fossil evidence that casts doubt not only on the manner in which "descent with modification" has occurred but also on the critical assumptions underpinning natural selection.

The buttress-builders have also pricked the pretensions of the high priests of Darwinism—the neo-Darwinists such as John Maynard Smith, Edward Wilson, and Richard Dawkins. While the paleontologists find common ground with these geneticists over the idea (if not the definition) of natural selection, they are dismissive of the neo-Darwinian focus on the gene rather than the organism as the unit of selection, and of their elevation of sexual selection to the status of primary Darwinian mechanism. The former they see as the absurdity of reductionism and the latter as a form of "madness" (Eldredge 1995: 213; Gould 1980).[1]

It is a great disappointment that the naturalists—as these paleontologists like to call themselves—with their historicist approach have been unable to sweep Darwinism aside and replace it with a realistic general dynamic model.

Books with promising titles such as Niles Eldredge's *Macroevolutionary Dynamics* (1989) offered much hope but ended up being no more than an untidy collection of ad hoc arguments designed to make natural selection compatible with the inconvenient modern fossil evidence. Unfortunately they have clung tenaciously to their Darwinist faith and have been unable to progress beyond a naive form of historicism which invests historical patterns with a lawlike nature. They also hold the surprisingly superficial view, shared by most natural scientists, that life forms are driven by exogenous forces such as climate, volcanic activity, and asteroid attacks.

Yet, at least the naturalists are aware of what needs to be explained. They wish to analyze the changing patterns of life on Earth. In contrast, neo-Darwinists are concerned with more abstract issues. Even in a book promisingly titled *The Major Transitions in Evolution* (1995), the neo-Darwinists John Maynard Smith and Eors Szathmary feel able to discuss the major stages in the emergence of life over the past 3,500 million years (myrs) without reference to the historical patterns of life. The neo-Darwinists have learnt little from the deficiencies of Charles Darwin's approach.

Exposing the Flaws in Darwin

Naturalists possess the data needed to provide a new and realistic theory of life, but they have opted instead to point out *some* of the flaws in Darwin's perception of the process of evolution and to support the core of his theory with a series of ad hoc arguments (or flying buttresses). The two main Darwinian flaws that the naturalists have exposed are: first, that "descent with modification" did not occur in a slow and continuous manner as Darwin and neo-Darwinists have maintained, but rather in relatively short bursts (of five to fifty thousand years) separated by long periods of stasis (of five to ten myrs); and second, that the transition between species was not a relatively smooth occurrence in which the new improved species eliminated the old species but rather a more calamitous affair in which the old species and orders collapsed, thereby providing the opportunity for other species to emerge and flourish.

What the naturalists have failed to expose is the totally unsuitable nature of Darwin's farmyard analogy and the factually incorrect assumption—the "doctrine of Malthus"—required to bring natural selection into play. Instead they have surreptitiously replaced Darwin's geometric population increase with climatic change as the driving force in natural selection. It does not seem to have occurred to them that this substitution is enough to demolish Darwin's farmyard theory, because it deactivates the essential struggle for existence and survival of the fittest on which natural selection depends. The naturalists attempt to justify this by claiming, quite wrongly, that Darwin's concept of natural selection is just a passive filter.

Punctuated Equilibria

As we have seen, a major prediction arising from Darwin's farmyard model is that transmutation or evolution occurs very slowly but continuously over vast periods of time. While the fossil evidence of the time seemed to contradict this prediction, Darwin was able to explain it away as being the product of the primitive state of paleontology in the mid-nineteenth century. Yet, despite the rapid development of this science over the following century, the evidence that Darwin needed to vindicate his predictions did not appear. Quite the contrary. The growing body of fossil evidence indicated that a large number of species, once they had emerged as separate entities, experienced very little directional change before they were finally extinguished. There was change but it was of an oscillating (or zigzag) rather than a directional nature, despite life histories of five to six myrs (Eldredge 1995: ch. 3). But, as paleontologists have been conditioned to think in Darwinian terms, they either discard this evidence as atypical, force it uncomfortably into a Darwinian straitjacket, or focus on non-evolutionary themes to avoid the growing contradiction between theory and evidence.

The first modern paleontologists to take the fossil evidence seriously and to insist that it contradicted Darwinian gradualism were Niles Eldredge and Stephen Gould (1972). In their famous "punctuated equilibria" paper they drew attention to what they called "stasis," which they contrasted with orthodox Darwinian "phyletic gradualism," and attempted to explain it in terms of "allopatric speciation," or the emergence of new species via geographic isolation from the main population. Darwin's view of life as a state of flux, they argued, was contradicted by the fossil record. Instead they painted a picture of a static world occasionally and briefly disrupted (or "punctuated") by the emergence of new species. Eldredge and Gould explain that

> The history of life is more adequately represented by a picture of 'punctuated equilibria' than by the notion of phyletic gradualism. The history of evolution is not one of stately unfolding, but a story of homeostatic equilibria, disturbed only 'rarely' . . . by rapid and episodic events of speciation. (Eldredge and Gould 1972: 82)

As with most other Darwinists, Eldredge and Gould tend to equate the "history of life" with "evolution" (directional genetic change) rather than with the total activity of life forms. (This is similar to equating the history of civilization with technological change.) Hence, while there are long periods in the history of life when there is little or no directional genetic change, life continues to be in a state of flux. Life is the sum total of the activities of life forms, whereas genetic change is merely one of four different dynamic strategies by which it is achieved. Life is dynamic even when there is no directional genetic change, because other dynamic strategies are being pursued by organisms to achieve their common objective of survival and prosperity. In other words, genetic change is not the objective but the occasional instrument of strategic pursuit. The naturalists have confused the ends and means of life. The importance of this distinction, which will become clearer in part III, has been overlooked by Darwinists of all types.

Debate in Darwinian circles has raged over the issue of punctuated equilibria versus phyletic gradualism over the past twenty-five years or so. The battle lines in the "great stasis debate" have been drawn between the naturalists and the neo-Darwinists. It is a debate they have yet to resolve. But, as I will show, resolution is no longer relevant because the "great stasis debate" is an outcome of Darwinian confusion. Life is never static; only particular dynamic strategies employed by life forms are put aside from time to time.

It seems fairly clear from the fossil evidence that there are very long periods, running into millions of years, during which emergent species experience little systematic anatomical change. Recently, Eldredge was able to confidently conclude:

> Evolution does not inevitably and irrevocably transform species as they persist through geological time. To the contrary, most species often seem to go nowhere, evolutionarily speaking. To be sure, some will accrue some evolutionary change over millions of years, but most of them hardly accrue any change at all. Stasis is now abundantly well documented as the preeminent paleontological pattern in the evolutionary history of species. (Eldredge 1995: 77)

Setting aside the confusion of "history of life" and "evolution," the question is whether this inconvenient evidence can be explained in Darwinian terms or whether it is necessary to scrap the farmyard model and develop a completely different dynamic theory.

Eldredge, a leading representative of the naturalist school, is quite clear on this issue:

> Stasis does not mean that we need an alternative to natural selection. But we do need to understand the circumstances in which natural selection works. On the one hand, how does natural selection keep species stable for long periods of time? And what circumstances obtain when natural selection does effect adaptive change? (Eldredge 1995: 77)

For a self-proclaimed Darwinist these are totally subversive questions, which would have panicked Darwin himself.

It is as if Eldredge is dealing with natural selection and punctuated equilibria in mutually exclusive compartments. It will be recalled from part I that Darwin's concept of natural selection involves the following causal sequence: geometric population increase → struggle for existence → survival or extinction based on physical/instinctual/intellectual comparative advantage → transfer of this advantage from survivors to their offspring. How, in the light of this sequence, is it possible for a Darwinist to ask: "How does natural selection keep species stable for long periods of time?" Charles Darwin's incredulous response to such a self-proclaimed disciple would be that natural selection and genetic stability are totally incompatible concepts. If natural selection is brought into play through geometric population increase for a species possessing heritable variation, there will be inevitable accumulation of beneficial variation. Darwin's concept of natural selection, therefore, is totally incapable of keeping species stable for long periods of time as Eldredge claims. *Stasis is, therefore, the best evidence we have that natural selection does not exist in reality.* The naturalists are only able to have their evolutionary cake and eat it too by abandoning the doctrine of Malthus and replacing it with an occasional exogenous driving force, such as climatic change.

Eldredge is particularly keen to rebut the neo-Darwinian accusation that the existence of punctuated equilibria implies the end of natural selection. At every opportunity he affirms his faith in natural selection, insisting that "naturalists such as myself completely agree [with the neo-Darwinists] that natural selection is the sole deterministic molder of adaptive evolutionary change" and that "we are merely dissatisfied with the lack of any cogent theory to explain why natural selection keeps species stable for so long—and what enables selection to trigger change when it does occur" (Eldredge 1995: 7, 77).

Naturalists are unable to consider the transmutation of species other than in the context of natural selection even when the predictions of Darwin's farmyard model are generally acknowledged to be totally inconsistent with the fossil evidence. Reality, they tell us, contradicts natural selection; hence all we need to understand is how natural selection can explain this contradiction! This response of faith begs the question of whether it is an appropriate stance for well-credentialed scientists. What, we might wonder, would it take for these naturalists to abandon their Darwinian religion? I suspect that no amount of evidence against natural selection will challenge their faith. As Darwin himself speculated, it is rarely possible to convince one's contemporaries to abandon the ideas on which they have built their scientific careers, despite overwhelming evidence and argument to the contrary:

> I by no means expect to convince experienced naturalists whose minds are stocked with a multitude of facts all viewed, during a long course of years, from a point of view directly opposite to mine. It is so easy to hide our ignorance under such expressions as the 'plan of creation', 'unity of design' [read "natural selection," "selected for this or that," etc.], &c., and to think that we give an explanation when we only restate a fact . . . but I look with confidence to the future, to young and rising naturalists, who will be able to view both sides of the question with impartiality. (*Origin*: 453)

Eldredge, and others like him, does not appear to realize—or refuses to acknowledge—the extent to which the naturalists have undermined Darwin's theory of natural selection. This is of particular interest because he has accused the neo-Darwinists, quite correctly, of perverting the master's doctrine. He regards natural selection as a "passive filter" that is activated by changes in the physical environment—particularly climate—rather than by automatic and rapid population change, despite the fact that Darwin believed that changing physical conditions had only a "slight and direct effect" (that is, not through natural selection) on variations.

Naturalists like Eldredge have substituted an exogenous physical driving force—both for the activation of adaptive change and for the extinction of species—for Darwin's automatic impulse of organisms to procreate at the maximum rate. One searches in vain in Eldredge's work for any discussion of high rates of population increase, the doctrine of Malthus, the struggle for existence, or the survival of the fittest. For the naturalist, natural selection has become a gateway through which, owing to some vague eccentricity, some organisms fail to pass. Darwin's "war of nature" has been abolished by our modern naturalists. And even those individuals that do pass through this gateway do so, for some mysterious reason, only at rare intervals.

How do the naturalists explain stasis? Essentially they view it as an outcome of two sets of forces, "habitat tracking" and isolation. The first of these appears to involve a suspension of Darwinian natural selection, whereas the second is merely a statistical artifact.

The "habitat tracking" argument states that a change in the physical environment is more likely to cause plant and animal species to migrate gradually to new areas possessing the conditions they prefer rather than holding on to old areas and evolving to exploit the changed conditions (Vrba 1985; Eldredge 1985). As Eldredge (1995: 64–65) says: "In the face of environmental change, organisms within each and every species seek familiar living conditions—habitats that are 'recognizable' to them based on the adaptations already in place." This is, of course, merely a *passive* response by organisms to a changing environment. In the naturalist vision of life, the driving force is the dynamic physical world, and individual organisms are merely its ciphers. But why is habitat tracking a far more likely outcome of environmental change than evolutionary adaptation? Eldredge (1995: 66) tells us that

> Naturalists . . . say that the most likely response of a species to environmental change is habitat tracking. The second most likely response to environmental change is extinction, which generally follows when suitable habitat cannot be found. The least likely outcome is wholesale, linear transformation of an entire species to meet the new environmental exigencies. Traditional evolutionists, including latter day ultra-Darwinians [neo-Darwinists] . . . have it the other way around: environmental change begets evolutionary transformation through natural selection; failing that, we expect extinction. (Eldredge 1995: 66)

But how can habitat tracking lead to the state of "stabilizing natural selection"? *In a Darwinian world* all individuals obey the doctrine of Malthus and procreate at a rapid, geometric rate, struggle for existence, and become subject to survival of the fittest. Migration in response to environmental change will not alter this "war of nature," particularly if the new regions are already occupied. This is the very reason that Darwin discounted the "changing physical conditions" as an important source of evolution. The essential condition required to bring natural selection into play is, as I have said before (but owing to the deafness of so-called modern Darwinians it is necessary to say so again, and again) is not environmental change but rapid and continuous population increase.

Habitat tracking makes sense only in a *non-Darwinian world*, which is not a world that should be inhabited by Darwinians of any stripe. In a non-Darwinian world—as is argued in part III—the migration of species is not a passive response to a changing environment but rather an outcome of the active dynamic strategies pursued by all life forms. Without realizing it the naturalists are struggling to find a non-Darwinian explanation within their Darwinian conceptual framework. The end result is massive confusion. In this respect the neo-Darwinists have greater claim to the mantle of Darwin, because they insist that the only alternative *in a Darwinian world* to extinction is genetic change. But, as I show in part III, *we do not live in a Darwinian world*.

The second reason for "stasis" advanced by Eldredge (1995: 82–84) is the impact of isolation on "the nature and structure" of species. It is based on an argument by Sewall Wright (1931; 1932)—later developed by T. Dobzhansky (1937) and E. Mayr (1942)—about the evolutionary effects of the geographical distribution of populations on species. Most species, it is claimed, are scattered over large regions, in which suitable habitats are "semi-isolated" from each other. This "disjunction" reduces the opportunity for genes to be exchanged between populations within a species. Hence, these populations, which occupy different ecosystems within the larger region of the species distribution, will experience different "evolutionary pathways," with "some going one way, some, another, others still yet another." The end result *for the species as a whole*, it is claimed, is "stasis," as these various pathways cancel each other out. This, Eldredge and others claim, is consistent with evidence that indicates some directional, gradual change at the population, but not at the species, level.

This is a statistical rather than a theoretical argument, which is designed to render the evidence of *both* change *and* stasis consistent with *both* natural selection *and* punctuated equilibria. It is a very fancy flying buttress indeed. Yet it is badly constructed. Why should the sum of genetic change at the population level lead to stasis at the species level? Clearly there is no good reason. Why would the doctrine of Malthus not lead to the positive accumulation of variations in a similar way in all populations owing to the common gene pool and similar population pressure (which is automatic in all organisms in all places)? Eldredge's hidden reason, of course, is that the naturalists have substituted ecological change for the doctrine of Malthus and, by doing so, have undermined rather than buttressed Darwin's farmyard concept of natural selection.

Having observed and "accounted for" stasis in the fossil record, the naturalists needed an explanation that could integrate it into Darwinian evolutionary theory. Why has this observed stasis been broken at rare intervals by abrupt and relatively rapid evolutionary change? The idea that Eldredge and Gould (1972) came up with is, as we have seen, "punctuated equilibria," in which stasis is "punctuated" by the relatively rare emergence of new species through a process known as "speciation." For the naturalists, therefore, directional genetic change is associated with the emergence of new species. Once a new species has fully emerged, which may take as little as five to fifty thousand years (thought by Eldredge to be "very generous"), it changes little until it is extinguished some five to ten myrs later (Eldredge 1995: 99). Despite the fact that punctuated equilibria presented an entirely different picture of life to that envisioned by Darwin—of stasis rather than flux—Eldredge (1995: 97) claims, rather naively, that "it should have been noncontroversial. It wasn't."

Punctuated equilibria shifts the evolutionary focus from the individual organism to the species, from the slow and gradual accumulation of "profitable variation" to the abrupt, brief, and rare process of speciation. And he expected it to be noncontroversial! Eldredge (1995: 106) continues: "speciation is obviously central to the fate of genetic variation, and a major shaper of patterns of evolutionary change through evolutionary time." He insists that genetic variation is the outcome of natural selection, but that the fate of variation depends on the process of speciation. In Darwin's *Origin* the fate of variation, of course, depends on natural selection.

The punctuated equilibria hypothesis consists of two main parts. The first part builds on Ernst Mayr's (1942) notion of "peripheral isolates," which refers to a population isolated on the margins of an ancestral species' range. This semi-isolated population occupies a suboptimal habitat and, through natural selection, it both adapts to that habitat and prevents crossbreeding (and, hence, loss of its genetic gain) through a change in its "specific mate recognition system" (which was taken from Paterson 1985), thereby becoming a new species. Somehow these adaptive changes are supposed to operate under the shaping influence of speciation, rather than the other way around as was traditionally thought. Eldredge (1995: 119) says "that's how we saw speciation as an actual cause of adaptive evolutionary change." But, he continues, "much as I still like our original [1972] suggestion explaining how adaptive change can be seen as a result (rather than a cause) of speciation, I confess I don't think it is the whole answer to the perplexing correlation between adaptive change and speciation . . . something else is afoot." This brings us to the second part of their explanation—to another flying buttress.

The second part is an even more obtuse discussion of the role of species in evolutionary change, which the naturalists call "species sorting" and even "species selection," an unfortunate name that implies direct competition with natural selection. Eldredge tells us:

> Species sorting is simply differential speciation or extinction of species within a larger group. Some lineages speciate at a higher rate than others, and some species are more prone to extinction than others. There are definite, repeated patterns in evolutionary history that reflect such differential speciation and extinction, patterns that had not really been introduced until we formulated the notion of punctuated equilibria. (Eldredge 1995: 119–20)

The question then becomes: What is responsible for these "differential speciation and extinction patterns"? Eldredge's answer, which is complex and appears circular, is that the success of a "fledgling species" depends on whether the process of speciation generates sufficient adaptive change to enable it to compete with sibling species. He asserts somewhat mystically that there's a built-in bias involving survival rates of fledgling species. Fledgling species are being sorted in a systematic fashion. Species sorting does not determine what new adaptations appear in evolution; rather, species sorting depends on whether or not such adaptive change occurs—through normal Darwinian processes. But the effect of that sorting is monumental in shaping the overall course of evolutionary history. It is that sorting that lies at the heart of the paradox: why so much of adaptive evolutionary change is tied up with speciation. (Eldredge 1995: 123)

And just in case you did not understand this, another passage from Eldredge may, or may not, help:

> Species are packages of genetic information—information pertaining to the functional properties of its organisms. It stands to reason, then, that if some species appear faster than others in a lineage, or if some become extinct more readily than others, there will be a juggling, a sorting, of species, and thus of the genetic information they contain. Anything that biases the births and deaths of species necessarily biases the fate of adaptations and the genetic information on which they are based. (Eldredge 1995: 128)

Adding to the complexity—the burgeoning of buttresses—is the further claim that speciation involves a directional bias. In other words, that fledgling species either automatically contribute to the genetic trend—for example, to the increasing brain size of hominids—or, at worst, have a neutral impact. They never reverse it. Why? We are not told. It appears to be one of the mysteries of life in the naturalists' world. The bottom line is that this is a supply-side theory—a theory in which the emergence and disappearance of species simply happen for reasons peculiar to those species. They certainly do not emerge in response to the demands of life. This accounts for the inherent mystery of naturalist theory.

Finally, how do the naturalists view the status of their theory? Eldredge explains that

> species sorting is not a new, or rival, theory of evolutionary adaptive stasis and change. It is *additional theory* that helps us to understand why adaptive change seems caught up with the origin of new reproductive communities: new species. . . . Species sorting affects the fates of adaptations once they are evolved. The two ideas are complementary—not rivals. (Eldredge 1995: 128; my emphasis)

This then is "additional" theory that is braced against existing, and clearly inadequate, theory. Whenever something appears in the fossil evidence that natural selection cannot explain, additional flying buttresses are erected. Is there any wonder that the outcome is mazelike? Although Darwin's theory is fatally flawed, it possesses a compelling simplicity that has enabled it to survive for almost 150 years. Had the acceptance of the idea of evolution depended upon naturalist theory it would never have made any impact on the world of ideas.

The lack of clarity in the naturalists' case arises from their inability to develop a general dynamic model centered on the role of speciation in explaining either directional genetic change or the differential success of species. As suggested already, the basic reason is that theirs is a supply-side model with no demand-side component. In part III a very different, realist, and far more straightforward model is employed to explain these issues. The differential success of species is largely an outcome of the dynamic strategies—of which only one involves genetic change—they employ. Hence, the reason that the genetic profile of a species consists of a series of long steps is that, once it has emerged, the component organisms turn from the pursuit of genetic to **nongenetic strategies**. While speciation is an outcome of the dynamic strategy of genetic change, the success of the species depends largely on the pursuit of nongenetic strategies by individuals. Accordingly a species is an outcome, not a cause, of the dynamic strategies of its individual membership. Species have no causal function because they (like genes) do not pursue dynamic strategies. Only organisms can be dynamic **strategists**. Of course it is true, as Eldredge (1995: 120) claims in attempting to justify his species-level approach, that "species have names; they have beginnings, histories, and ends. They are discreet historical entities." What species do not "have" are dynamic strategies—they are merely the *outcome* of individual strategies—and, hence, they do not play any causal role in the dynamics of life as claimed by the naturalists. This is because, unlike human society, animal species do not provide strategic leadership; their overall strategies are merely the aggregation of individual strategies. The flying buttresses constructed by the naturalists are just as flawed as the structure they have been designed to support.

The Rise and Fall of Species

The naturalists, quite properly, wish to explain the patterns they observe in the fossil evidence, not only at the level of the individual organism but also at that of species and groups of species. They wish to explain macroevolution as well as microevolution. Indeed they are determined to explain "the entire history of life" (Eldredge 1995: 126), involving the rise and fall of species and groups of species—something the neo-Darwinists are unable to do.

The examination of long-term fossil evidence contained within selected ecosystems has suggested to the naturalists that "the entire fauna appears rather abruptly, persists with great monotony (and much habitat tracking), and eventually disappears with the same abruptness with which it first appeared" (Eldredge 1995: 147). Their reading of the evidence suggests that groups of species, after persisting for millions of years, are extinguished, allowing—even forcing—new species to emerge. These extinctions, which are discussed more fully in chapter 9, range from single species, entire dynasties, the flora and fauna of small localities, to global extinctions that have occurred five or six times. Global extinctions appear to have occurred over periods of up to five hundred thousand years, with subsequent recoveries taking some five to ten myrs.

Clearly this is very different to Darwin's explanation, which is a story about species gradually succumbing to new, more highly evolved species. Darwin states quite clearly in *Origin* that species are produced and exterminated by slowly acting and still existing causes, and not by miraculous acts of creation and by catastrophes; and as the most important of all causes of organic change is one which is almost independent of altered and perhaps suddenly altered physical conditions, namely, the mutual relation of organism to organism [that is, struggle for existence]. (*Origin*: 457–58)

How then do naturalists explain the rise and fall of species that they quite correctly see in the fossil evidence? Essentially they surreptitiously discard Darwin's farmyard hypothesis and, in direct contradiction of the master, explain it in terms of catastrophic changes in the physical world, such as massive volcanic eruptions, huge asteroid impacts, and major climatic change. They argue that species in a state of stasis will continue that way forever unless some random physical event causes extinction and further speciation. The rise and fall of species and groups of species, therefore, are driven by exogenous physical events: a direct response to the dynamics of the Earth and universe. Presumably they would attempt to explain the rise and fall of human society in the same way. Certainly Eldredge (1995: 157) discusses the emergence of *Homo habilis* and the extinction of *Australopithecus africanus* as the outcome of the habitat destruction at the onset of an ice age some 2.8 myrs ago.

A growing number of paleontologists appear to favor climatic change as the driving force in both the extinction of old species and the emergence of new species (Cowen 2000: ch. 18). Eldredge, for example, asserts:

Climate change induces extinction *and* speciation simultaneously, along with habitat tracking. All components of the 'turnover'—the abrupt disappearances of older species, and their rapid replacement by others—can be traced back to a single climatic event. (Eldredge 1995: 150; emphasis in original)

This fall and rise process has been called, by Elizabeth Vrba (1985), the "turnover pulse" model. But, as Eldredge notes, there is a slight problem of timing with this model—the emergence of new species is not simultaneous with the extinction of old species, often taking millions of years to occur. It is not difficult to anticipate Eldredge's response to this inadequacy of existing theory to explain the real world—"we need some additional theory!" A further flying buttress to be built in support of the failing Darwinian structure. Speculation piled upon speculation with no overarching explanation, no general dynamic theory.

And what is that "additional theory"? None other than our old incomprehensible friend, "species sorting." Quite cheerfully Eldredge asserts:

Once again, we must consider fledgling species. In a world suddenly devoid of a majority of species that used to staff the ecosystems, we can imagine the probable fate of the few new species that might, in the course of time, appear through normal speciation. . . . Survival rates would be far higher than usual. It is the rate, not of speciation per se, but of successful speciation, that goes way up after a major extinction event. (Eldredge 1995: 151)

But, *in a Darwinian world*, how do new species emerge when there is a sudden massive *reduction* in population pressure? Darwin, of course, would argue that it is just not possible; which was exactly why he assumed that population growth was always rapid and continuous and that new species replaced old ones in a slow but steady fashion. Apart from chanting incantations about species sorting—such as "species 'speciate'" (Eldredge 1995: 181) or "species give rise to species, species multiply" (Alexander and Borgia 1978)—the naturalists fail to explain how this could occur.

There exists, however, a general dynamic theory that can explain not only the action of individuals but also the rise and fall of species and even of dynasties. This is the dynamic-strategy theory presented in part III, which treats the fluctuations of life in terms of the exploitation and exhaustion of dynamic strategies. This simple and elegant theory, which can explain both microdynamic and macrodynamic processes, dispenses with the need for the fanciful adhockery of these naturalist arguments, and does away with the farmyard theory of natural selection.

The Buttresses Also Have Flaws

In attempting to explain the patterns in the fossil record within a Darwinian framework the naturalists have built a large number of disfiguring buttresses. The simplicity and elegance of Darwin's evolutionary structure have been distorted beyond recognition. The original, pure architectural forms have been transformed into a bizarre Gothic construction. And the outcome is one not of greater clarity but of greater confusion. In the end these flying buttresses merely draw attention to the fact that, elegant as it once was, natural selection is not, and never could be, a general theory of life. A realist general theory would not require any ad hoc support. A better strategy for the naturalists would have been to reject natural selection at the outset and to construct a new, internally consistent general dynamic model that could explain the historical evidence. Clearly this is beyond their ambitions.

The much heralded concept of punctuated equilibria cannot be regarded as a dynamic theory. It is merely a stylized genetic profile supported by a number of ad hoc arguments. Eldredge discusses the method employed in developing this concept in the following terms:

> It is a two-step process: first, establish the very existence of a recurrent pattern in evolutionary history. Then, using what we know about the evolutionary biology of living organisms, formulate a combination of factors that is most likely to yield those very kinds of evolutionary patterns we see in the fossil record. (Eldredge 1995: 58)

The first step is correct, but thereafter this exercise in historicism runs aground.

Eldredge and Gould attempt not only to describe the recurrent pattern in the fossil record—the stepped genetic profile—they also invest it with lawlike significance. The point is that historical patterns are something that require explanation, they are not a source of explanation (Snooks 1998a: ch. 7). The latter is the old historicist fallacy. The naturalists, however, have transformed the historical stepped genetic profile into the theoretical "punctuated equilibria." Compounding this problem, their second methodological step is designed not to generate a new and more appropriate general theory to explain the recurrent historical pattern but rather to employ natural selection, together with a number of ad hoc supporting arguments, to "explain" the "theory" of punctuated equilibria.

This is just not good enough. A new beginning is required. In part III this is achieved through what I call the **quaternary inductive method**—originally developed in my global-history trilogy (Snooks 1996; 1997; 1998a)—involving the identification of **timescapes** (or recurrent patterns), the development of a new general dynamic theory to explain these timescapes, the derivation of specific (historical) dynamic mechanisms by applying the general theory to the timescapes, and uncovering the laws underlying these general and specific mechanisms (but *not* the timescapes). The resulting dynamic-strategy theory, which has an elegant simplicity, succeeds where the buttressed Darwinian model fails.

The naturalists' account of the rise and fall of species and groups of species is also fatally flawed. As we have seen, this process is driven by random changes in the physical conditions of life including climate, volcanic action, and asteroid impacts. Life, in other words, is at the mercy of the physical world, to which it responds passively. Life is but a cosmic lottery! This hardly tallies with what we know about the history of mankind. All physical and biological explanations of life have foundered on these rocks. It is merely a cop-out to say, as many natural scientists do, that our intellect makes us "unique" and our society is "artificial." This implies that mankind must remain unexplained when we write the history of life.

But, as I argue in part III—and throughout my global-history trilogy—human beings attempt to maximize the probability of survival and prosperity in the same way as other species: desires drive, ideas merely facilitate. Differences in intellect between animal species are a matter of degree rather than of kind as Darwin and Wallace were keen to demonstrate. Accordingly, all of life can be explained by the same general dynamic model. Further, the systematic nature of the great waves of life (and of human society) over the past three to four billion years belie the randomness of those physical forces favored by the naturalists as the drivers in life.

The ad hoc arguments introduced by the naturalists that are least persuasive are those about the role of species. It is understandable that the naturalists should object to the neo-Darwinian idea that natural selection on its own can account for the fluctuating fortunes of life, such as the rise and fall of species. It clearly cannot. This is precisely why Darwin had to picture evolution as a gradual and continuous process—of old species gradually being replaced by new species—rather than as a series of great waves of life.[2] Reality can only be explained by developing a general dynamic model that encompasses not only the driving force at the individual level but also the mechanism by which this is translated into the observed wavelike development path of life and the steplike profile of genetic change. A buttressed Darwinian model is no more able to achieve this than Darwin's original farmyard theory or the current neo-Darwinian models.

Of course, species are "real" in the sense that they exist. But they do not possess objectives as the naturalists insist, nor do they attempt to achieve them. Their insistence that "species speciate" or "species multiply" has a metaphysical ring to it—almost of religious incantation. The reality is that only individual organisms have objectives and strategies. Species are not strategic organizations as are human societies that coordinate the **strategic pursuit** of individuals through the provision of strategic leadership. Species are merely the outcome of successful dynamic strategies, both genetic and nongenetic, pursued by individual organisms. The more successful are the individual strategies, which are coordinated through **strategic imitation** (mimicking successful individuals), the larger and more dominant the species that they collectively define. Also, as will be argued in the next chapter, genes are merely associated with the building blocks of organisms and are subject to their dynamic strategies.

Like all self-proclaimed Darwinists, the naturalists have a tendency to equate genetic change leading to the transformation of species with the dynamics of life in general. They share this with the master. No doubt, if questioned on this issue, most Darwinists would agree that there is more to life than genetic change, but would probably argue that for all practical purposes they are the same. The important point is that the Darwinists, both naturalist and geneticist, by proceeding as if the transformation of species is the same as the dynamics of life have developed models that are incapable of explaining not only the dynamics of life in general but also of genetic change in particular. As the dynamic-strategy model presented in part III shows, genetic change is merely one of the dynamic strategies employed by individuals to survive and prosper—merely one fundamental strategy impacting on the dynamics of life. It is essential to realize that individuals attempt to maximize not genetic change but rather the probability of achieving the universal objective of survival and prosperity.

Finally, the focus on "stasis" by the naturalists is doubly misleading. This focus is, as we have seen, an outcome of observing in the fossil record the genetic stability of species once they have fully emerged. But it is misleading for at least two reasons. First, because of the tendency to equate genetic change with the dynamics of life, the naturalists are responsible for giving the impression that *life* is in stasis rather than flux. Second, the "stasis" concept is not even applicable to the issue of genetic change. Directional genetic change is not static in the sense claimed by the naturalists: it is just that as a dynamic strategy it is not as "economical" for organisms as other nongenetic strategies over long periods of time. The naturalists have focused on the wrong issue. In the real world, therefore, the *concept* of stasis is totally irrelevant. It is not something that has to be explained, as such. Hence the usual theories about convergence to equilibrium following some sort of displacement are figments of the naturalists' imaginations. What does need to be explained, however, is why organisms alternate between genetic and nongenetic dynamic strategies during the lifetime of the species of which they are members. The outcome of this insight is that the Darwinists of all varieties have wasted a great deal of time and energy in the so-called "great stasis debate" of the last quarter of the twentieth century. It has led to many unnecessary flying buttresses.

Conclusions

The naturalists sense that the traditional Darwinian approach to biological change is not entirely satisfactory because its predictions do not tally with what they see in the fossil record. As paleontologists they wish to develop a theory that can explain the "recurrent patterns" recorded in stone. This is the essential starting point in explaining dynamics, and it held out promise that they might develop an effective dynamic model. Unfortunately this promise was not fulfilled, largely because they fell into two rather obvious traps, one after the other.

The first trap is that of naive historicism, which is an outcome of treating historical patterns as if they are dynamic processes rather than dynamic outcomes. Instead of treating the long-stepped genetic profile of each species as a pattern of *outcomes* that had to be explained, they anointed it with an evocative name—"punctuated equilibria"—and invested it with unwarranted dynamic characteristics.[3] The second trap was the temptation to retain their Darwinian faith. What was required was not just to recognize the flaws in the Darwinian structure but also to raze it to the ground in order to build a new theoretical structure capable of explaining the "recurrent patterns" in the fossil record. They refused to face the fact that "stasis" is the best evidence we have that natural selection is an illusion. By retaining the theory of natural selection, the naturalists have been forced into developing supporting arguments that are ad hoc, fanciful, and even metaphysical to justify this contradiction. Ironically this has only drawn attention to the deficiencies of Darwin's farmyard theory. Once the supporting arguments—the flying buttresses—fall away owing to their lack of reality, the whole structure collapses. What we are experiencing is nothing less than the collapse of Darwinism.

Chapter 6

The Cathedral-Reconstructionists:
Neo-Darwinism

The simple structure built by Darwin was relatively neglected for two generations after his death in 1882. There was even the threat of its demolition to make
way for an entirely new edifice based on Mendelian genetics, which was rediscovered by the scientific world around 1900. Some claimed that, now they had
discovered the source of mutation and replication, there was no further need for
Darwin's amateurish theories. This was real twentieth-century science, not crude
nineteenth-century experimentation with domesticated animals.

Once the euphoria had passed, however, population geneticists realized that
a knowledge of genetics could still not explain the transformation of species.
That required a dynamic theory of life. And at the time there were only two
theories of this nature that enjoyed substantial support. One was the idea of a
supreme creator, and the other was Darwin's theory of natural selection: two
great cathedrals standing side by side. For the twentieth-century scientist this
was not a real choice. What made Darwinism even more attractive to the new
geneticists was that it was unfinished and possessed a degree of ambiguity.
There was, in other words, considerable scope for reconstruction—of rebuilding
the Darwinian cathedral for their own very different purposes. In particular
Darwin's theory could not explain mutation and the process of biological inheritance, and there was a degree of ambiguity concerning the relationship between natural selection and sexual selection.

When the cathedral-reconstructionists closely examined Darwin's original
plans, they found themselves unhappy with some important architectural details.
In particular they felt that Darwin had been unduly influenced by the political
economists of his day. They were inclined to take a more sociological approach,
which emphasized sex rather than materialism. Not only was this more compatible with their views on population genetics, but Sigmund Freud and Carl Jung
rather than Adam Smith and Malthus were the popular prophets of their day.

On scrutinizing Darwin's theory, the cathedral-reconstructionists were re-
lieved to find that they were able to exploit the ambiguous relationship between

natural and sexual selection. This enabled them to seal off that major part of Darwin's original structure devoted to competition for scarce resources and to enlarge and embellish that minor part devoted to reproductive success to make it appear to be the main architectural feature of Darwinism. This would mean ignoring the junior partner in the architectural practice of Darwin and Wallace—Wallace had expressed reservations about Darwin's sexual selection hypothesis—and focusing on a minor interest of the senior partner. In this way the population geneticists were able to hijack the mantle of Darwinism. Henceforth the reconstructionists would become known as neo-Darwinists rather than neo-Wallacians, and they would rebuild Darwinism from genetic building blocks.

The Rise of Neo-Darwinism

To call the reconstructionists neo-Darwinists (or even ultra-Darwinists) is misleading. While the Darwinian mantle has provided the legitimacy that the population geneticists were looking for, it has also disguised the extent to which they have distorted Darwin's natural selection theory. The early reconstructionist challenge to Darwinism came from two sources on the continent of Europe: the German biologist August Weismann (1885), and the Austrian monk Gregor Mendel (1866).

Weismann challenged Darwin's Lamarckian theory of heredity called "pangenesis," propounded in *The Variation of Animals and Plants Under Domestication* (1868). He did this by conducting mutilation experiments (most famously by removing the tails of rats soon after birth) to discover whether these bodily modifications could be passed on to their progeny. While this is a rather naive test of Lamarckian theory—which requires organic change in response to the environment—it was taken seriously at the time. It led Weismann to formulate the idea that inheritance is confined to the "germ" cells—the ovaries and testes—and not to other parts of the body—or somatic cells—as Darwin had argued. The hypothesized barrier between somatic and germ cells became known as Weismann's barrier, and it has become part of the central dogma of modern biology. Richard Dawkins, for example, admits to being an "extreme Weismannist" and has offered to eat his hat if anyone is able to demonstrate that this part of the central dogma is not true. There are some scholars, such as the Australian biologists Steele, Lindley, and Blanden (1998), who are determined to see him do so. Whatever the outcome, it is a fact that Darwin's speculations about the sources of mutation were rejected by the reconstructionists soon after they were propounded.

Mendel's now famous experiments with the breeding of peas, largely ignored until the turn of the century, became the basis for the modern study of genetics, which was finally able to explain the source of, if not the reason for, Darwinian variations. The first two decades of the twentieth century witnessed a revolution in the understanding of heredity and mutation. With this the study of biology was transformed and placed on a more scientific basis. To many at the time, Darwin's farmyard thesis appeared to have been displaced by the scientific study of genetics. The dynamics of life, they thought, could be understood solely by an analysis of genetics.

It became apparent to the more perceptive population geneticists in the 1920s, however, that while a knowledge of genetics could clarify the nature of heredity it could not provide an understanding of the dynamics of life. Conversely, while Darwin did not understand the principles of heredity he did claim to possess a dynamic theory. In response to this impasse a number of American and British biologists including Ronald Fisher (1930), J. B. S. Haldane (1932), Sewall Wright (1931; 1932), and Ernst Mayr (1942) attempted to integrate Mendelian genetics and Darwinian dynamics. This was the beginning of the neo-Darwinian synthesis, which was further developed by the Russian-born biologist Theodosius Dobzhansky (1937; 1970), and the Englishman John Maynard Smith (1972). Neo-Darwinism, therefore, has international origins.

As the neo-Darwinists were geneticists of one type or another, this synthesis began to look like a microbiological takeover of Darwinian macroevolutionary dynamics. Their interest in the role played by genes in heredity led to a focus on Darwin's concept of "sexual selection" as if it were "natural selection." In turn this enabled the neo-Darwinists to develop what they saw as dynamic models of life by using genetic building blocks. In other words they attempted to link the dynamics of life exclusively to the genetic principles of heredity. This takeover of Darwinism by the geneticists appears to have begun with George Williams, who in 1966 argued that individuals attempted to maximize the survival of their own genes in the gene pool, and was driven to its logical conclusion by William Hamilton (1964), Edward Wilson (1975), Robert Trivers (1971; 1972; 1974), and Richard Dawkins (1976; 1982).

Neo-Darwinism, therefore, shifted the focus in evolutionary theory from Darwin's concept of the struggle by individuals for scarce natural resources, and hence for survival, to the concept of "reproductive success" at first of individuals and later of genes. Reproductive success was defined initially in terms of the survival of offspring (Fisher), then in terms of the survival of an individual's genes (Williams), next in terms of the survival of genes similar to those in the host individual no matter where they might reside (Hamilton), and finally in terms of a gene's efforts to propagate copies of itself in the gene pool by manipulating not only its host but other individuals as well (Dawkins). At the same time there was a shift in emphasis from evolution being for the good of the individual to being good for the gene. This was a shift in evolutionary focus from the selfish organism to the selfish gene (and the altruistic organism).

A central question in this chapter is whether the neo-Darwinian dynamic model constructed from genetic building blocks works any better than Darwin's theory of natural selection. The neo-Darwinian model of the dynamics of life can be characterized as follows.

- The motive force in life is provided by genes, which act "as if" they are striving to increase their influence in the gene pool. This is the essence of the "selfish gene." Why this should be so we are not told. Nor are we provided with any supporting evidence. In effect this *assumption* about gene "behavior" internalizes the *assumption* made by Darwin about the automatic tendency for all organisms to procreate geometrically. But it relocates it from the organism to the gene.

- To achieve this outcome, each individual is manipulated by some combination of its own genes and those hosted by other individuals to procreate in a manner that maximizes the presence of the controlling genes in the gene pool.

- Those individuals who are controlled by the fittest genes, and hence get to acquire the most profitable variations, are able to procreate most often and most effectively. These genes are carried by their successful propagating machines into the future of the species. The other genes are exterminated along with their luckless "lumbering robots." And in the process, the network of gene control over their hosts, other individuals, and physical infrastructure—the "extended phenotype"—is transformed.

This is without doubt the most bizarre dynamic model I have ever encountered. Not only does it exude an atmosphere of science fiction, it cannot be tested. How do we test whether genes act "as if" they are trying to increase their influence in the gene pool, whether genes are able to manipulate organisms, or whether genes are able to influence the physical world? Clearly we cannot. The neo-Darwinian dynamic model is, therefore, a metaphysical construct which, like the creationist model, must be accepted as a matter of faith. Little wonder the neo-Darwinists show little or no interest in the rise and fall of species. Their focus is on the underworld of gene activity, not the real world which the rest of us inhabit.

In the remainder of this chapter I focus on the way that Richard Dawkins has taken neo-Darwinism to its logical extreme, and consider its implications for Darwinism as a whole. And in the following two chapters I examine the way neo-Darwinism has been used in sociobiology to interpret animal and human behavior.

From the Selfish Organism to the Selfish Gene— Reductio ad Absurdum

The final touches to the neo-Darwinist reconstruction were added with enthusiasm and flair by the British ethologist (concerned with animal behavior) Richard Dawkins. In the mid-1970s Dawkins set out to take the neo-Darwinist version of life to its logical conclusion. By doing so he not only painted neo-Darwinism in brilliant highlights that have dazzled the expert and layperson alike, but he also demonstrated more clearly than any critic could exactly how bizarre that vision of life really is. The last quarter of the twentieth century, therefore, was the baroque period in the reconstruction of the cathedral of Darwinism—a time of extravagance and virtuosity before the fall.

Dawkins's main contribution to the final flourish of neo-Darwinism is found in two of his earliest books, *The Selfish Gene* (1976) and *The Extended Phenotype* (1982). More recently Dawkins (1998) has become an apologist for science in general as well as neo-Darwinism in particular. The basis for the early books is the conventional neo-Darwinist distortion of Darwin's theory of natural selection. Some sociobiologists such as Robert Trivers (1985) claimed that Darwin would have approved of the neo-Darwinist reconstruction because he regarded sexual selection as more important than natural selection. This is not correct. As we have seen, while Darwin recognized that natural selection was unable to encompass all forms of descent with modification, he was clear that sexual selection was only an "aid" to natural selection. Indeed, sexual selection was the first flying buttress of Darwinism. As I will show, the neo-Darwinists have attempted to transform natural selection from a materialist to a sociological theory.

As his point of departure, Dawkins used the work of J. Maynard Smith (1972) and W. D. Hamilton (1964). They cleared the way for the "selfish gene" by arguing that reproductive success should be thought of in terms of the survival of an individual's genes in a wider kinship group rather than of its offspring. This was Hamilton's concept of "inclusive fitness" which, by the late 1970s, was widely regarded as a more sophisticated measure of reproductive success. While Hamilton and other forward thinkers of the time still preferred to interpret inclusive fitness from the standpoint of the individual, it was but a short step to view it in terms of the gene. Richard Dawkins (1976) and Edward Wilson (1975) were the first to take that short step. What attracted so much attention was not the distance traveled but the panache with which this ground was covered.

The Selfish Gene (1976) is a tour de force of scientific explanation. It is just a pity that its basic thesis—the thesis of neo-Darwinism—is wrong. Dawkins, nevertheless, adopted this flawed idea and pushed it to its ultimate absurdity. What is remarkable is that, throughout this rhetorical triumph, Dawkins's faith in this exercise in reductio ad absurdum did not waver. The absurdity lies not in the lengths to which he took this idea—the step was a small one—but in the original concept of "inclusive fitness." We are invited by Dawkins to cheerfully accept a world in which the motive force and behavior of organisms is directed by and for the good of their DNA. Wallace was right to warn Darwin of the dangers in employing the sexual selection concept.

In *The Selfish Gene* Dawkins develops hypotheses about genes as "replicators," individuals as "survival machines" for the genes they contain, the ways in which genes attempt "to get more numerous in the gene pool" (in particular through Hamiltonian "kin selection" whereby those individuals containing copies of some of its own genes are favored), and the extension of the "replicator" concept to the world of ideas through the concept of "memes." Essentially this book, which I have discussed at length elsewhere (Snooks 1996), was intended to persuade us that we should switch our view of life from the selfish organism pursuing reproductive success to the selfish gene attempting to maximize copies of itself in the gene pool. As is well-known, *The Selfish Gene* was a great popular success and went a long way to achieving its objective, even making its way into the discipline's textbooks. Some, however, managed to resist the seductiveness of this final phase of cathedral-building. Most prominent among these are the paleontologists, who literally have their feet, and eyes, fixed firmly on the ground. Some of these even regard Dawkins's thesis as a form of intellectual madness.

The Extended Phenotype

But it is Dawkins's later book, *The Extended Phenotype* (1982), that I want to discuss in detail here. It is a more serious work than *The Selfish Gene* and the author himself regards it as containing his most original contribution to neo-Darwinian theory. Others might regard it as his most Gothic work.

Dawkins's Objectives

Dawkins's stated objective is to change our view of life from the standpoint of the individual to that of the gene. He argues that it is the gene and not the individual that maximizes reproductive success or inclusive fitness. The selfish gene does this by manipulating not only its host organism, or "survival machine," but also other organisms and their built structures ("artifacts"). This in a nutshell is the idea of the extended phenotype. While the conventional phenotype is a set of observable characteristics possessed by an organism and determined by its genotype and external environment, the extended phenotype includes characteristics of all organisms and their built structures that are manipulated by the genes in a given organism. Dawkins calls this "action at a distance."

At first glance this shift in focus from the individual to the gene and its network of biological and physical influences may appear to constitute a radical change in the neo-Darwinist position. It is not. All that Dawkins has done is to take the existing (1970s) neo-Darwinist argument to its logical conclusion, something that Dawkins openly admits. This is why his viewpoint was so quickly accepted by the profession. Radically different views take generations to gain acceptance.[1]

What status does the extended phenotype have in the world of science? Dawkins is quite candid on this issue:

> The vision of life that I advocate, and label with the name of the extended phe-
> notype, is not provably more correct than the orthodox view. . . . But I doubt
> that there is any experiment that could be done to prove my claim. . . . The ex-
> tended phenotype may not constitute a testable hypothesis in itself, but it so far
> changes the way we see animals and plants that it may cause us to think of test-
> able hypotheses that we would otherwise never have dreamed of. (Dawkins
> 1982: 1–2)

This is the status that the extended phenotype shares with all evolutionary theory. Being untestable, Darwinism occupies the realms of metaphysics—of what Karl Popper called the "psychology of knowledge" rather than the "logic of knowledge." Dawkins (1982: 181) is quite candid about this issue as well, admitting that "like God, natural selection is too big a theory to be proved or disproved by word-games. God and natural selection are, after all, the only two workable theories we have of why we exist." Of course, metaphysics can be endlessly fascinating, but it is a matter of faith rather than science. In the end, religion and Darwinism are two great cathedrals competing for the attention of the true believers. Science is not a matter of "word games" but of experiment and empirical verification/falsification.

Dawkins's Method

Dawkins' intellectual party trick is to pursue existing arguments to their logical conclusion. In the process it is remarkable what can be revealed about the original arguments, particularly what was unanticipated by their authors. While Dawkins believes that he has provided important insights into orthodox theory, unwittingly he has exposed the absurdity of the entire neo-Darwinian position.

Essentially Dawkins employs a deductive method that proceeds via a number of small logical steps from what is allegedly "known" to what is unknown. This method relies very heavily on the truth of the assumptions and the realism of the intellectual framework from which you begin. In a number of places in his book Dawkins (1982: 208, 227) says revealing things like: "the logic of the argument now seems to compel us to contemplate the possibility of . . . ," and "the concept . . . pushes our idea . . . out to its logical culmination." He also engages in what he calls "thought experiments" (rather than reality experiments) even when he rates them as "wildly improbable" (Dawkins 1982: 3).

His test of scholarly veracity is whether a writer employs "good [that is, logical] argument" rather than how realistic their argument might be (Dawkins 1982: 116). Dawkins even claims that his "belief in the inviolability of the central dogma is not a dogmatic one! It is based on reason" (Dawkins 1982: 174). Ironically, one of the dictionary (*OED*) meanings of "dogmatic" is: "based on *a priori* principles, not on induction"—that is, the deductive method employed by Dawkins! The setting forth of opinion (dogma) has always been associated with deductive argument (dogmatism) that makes no appeal to reality. Dawkins has merely trumped himself.

What Is the Extended Phenotype All About?

In the jargon of *The Extended Phenotype*, Dawkins wants to persuade his colleagues, together with the interested bystander, that life should be viewed from the standpoint of the "replicator" (gene or "meme") rather than the "vehicle" (organism, group, or society), and that the influence of the replicator should be thought of as the extended phenotype (various organisms and their artifacts subject to a gene's manipulations) rather than the conventional phenotype (host organism). He attempts to do so by "undermining the reader's confidence, in the central theorem of the selfish organism" (Dawkins 1982: 80) by developing a general theory about replicators—genes in biological life and memes in human society—by illustrating his selfish replicator argument in terms of the way replicators are supposed to manipulate both their host organism and other organisms (involving parasitic and symbiotic relationships), and by arguing that the evolutionary impact of replicators is experienced through the extended phenotype. Before beginning it is worth noting that *The Extended Phenotype* is not well structured, probably because it is a compromise between extending and defending his early work in *The Selfish Gene*.

Dawkins believes that the best way to undermine confidence in the central theorem of the selfish organism is to show how organisms allow themselves to be systematically manipulated. If some organisms "consistently work in the interests of other organisms rather than of themselves" this would amount to a "violation" of the central theorem. And it would open the door for the selfish gene. To demonstrate this possibility, Dawkins takes us through a number of examples of animal species (particularly of birds and plants) in which, he claims, some individuals work for the benefit of other individuals and, thereby, against their own interests. "Altruism" at the individual level, therefore, is a prerequisite for "selfishness" at the genetic level.

His main examples are the cuckoo bird and various types of ant colonies. It is well-known that the female cuckoo is able to avoid the effort of rearing her own young by laying eggs in the nest of another species of bird after removing the host's eggs. This is interpreted by Dawkins as the outcome of cuckoo genes manipulating the behavior of other birds to maximize the survival of copies of themselves at the expense of copies of the host's genes. It is, he argues using cold war jargon, the outcome of a genetic "arms race" between the cuckoo and its adversary. One species experiences adaptations that improve its ability to manipulate the other species, who in turn may experience adaptations that improve their resistance. In this case it is the cuckoos that have won the genetic arms race.

Dawkins also uses ants to illustrate his gene-manipulation thesis. We are told about some ant species in which queens have no workers of their own. These queens invade the nests of other ant species, kill the resident queens, and use the host workers to raise their alien young. In other species the workers respond to an invading queen by killing their own queen, who is also their own mother. Naturally this is difficult for a neo-Darwinist to explain, because it reduces the production and survival of copies of the workers' genes. Even Dawkins (1982: 71) refers to this as "genetic madness," and seems to be at a loss to explain it:

> Why do the workers do it? I am sorry I can do no more than, once again, vaguely talk about arms races. Any nervous system is vulnerable to manipulation by a clever-enough pharmacologist. (Dawkins 1982: 71)

In a third type of ant species, the queen remains in the original nest while her workers raid other nests to bring back the larvae and pupae of what in the fullness of time become slave workers.

How is this systematic manipulation of some individuals by the genes of others carried out? In the case of cuckoos, Dawkins talks about "egg mimicry," but for other birds and animals we are told that manipulation is exerted by one side entering the minds and nervous systems of the other side in a druglike way through songs and visual displays during courting, or through the cries of the young for food. And in ant species the invading queens somehow subvert the foreign workers through some sort of "mind control," possibly by chemical secretions. Or so we are told! Once again no evidence is or can be provided.

Dawkins discusses other examples of manipulation, but this is enough to demonstrate the type of argument he employs. While the intention is to undermine our confidence in the idea of the selfish organism, Dawkins merely undermines our confidence in the ability of neo-Darwinism to explain animal behavior at all. More will be said on this matter when we review the work of sociobiology in the next chapter. Only neo-Darwinists who had previously accepted the earlier "central theorem" of inclusive fitness are likely to embrace these strange explanations of animal behavior. They are certainly not shared by those biologists who have made a career out of examining the historical record—paleontologists such as Stephen Gould and Niles Eldredge and evolutionary geneticists such as Richard Lewontin and Steve Jones. And even less by the intelligent layperson. One has the feeling that Dawkins and his followers (for example, Dennett 1995) are only able to perpetrate this elaborate practical joke because so little is known about nonhuman species. As we shall see, these unrealistic ideas about "mind games" always unravel when they are applied seriously to human society, about which we know a great deal more.

The relationships between the organisms to which Dawkins refers can be explained quite simply in economic terms. There is no need to invoke complex explanations about "alien" control of our minds and bodies. What Dawkins describes is merely cooperative and uncooperative behavior that can be explained in terms of the dynamic-strategy model presented in part III. Individuals cooperate with each other to maximize their individual and joint material interests. This is not a case of "altruism" and has nothing to do with kin selection. One form of cooperation is symbiosis, in which both individuals and groups of individuals gain from the association. The matter of "anting," where birds allow ants to run through their feathers, is a case in point. Dawkins sees "anting" as an example of ant genes manipulating the behavior of birds. In reality, both sides appear to gain materially from the association. There is absolutely no need to invoke science fiction stories about mind and body invasions by unseen "aliens." In the dynamic-strategy model in part III, symbiosis is a dynamic strategy similar to that of commerce in human society.

Uncooperative activity often involves predatory behavior. In the dynamic-strategy model this is called the conquest strategy. It is a strategy that enables some individuals and societies to increase their probability of survival and prosperity by hijacking resources held by other individuals and societies. This is clearly the case with the ant examples employed by Dawkins. Ant species that take "slaves" are pursuing a conquest strategy, as are queens who "invade" other nests and kill the resident queen. In human society we are familiar with cases where citizens depose their own ruler in favor of a foreign monarch (for example, the "glorious revolution" in England in 1688 when William of Orange was invited by the English parliament to drive out their king, James II) who they are convinced will be better able to provide strategic leadership. That is what democratic elections are all about in modern human society.

Cuckoos also pursue a specialized form of the conquest strategy. They do not kill their host or take her captive; instead they make her an unwitting slave through deception. Once again there is no need to invent fantastic stories about alienlike genes, which even Dawkins is prepared to admit are not always very convincing. One always has the feeling he is less interested in reality than seeing how far he can bend our credulity by playing extreme games of a neo-Darwinian kind. The dynamic-strategy model, by contrast, can explain all the Dawkins examples much more simply, realistically and, hence, convincingly.

Having attempted to undermine our confidence in the idea of the selfish organism, Dawkins presents his argument for the selfish gene and its network of influences—the extended phenotype—in "the world at large." Essentially the argument is based on the role of the "replicator" and its special relationship with the neo-Darwinian version of natural selection (reproductive success). The replicator is the active force—the manipulator of minds, bodies, and artifacts—as well as the beneficiary of evolution. Organisms are merely expendable vehicles exploited by replicators.

It is an argument that depends critically on the truth not only of Darwin's original concept of natural selection but also of the neo-Darwinian reinterpretation. A successful challenge to either will cause Dawkins's argument to crash to the ground. Dawkins expects us to accept his extreme view of life on the grounds that it is self-evident. He expects us to transfer our allegiance from the organism to the gene because the latter is a "replicator" and the former is not. It is as simple as that. With Dawkins an appeal to the idea of the replicator is a substitute for both realistic evidence and argument. We are just expected to accept the special status of replicatorhood!

In biology the replicator is the gene, or what Dawkins calls the "active germ-line replicator." He claims that it is the unit of selection in the evolutionary process. Dawkins tells us that

> The reason active germ-line replicators are important units is that, wherever in the universe they may be found, they are likely to become the basis for natural selection and hence evolution. (Dawkins 1982: 84)

This replicator is "active" in that it can influence its own replication and it is "germ-line" as it will be the ancestor of an indefinitely long line of descendent replicators. It is, in other words, potentially immortal. In contrast, the organism is passive in that it is genetically programmed to operate for the benefit of the replicator and it is mortal. Like the advocates of religion, Dawkins rejects the ephemeral and places his faith in the eternal.

Not only is the replicator, or selfish gene, the active force in life, but it is also the primary beneficiary. Dawkins explains:

> When we ask *whose* survival they [the phenotypic effects] are adapted to ensure, the fundamental answer has to be not the group, nor the individual organism, but the relevant replicators themselves. (Dawkins 1982: 84; emphasis in original)

While we can only see the phenotypic effects as adaptations to survival, it is the selfish gene rather than the organism that gains most because the "successful" organism ultimately dies, whereas the "world tends automatically to become populated by germ-line replicators whose active phenotypic effects are such as to ensure their successful replication" (Dawkins 1982: 84). What Dawkins does not make clear, however, is that gene "immortality" is of a rather peculiar kind, because what is being passed on is not life but genetic information. Hence, it is genetic information rather than the selfish gene that is immortal, in the same way that it is Plato's *Republic* rather than Plato that lives on.

We also need to question whether it is the replicator that is the driving force in life. Dawkins is not very clear about this. While his use of the selfish gene metaphor suggests that it is the driving force, in his less rhetorical and more reflective moments Dawkins appears to contradict this impression. In considering the question of whether genes do actually "strive," Dawkins says:

> Genes manipulate the world *as if* striving to maximize their own survival. They do not really 'strive', but my point is that in this respect they do not differ from individuals. Neither individuals nor genes really strive to maximize anything. Or, rather, individuals may strive for something, but it will be a morsel of food, an attractive female, or a desirable territory, *not inclusive fitness*. (Dawkins 1982: 188–89; my emphasis)

This is very revealing. Dawkins admits to believing that neither genes nor individuals strive for inclusive fitness, but that individuals (but not genes) "may strive" for survival and consumption. It would seem that beneath all the misleading rhetoric we are in agreement: individuals, but not genes, "strive" to maximize not the multiplication of their genes but their own survival and prosperity. It is this "striving" by individuals for survival and prosperity that is the real driving force in life. It does not come from genes. Certainly Dawkins is unable to explain why genes act "as if" they are selfish replicators.

Why then does Dawkins grant special status to replicators? What distinguishes genes from all other forms of life? Dawkins appears to have no doubts, as can be seen from the following extract:

> The special status of genetic factors rather than non-genetic factors is deserved for one reason only: genetic factors replicate themselves, blemishes and all, but non-genetic factors do not. (Dawkins 1982: 99)

Accordingly life can be divided into two groups, the replicators and the vehicles. In the biological world, replicators are genes whereas vehicles can be either individuals, populations, or species. Dawkins is not fussed about what form the vehicle takes, because to him the central aspect of life is the replicator. Vehicles are only important because they are useful in facilitating the objectives of the replicators. He tells us that

> a vehicle is an entity in which replicators (genes and memes) travel about, an entity whose attributes are affected by the replicators inside it, an entity which

may be seen as a compound tool of replicator propagation. (Dawkins 1982:
112)

Further we are told that "the neo-Weismannist view of life which this
[Dawkins's] book advocates lays stress on the genetic replicator as a fundamental unit of explanation" (Dawkins 1982: 113). And, in this view of life, the
vehicle is merely "an integrated and coherent 'instrument of replicator preservation'" (Dawkins 1982: 114).

There can be no doubt that Dawkins has no interest in "vehicles," or individual organisms. Indeed, his primary objective is to undermine our interest in
and admiration for these "survival machines." "The main purpose of this book,"
he tells us (Dawkins 1982: 115), "is to draw attention to the weaknesses of the
whole vehicle concept," particularly as employed by the group-selectionists such
as V. C. Wynne-Edwards. It is Dawkins's intention to replace the orthodox focus on the vehicle with a new focus on a wider set of phenotypic characteristics—"the extended phenotype." For Dawkins (1982: 117) "the doctrine of the
extended phenotype is that the phenotypic effect of a gene (genetic replicator) is
best seen as an effect on the world at large, and only incidentally upon the individual organism—or any other vehicle—in which it happens to sit."

This brings us to the final section of Dawkins's magnum opus, which deals with
the theory of the extended phenotype. He regards these final four chapters not
only as the most important in the book but also the most original work he has
done or probably will ever do.[2] His method in this section is to take a series of
small logical steps to arrive at the extended phenotype. If you accept the initial
premise of neo-Darwinism the first step is not so hard to swallow, nor the second, and so on. While Dawkins views this method as persuasive and unassail-
able, I see it as leading us into total absurdity by degrees. While hiding behind
the acceptable idea of the selfish organism, the neo-Darwinian notion of individuals attempting to maximize reproductive success did not seem too unrealistic to the educated nonspecialist. But when the logical implications of that
worldview are made crystal clear, as Dawkins has done, the fantasy of the entire
enterprise is stripped bare for all to see. Dawkins exposes the fatal flaws in the
entire structure of neo-Darwinism.

What Dawkins does in these final chapters is to explain the concept of the
extended phenotype and then to apply it to some of the issues—namely the activities of parasites and symbionts—that had troubled him earlier. Briefly his
thesis is that genes attempt to maximize copies of themselves in the gene pool
by manipulating the behavior of both the host and neighboring organisms. This
behavior together with the physical structures that these organisms build are all
part of the extended phenotype. The examples provided of these "artifacts" are
caddis houses, beaver dams, and termite mounds.

In the case of beaver dams, the genes carried around by beavers are "selected" to manipulate the behavior of its host and neighboring beavers to build
dams that extend over several miles in order to increase access to food. Dawkins
(1982: 200) tells us that "the lake may be regarded as a huge extended phenotype, extending the foraging range of the beaver in a way which is somewhat

analogous to the web of a spider." He suggests that if beaver dams could be fossilized, beaver-lake paleontologists would discover evolution in their phenotypic characteristics such as increased size! This idea, he argues, is merely a logical extension of the conventional neo-Darwinian view that genes are responsible for the phenotype characteristics of the host organism.

The extended phenotype is then employed to cast light on the problem raised earlier in his book about the role of parasites and symbionts. Dawkins argues that the genes of a parasite work to manipulate the behavior of its host to promote the propagation of copies of the parasite's genes. This is achieved through both host behavior and the impact of the "artifacts" that it creates. The distinction he introduces between parasites and symbionts involves the degree to which their objectives overlap. This is defined in terms of the way the parasite leaves the host: at one extreme it involves the use of the host's "propagule" (reproductive particle) for its own reproduction, and at the other it involves the departure of the parasite "in the host's exhaled breath" (Dawkins 1982: 225). In other words, in the case of parasites and symbionts an extended phenotype is jointly manipulated, not necessarily cooperatively, by genes from distantly related individuals, individuals of different species, even different animal and plant kingdoms.

The lengths to which Dawkins is willing to take this argument is reflected in statements of the following representative type: "If fluke genes can be said to have phenotypic expression in a snail's body, there is no sensible reason why cuckoo genes should not be said to have phenotypic expression in a reed warbler's body" (Dawkins 1982: 227). In other words, the reed warbler's body (or snail's body) is part of the extended phenotype of the cuckoo's (or fluke's) genes, in the same way that a beaver dam is part of the extended phenotype of a beaver's genes. In each case, according to Dawkins, it is a matter of genes manipulating at a distance the world that you and I populate. He is even prepared to speculate that this "action at a distance" could, under special circumstances, extend beyond the few miles of a beaver's dam, even to a different continent.

When Dawkins wrote *The Extended Phenotype* about 1980, most field biologists, we are told, subscribed to Hamilton's view "that animals are expected to behave as if maximizing the survival chances of all the genes inside them" (Dawkins 1982: 248). This was the orthodox central theorem of inclusive fitness. It is Dawkins's claim that he has "amended" this outlook to create a new central theorem of the extended phenotype, which states that "an animal's behaviour tends to maximize the survival of the genes 'for' that behaviour, whether or not those genes happen to be in the body of the particular animal performing the behaviour" (Dawkins 1982: 248). The implication of Dawkins's new theorem is that "the individual performing the behaviour is not the entity for whose benefit the behaviour is an adaptation. Adaptations benefit the genetic replicators responsible for them, and only incidentally the individual organisms involved" (Dawkins 1982: 249).

But if organisms are merely vehicles, why have they developed such complex forms and behavior patterns? Dawkins responds to this inevitable question by arguing that it is the way genetic replicators are able to fashion better tools or instruments for survival. He reminds us that "an organism is the physical unit associated with one single life cycle," and that each cycle starts with a single cell. We are also told that "the fact that each cycle restarts in every generation from a single cell permits mutations to achieve radical evolutionary changes by going "back to the drawing board" of embryological engineering" (Dawkins 1982: 259, 264). Sexual reproduction enables this endless cycle to occur. Mere growth of existing life forms rather than reproduction would not generate systematic change.

Dawkins is, however, uneasy about this biological-complexity argument. He admits that it is not "a completely satisfying answer to the question of why there are large multicellular organisms" (Dawkins 1982: 263). For reasons to be discussed in part III, his unease is well-founded, because, in reality (rather than in neo-Darwinian fantasy), it is the organism that is both the driving force for, and the beneficiary of, the eternal strategic pursuit, and it is the "genetic replicator" that is but *one* of the instruments employed by the organism in this pursuit. The neo-Darwinians have perversely reversed reality.

Why complex organisms? Because they are better able to succeed in the strategic pursuit, as is demonstrated by the domination of this planet by mankind. Why organisms at all? Because only organisms have the desire and the ability to pursue dynamic strategies. Certainly genes cannot. Even Dawkins admits that in reality, as opposed to his word games, genes "do not strive." Why sex? Because it not only increases the number and range of "profitable variations" that can be used when necessary in the organism's strategic pursuit, but is central to the process of **strategic selection**. As we shall see, genetic change is one of the four dynamic strategies employed by organisms struggling to survive and prosper. All of this is explained in part III.

What, we need to ask, is the purpose behind Dawkins's quest for logical extremism? Could it be an attempt to overcome the obvious deficiency in the theory of natural selection—either the original Darwinian or distorted neo-Darwinian version—of being unable to explain the societal structure of animal populations, particularly of the human race? If so, it is a complete failure. In the theory of the extended phenotype we have a bizarre attempt to explain changes in "the world at large" through genetic "action at a distance." Any attempt to explain human society—the ultimate test for any theory of life—in terms of gene manipulation at a distance would end up resembling a science fiction story about body-snatching aliens who are able to infiltrate the minds and behavior of "earthlings." Could it be that science fiction is the logical outcome of the development path of neo-Darwinism? Certainly Dawkins's story line would be perfectly acceptable to science fiction publishers.[3]

What is certain is that neo-Darwinism amounts to a microbiological takeover of macroevolutionary dynamics. Whatever its failings, Darwinian dynamics takes a macroevolutionary approach, an approach that has been continued by the naturalists (or paleontologists) like Eldredge and Gould to this day. Neo-Darwinism, especially in the hands of extremists like Dawkins, is an attempt to construct a dynamic model of life from the microbiological building blocks of genetics. Life to the neo-Darwinist is a giant genome dominated by the biological replicator rather than a strategic pursuit dominated by the individual organism.

In this, modern biology is not unique. The same thing has happened in the social science discipline of economics. Macroeconomics was mapped out by John Maynard Keynes in *The General Theory* (1936) without reference to the prevailing neoclassical theory that was, and is, based on microeconomic production theory. From the 1950s there was an attempt by the neoclassicists to rework Keynesian macroeconomics by reinterpreting it in microeconomic terms. This was called the neoclassical synthesis, which led to the distortion and eventual rejection of Keynesian macroeconomics. In its place, microeconomic production theory was used, unsuccessfully—see my *Global Crisis Makers* (Snooks 2000)—to reinterpret macroeconomic relationships and to construct their totally unsuccessful "growth" models. In a fascinating parallel with modern evolutionary biology, neoclassical economists view human society as a giant factory dominated by the machine, rather than a strategic organization dominated by the dynamic strategist.

It is significant that in both biology and economics the technicians, highly trained in mathematics and statistics, take a reductionist approach to dynamics. They both seek the smallest unit possible to explain their respective life systems. Why? In my opinion it is because this is easier and generates quicker results (Snooks 1998b: 70–74; 1999: 107–14).

The Neo-Darwinian View of Social Systems

In part I it was shown that Darwin was unable to explain the dynamics of human society using the theory of natural selection. What we need to discover here is whether the neo-Darwinists can do any better with their very different evolutionary model.

The neo-Darwinists have offered a number of theoretical explanations for social interactions between individuals. These include kinship theory pioneered by W. D. Hamilton and the theories of the extended phenotype and of "memes" developed by Richard Dawkins. Can these theories explain the dynamics of human society and if not what does this imply for their value in explaining the dynamics of animal society? In this chapter I focus on the theories of Richard Dawkins and in the following two on those of Bill Hamilton and the sociobiologists (such as Edward Wilson).

It is interesting that Dawkins has anticipated this type of critical exercise. He has done so, rather lamely however, by claiming that while humans act in a way that clearly violates the central theorem of neo-Darwinism—in the form of homosexuality, contraception, falling birthrates in rich countries, child adoption, and so on—it is not a real issue because human society is "artificial." There is nothing unexpected or original here. This is the usual defensive approach employed by Darwinists when faced with the unaccommodating facts of human society (see also Jones 1999: 435–37). Dawkins, for example, asserts:

Adoption and contraception, like reading, mathematics, and stress-induced illness, are products of an animal that is living in an environment radically different from the one in which its genes were naturally selected. The question, about the adaptive significance of behaviour in an artificial world, should never have been put; and although a silly question may deserve a silly answer, it is wiser to give no answer at all and to explain why. (Dawkins 1982: 36)

In fact this is a very silly answer. All animals at some stage live in very different environments to those in which they emerged as new species. This is one of the reasons, according to Darwin and his followers, behind "descent with modification." In any case why should we regard human society as any more artificial than the society of other animal species? Civilization is something that has emerged from the strategic pursuit that we and all other forms of life have been engaged in from the very beginning. It is not something that has been imposed on life by some alien force. In particular it is something that has been constructed in response to the dynamic strategies we have been employing for the past 1.6 million years (myrs), not just the past few hundred thousand or even few thousand years (Snooks 1996; 1997). Even pre-urban hunter–gatherer societies practiced behavior—such as birth control through infanticide, adoption, and homosexuality—that violates the central theorem of neo-Darwinism. In what fundamental ways are these simple hunter–gatherer societies more "artificial" than or "radically different" to those of the early hominids or pongids? The silence is deafening!

Dawkins's answer is also rather silly in that it is a clear admission that neo-Darwinism is unable to explain the behavior of either modern mankind or the early hominids. This is a damaging admission from the supporters of such a widely accepted theory of life. Could it be that neo-Darwinism is no more relevant to nonhuman life than it is to human society? Could it be that we have been deluded by the cathedral-reconstructionists into thinking that nonhuman life is more like some alien form of life than our own? Could it be that evolutionary theory is more like science fiction than real science? Could it be that we have all been taken in by a good story well told?

Exactly what does the theory of the extended phenotype tell us about the dynamics of human society? Let us review the evolutionary troops. "Artifacts," part of the extended phenotype, could be interpreted as capital infrastructure and, possibly, even as institutions (laws and customs) and organizations, although Dawkins does not do so. His theory tells us that the interaction between individuals in society, together with society's infrastructure and, possibly, its institutions and organizations, are an outcome of genes acting as if they are striving to maximize copies of themselves in the gene pool. Those individuals and artifacts that survive are an outcome of natural selection operating at the level of the genetic replicator. Dawkins would say that, if human societies (not their ruins) could be fossilized (as he suggests for beaver dams), paleontologists would detect ecosociopolitical evolution.

Dawkins, however, does not extend the extended phenotype quite this far. I am just adopting his approach of taking accepted ideas to their logical conclusion. Surely Dawkins could not object. Surely he would agree that if we can accept the idea of the extended phenotype as an explanation of the evolution of nonhuman life forms and their artifacts, then it is but a small logical step to apply it to humanity and its civilizations as well.

But when we do extend the extended phenotype, what do we find? We find that it is impossible to test the theory—of genes "at a distance" manipulating the "world at large"—using historical evidence. To do this we would need to access evidence from the underworld of the selfish gene. Of course, Dawkins recognizes this problem and, as we have seen, claims to be comfortable with an untestable and hence metaphysical theory. If the extended-Dawkins approach was generally accepted what would it imply for schools of life (both biology and social) sciences? They would become de facto schools of theology. But then I know some philosophers and their fellow-travelers in my own school of social sciences who are making that fruitless journey.

Even if we switch back to older versions of neo-Darwinism in which selfish organisms attempt to maximize their "inclusive fitness" or just their own genes, or even just their own offspring, what can we make of human society? The answer is very little. As we shall see in the next chapter, the inclusive fitness idea attempts to explain relationships between individuals in terms of whether or not—or the extent to which—they share copies of some of the same genes. I show that this is not a satisfactory explanation of our complex societies based as they are on interactions between not only family members but also friends, associates, enemies, and even total strangers. And the older and "cruder" concept of organisms attempting to maximize the number of their offspring has never been true for human society even in its earliest and simplest forms. Right from the beginning, and probably before that, our species has subjected its fertility to the control of its dynamic strategies. Like Darwin's original theory of natural selection—for which at least a superficial case can be made—the modern theory of evolution cannot explain the dynamics of human society. It cannot even explain the behavior of humans at the individual level. What chance, therefore, does it really have of explaining the dynamics of any other species? None. As I will show in the next chapter.

The other neo-Darwinian theory that some might expect to have relevance to at least aspects of our human story is Dawkins's general theory of the replicator. As is well-known, Dawkins has extended his ideas about germ-line or genetic replicators to a general theory about all replicators. The replicator that he believes has relevance to human society is human "ideas," which he calls "memes." A meme, we are told, is "a unit of information residing in the brain." It can be a concept, a tune, a poem, a design in clothes (or anything else), or "ways of making pots or building arches" (Dawkins 1989: 192). Dawkins explains his meme theory as follows:

> Just as genes propagate themselves in the gene pool by leaping from body to body via sperm or eggs, so memes propagate themselves in the meme pool by leaping from brain to brain via a process which, in the broad sense, can be called imitation. . . . When you plant a fertile meme in my mind you literally parasitize my brain, turning it into a vehicle for the meme's propagation in just

the way that a virus may parasitize the genetic mechanism of a host cell. (Dawkins 1989: 192)

But, unlike genes, meme copies are rarely literally faithful to the original. There is considerable copying error.

Dawkins asserts, without any evidence, that the selection process among memes is Darwinian. He tells us that "those memes that spread do so because they are good at spreading" (Dawkins 1998: 304). And as a result, the world becomes filled with those memes that are good at getting themselves copied in brain after brain. This is because the meme is "catchy" or "trendy" or for some other reason ingratiates itself into our minds.

Like genes, memes are "active replicators" and they are able to "manipulate the behaviour of living bodies" (Dawkins 1998: 306). And like genes they occupy a replicator pool which is run along the lines of a cooperative! Dawkins explains:

> Just as a species gene pool becomes a cooperative cartel of genes, so a group of minds—a 'culture', a 'tradition'—becomes a cooperative cartel of memes, a memeplex, as it has been called . . . of mutually assisting memes, each providing an environment which favours the others. Whatever may be the limitations of the meme theory, I think this one point, that a culture or a tradition, a religion or a political complexion grows up according to the model of 'the selfish cooperator' is probably at least an important part of the truth. (Dawkins 1998: 306)

At the end of the 1990s, Dawkins appears less cautious than twenty years earlier about the value of his meme theory in explaining human civilization. In *The Extended Phenotype* (1982: 112) he said of his meme theory: "my own feeling is that its main value may lie not so much in helping us to understand human culture as in sharpening our perception of genetic natural selection." Which Richard Dawkins is correct? The answer is neither. The meme theory can tell us nothing about that dynamics of human society, because our culture is driven not from the supply side by self-propelled "ideas" but from the demand side by the unfolding dynamic strategies that are driven by the **strategic desire** of mankind for survival and prosperity. And this failure as a model of societal change exposes the flaws in the neo-Darwinian theory of life. For what is wrong with the meme-replicator theory of human society is wrong with the gene-replicator theory of nature. The rest of his tribe will come to regret his unwitting exposé.

Essentially Dawkins's general replicator story is a supply-side theory, which asserts that ideas are responsible for driving ecosociopolitical structures. Ideas compete with themselves in the brains of all individuals in society. This is made possible by the well-known fact that humans (as well as animals) engage in mimicry. It is a matter discussed in biological circles at least since Alfred Wallace (1871: chs. 3, 6) and in the social sciences since Herbert Spencer (1851) and Joseph Schumpeter (1912). But in contrast to these earlier theories, Dawkins claims that it is the ideas themselves rather than the individuals that are responsible for meme propagation. Individuals are only vehicles for these active replicators. Those ideas that survive are, in Darwinian fashion, those that are best at leaping from brain to brain. And these successful memes manipulate the minds and behavior of the individuals they invade.

Once again, in evaluating this extension of neo-Darwinian theory, we encounter the problem of evidence. Clearly the meme theory is not *directly* testable. How could we test the hypotheses that ideas compete to propagate themselves; that ideas are responsible for leaping from brain to brain, transforming them as they go; that memes are responsible for manipulating the behavior of individuals to ensure their (the memes') survival; or that memes are responsible for actively shaping our ecosociopolitical institutions and organizations? On the other hand, this hypothesis about mind-snatchers does seem to defy common sense—to defy what we know about ourselves and our society. Once again Dawkins leaves us with the feeling that we have read some fairly lively and ingenious science fiction in which unseen alien beings are taking over our minds and bodies. While some might believe this theory about aliens when it comes to other animals, few are likely to believe it about ourselves. Those that do have real problems in perceiving reality. Ultimately it is at the level of belief that we are asked to accept the meme theory.

But it must not be left at the level of metaphysics. We possess a massive body of reliable information about human society that cannot be ignored as Dawkins and his followers have done. Here is the challenge that they must face before writing further speculative works on meme theory: Use it to explain the details of, for example, the rise and fall of ancient Greek civilization or, more specifically, the rise and fall of Athenian democracy or of Greek religion; or, more topically, explain the rise and fall of the U.S.S.R., or the transition of East Asia from "miracle" to "meltdown" and back again. A general theory of life must be able to explain all these changes, and more. Impossible? Not at all. The dynamic-strategy theory presented in part III can do so (Snooks 1996; 1997; 1998a).

While it is not possible to test meme theory directly, we can do so indirectly by empirically establishing a diametrically opposite *realist* theory. The dynamic-strategy theory tells us that in the long-run ideas do not propagate themselves but are a response to **strategic demand**. It is society's unfolding dynamic strategy, driven by pioneering strategists (that is, organisms) who are exploiting available materialist opportunities in the pursuit of survival and prosperity, that generates a strategic demand for a range of essential inputs including ideas of all types. Those individuals that follow the pioneers adopt ideas (among other strategic instruments) not because these ideas are inherently successful or compatible but because they are employed by successful pioneers. What is being imitated is not the success of the ideas—in any case they are only a small part of a raft of strategic instruments—but the success of the pioneering strategists (that is, organisms). This is a demand-side theory that focuses on individuals driven by strategic desire, and it is a theory that has been verified in considerable detail. It is also a theory capable of falsification—and hence is not metaphysical—and it has already survived numerous tests of falsification.

This is all very well at the societal level, you might say, but what about the rise and fall of scientific ideas, or of fashions in clothes or popular music? Surely these are not the outcome of a supply response to changes in strategic demand? My response is that in reality they are.

First, take the case of the development of scientific ideas. Young scientists choose their area of research not by adopting successful ideas that are supposedly leaping from mind to mind like a virus but rather by imitating the work of successful and eminent scientists in their fields. It is for this reason that scientists form schools that take their lead from famous scholars and call themselves Darwinists, Weismannists, Lamarckians, Marxists, Keynesians, and so on. The successful pioneers in any scientific field are busily exploring the opportunities inherent in their intellectual paradigm, which are an outcome of society's unfolding dynamic strategy. Research staff and young graduates respond to this strategic demand by supplying those ideas, usually variations on existing themes, that generate research funds, Ph.D. scholarships, postgraduate scholarships and, ultimately, research jobs.

This argument contrasts with Dawkins's supply-side explanation that successful scientific ideas are those that are most compatible with existing ideas. Clearly his hypothesis cannot explain those ideas that generate new intellectual paradigms, because they are incompatible with conventional ideas. Conversely the dynamic-strategy theory can explain intellectual paradigm shifts: they are the outcome of major changes in strategic demand. They are the ideas of genius that are later embellished by lesser minds.

The argument is similar in the case of popular music or clothes fashions. Successful tunes and designs do not enter our heads of their own accord, nor do they manipulate our behavior. Rather, because the followers in society want to be associated with material success, even if only vicariously, they imitate not the fashions per se but the habits and lifestyles of rich and famous *individuals*. Of course, catchy tunes and trendy designs may temporarily take our fancy, but we will only persist with them if they continue to be associated with the "movers and shakers." Needless to say only a few of the followers will achieve the success of their heroes, but the rest will at least derive a vicarious satisfaction from doing some of the things that rich and famous *individuals* do.

Conclusions

In its early form, neo-Darwinism—a synthesis between Mendelian genetics and Darwinian sexual selection—was able to disguise not only the extent to which it had distorted Darwin's farmyard concept of natural selection but, more importantly, the absurdity of its central theorem. The plot began to unravel in the mid-1960s and early 1970s when Hamilton and Trivers set off to explore the logical implications of the discipline. Yet the full extent of the neo-Darwinist folly was not clear because of the technical nature of their work and their focus on the organism despite their insistence that it was acting in the interests of genes similar to its own wherever they were to be found.

It was not until Richard Dawkins pursued the neo-Darwinian view of life to its ultimate conclusion in the mid-1970s and early 1980s that the full extent of its inherent absurdity was fully exposed: of genes establishing networks of influence by which they manipulated not only their host organism but also other organisms and their "artifacts." The story he tells is more akin to science fiction than science reality, in which unseen aliens infiltrate the minds and bodies of human and nonhuman organisms in order to perpetuate their influence. In the process, mankind is liberated from the dominance of God only to be delivered into the hands of another metaphysical being called the replicator. It is as if the Renaissance had never been. Of course this is not why we reject this neo-Darwinian absurdity: we reject it because it is totally unrealistic. Part III will show that there is a nonevolutionary theory that can not only explain the dynamics of life but also reinstate mankind's dignity.

Chapter 7

The Image-Makers and Evangelists I: Sociobiology and the Animal World

The sociobiologists are the image-makers in the cathedral of neo-Darwinism. It is their role to embellish the structure built by the neo-Darwinian architects. They translate the Darwinian gospel into the visual stories that adorn the walls and ceilings of this great building. They tell a simple story about the behavior of animals in the natural world, the emergence of mankind, and the "evolution" of human culture. They attempt to translate the mystical abstractions of the cathedral-reconstructionists into a medium that the people can understand and with which they can empathize. They can also be seen plastering over the growing cracks in the cathedral walls and ceilings.

Some sociobiologists are also evangelists. Taking their own stories seriously, they are driven to be missionaries. They try to persuade the people of the literal truth of their simple stories. In doing so they make the neo-Darwinian theology more palatable to the common taste. They even claim that their gospel can integrate all aspects of the penitent's life. If only we have faith. In this chapter we will examine the way sociobiology treats the animal world and, in the following chapter, the world of mankind.

Sociobiology, which is an applied branch of neo-Darwinism, has two central objectives and two methods of achieving them. In the past it has focused largely on explaining animal behavior using the relatively recent concept of "kin selection." Increasingly, however, attention is being devoted to explaining the nature of human culture using the new concept of "coevolution," which involves the alleged interaction between genes and culture under the shaping influence of natural selection. The latter is meant to reverse the failures of earlier attempts by Darwin and neo-Darwinists to explain the dynamics of human society. Animal behavior is discussed in this chapter and human culture in the next.

Sociobiology's Model of Animal Behavior

In outlining the sociobiological model of animal behavior we need to consider the forces underlying their motivation and decisionmaking. Individuals in this model do not make choices freely within existing environmental constraints, such as natural resource or budget limitations, but are genetically programmed, or "selected," to act as they do, following the preordained objective of maximizing reproductive success.

Motivation

Sociobiologists begin with the assumption that organisms attempt to maximize their "reproductive success." This involves, as we have seen, an attempt to maximize the survival of their genes, or, as Richard Dawkins (1989: 19) has put it, as an attempt by the immortal genes locked within "gigantic lumbering robots" to maximize their presence in the gene pool. Why? Because they are "selected" to do so. It is an outcome of the evolutionary process.

The credo of the sociobiologists is that life is basically social rather than economic and that evolution is a sociobiological rather than a materialist process. In this way they reinterpret Darwin's theory of natural selection to be a process of "differential reproductive success" rather than a highly competitive struggle for scarce resources. By a subtle juggling of concepts, the very substance of Darwinism has been discarded. Only his name has been retained to provide the rationale and authority that sociobiologists crave.

Robert Trivers, one of the pioneers of sociobiology, asserts that

> In Darwin's view, evolution resulted from two factors: heritable variation and differential reproductive success. By *heritable variation*, Darwin meant that in each living species there is variability, some of which is inherited by the offspring. By *differential reproductive* success, Darwin meant that in each species some individuals leave many surviving offspring, some leave few, and some leave none at all. If these two components are coupled, we are likely to see changes in the heritable constitution of a species. (Trivers 1985: 12; emphasis in original)

But what is it that determines "differential reproductive success"? Trivers (1985: 13–15) claims that there are two explanatory factors: "mortality selection" and "sexual selection." By mortality selection (a term—as with reproductive success—not used by Darwin) Trivers means prereproductive mortality caused by high rates of fertility and population growth that generates "intense competition to survive, a competition that would result in non-random differential mortality." Those with a heritable (genetic) advantage survive, the rest do not. By "sexual selection" (a term, as we have seen, that Darwin did use) Trivers means intrasexual competition based on differences in fecundity and in the ability to gain access to members of the opposite sex. He completes his critique by asserting that "so important did Darwin consider intrasexual competition that he called it by a separate name, *sexual selection.*"

It is on this type of critique of Darwin that the sociobiologists base their belief that individuals strive to maximize their reproductive success. As we know, this is a more widely acceptable form of the neo-Darwinian argument that genes, by manipulating organisms and their artifacts, attempt to maximize their presence in the gene pool. In turn, this critique has given rise to the further proposition that social organization in the animal world is based on the costs and benefits of reproductive success. Sociobiologists believe, therefore, that the principal outcome of life on Earth—population increase—has become the sole "motivation" for animals. It is, of course, a motivation over which they have no control. They are "selected" to do what they do. As Trivers (1985: 20) says: "a shorthand way of describing natural selection is to say that it favors the individual that leaves the most surviving offspring, or favors the traits that permit an individual to leave the most surviving offspring."

As can be seen from part I this critique bears little resemblance to Darwin's theory of natural selection. Sociobiologists have not only changed the functional definition of natural selection to *include* sexual selection—which Darwin kept separate as a default mechanism—but have reversed the importance of both concepts. Darwin's natural selection is demoted to "mortality selection" and, in the process, becomes less important than "sexual selection," which is redefined to include fecundity as well as male access to females—redefined to become something quite new, namely "reproductive success."

Part I shows that Darwin used the separate name "sexual selection," not because he thought it "so important" but because he thought it less important, "less rigorous," more partial in its impact, and merely an "aid" to his key concept of natural selection, which was a materialist process of struggle for scarce resources. He never attempted to integrate sexual selection into a general theory of evolution. And the codiscoverer of natural selection, Alfred Wallace, argued that the "unimportant" issues identified by Darwin as candidates for sexual selection could actually be better explained by their concept of natural selection. In essence, the sociobiologists have distorted Darwin's evolution concept by replacing his materialism—the struggle for scarce resources—with modern sociology—the interaction, largely sexual (and hence genetic), between gene-directed organisms. The influence of Adam Smith and Malthus, therefore, has been replaced by that of Sigmund Freud (1856–1939) and Carl Jung (1875–1961).

There is no extended discussion of animal motivation in the sociobiological literature. It is taken as given. Individuals do what they are "selected" to do. Trivers, for example, tells us:

> Instead of a disorganized list of items that we may care to invest ourselves in, such as children, leisure time, sexual enjoyment, food, friendship, and so on, Darwin's theory says that all these activities are expected to be organized eventually toward the production of surviving offspring. (Trivers 1985: 21)

While Trivers talks in terms of maximizing the number of offspring, we should remember that this is code, by someone who is supposed to be explaining the behavior of animals—organisms we recognize and love—for genes manipulating organisms to maximize their presence in the gene pool. Edward Wilson, the other pioneering sociobiologist, is more forthright in this matter, asserting that

> In a Darwinist sense the organism does not live for itself. Its primary function
> is not even to reproduce other organisms; it reproduces genes, and it serves as
> their temporary carrier. Each organism generated by sexual reproduction is a
> unique, accidental subset of all the genes constituting the species. (Wilson
> 1975: 3)

Interestingly this was published a year before Dawkins's *The Selfish Gene*. Clearly Wilson was preempted by his more flamboyant English colleague.

Having stripped Darwinism of its exogenous materialist driving force—the life and death struggle for scarce resources—and replaced it with a passive genetic sorting device, sociobiologists like Robert Trivers (1985: 19) are forced to consider whether "selection was too weak a force to bring about substantial evolutionary change." To provide his devitalized evolutionary system with some credibility, Trivers introduces the concept of "intensity of selection," which is inversely related to the time required for genetic change. The weaker the selection pressure, the greater the time needed to generate a given amount of genetic change. "The intensity of selection," he tells us, "is a function of the variability in reproductive success. Where variability is high, selection is strong" (Trivers 1985: 24).

The variability in reproductive success in the sociobiological model depends on the social behavior of animals. In herring gulls, for instance, it involves the distance between nests, the dates of egglaying, whether camouflage techniques are used, and so on. And social behavior is determined by genetic variation. This is, therefore, a circular system in which genetic change is alleged to be driven by genetic variation, while the "intensity of selection" is no more than a measure of the range of genetic variation in a species.

No attempt has been made in sociobiology to really test the central assumption that individuals in all species are "selected" to maximize the number of surviving offspring, let alone to maximize copies of their genes in the gene pool. Instead, innumerable case studies of animal populations have been employed merely to *illustrate* its plausibility. As we shall see, the evidence marshaled by the sociobiologists can be better explained by the materialist model presented in part III. Some of these data even critically challenge the sociobiological model.

To illustrate the credibility gap for sociobiology I will briefly review a number of the main behavioral issues in this discipline, including parent–offspring conflict, parental investment, gender mortality rates, and female choice. The conventional approach in sociobiology is to examine these issues in terms of the costs and benefits involved in reproductive success. These issues are of widespread interest to both the natural and social sciences. My objective in reviewing them is to provide (in part III) a more realistic model of animal behavior than is currently available.

First, we consider parent–offspring conflict. It is well-known to observers of animal behavior that parents, particularly females, invest considerable time and effort in their offspring. This, sociobiologists argue, is an outcome of their genetic structure, which has evolved under natural selection over very long periods of time. Robert Trivers (1985: 148), who pioneered work on this subject, tells us: "The parent has been selected to invest in its offspring in such a way as to maximize the number eventually surviving." This investment is interpreted not in economic terms involving *material* benefits and costs, but in biological terms involving *reproductive* benefits and costs. In the sociobiological model, parental investment generates a benefit in terms of the increased probability of survival of present offspring, but also involves a cost in terms of the opportunity forgone to raise additional offspring. Also in contrast to economic cost–benefit calculations, sociobiology does not allow for freedom of individual choice. "Choice" is genetically determined. Trivers attempts to explain:

> The parent is naturally *selected* to maximize the difference between the benefit and the cost. In particular, it is *selected* to avoid any investment in the offspring for which the cost is greater than the benefit, since such investment would decrease the total number of its offspring surviving. (Trivers 1985: 148; my emphasis)

While the content is different, the language of sociobiology is the same as that of economics. The influence on sociobiology is obvious.

Unsurprisingly, offspring have a different objective to that of their parents. They are "selected" to encourage parents to invest more than the parents are "selected to give." This produces a conflict between parents and offspring over the amount of parental investment. Offspring are, accordingly to this theory, "selected" to use psychological manipulation—such as temper tantrums in chimpanzee infants and convulsions in pelican chicks—to induce greater investment from parents. Trivers (1985: 149) explicitly rejects any idea that conflict is based on any clash of self-interest—a sure indicator that real choice is involved: "conflict does not occur because of the innate selfishness of the offspring. Indeed, it would be just as accurate to say that conflict occurs because of the innate selfishness of the parent." Indeed!

This conclusion would seem to defy the commonsense position that, indeed, *both* parents and offspring pursue self-interest. Nevertheless, Trivers believes that all the studies he describes concerning parent–offspring conflict—monkeys, baboons, chimpanzees, pelicans, and herring gulls—support his genetic hypothesis of behavior rather than any alternative hypothesis. But he does not attempt to refute his beliefs or test the alternative materialist hypothesis.

The second issue concerns parental investment. Sociobiological studies of parental investment in offspring usually focus on female investment. As Trivers (1985: 237) says: "In general, males invest little or nothing parentally in their offspring, but instead compete among themselves for access to females." Why, if males are "selected" to maximize their reproductive success, do they spend so much time fighting and so little time helping their offspring to survive? This obvious question is ignored. We are only told that males are "selected" to maximize the spread of their genes by competing with each other for access to the females.

Once again a variety of animal studies, including pigeons, seals, red deer, reptiles, and a variety of insects are employed to *illustrate* the way the reproductive-success hypothesis is supposed to work in nature. Yet Trivers does not consider whether this evidence provides a more persuasive case for a very different argument that the males are competing with each other not directly for access to the females but for territory—scarce natural resources—and only then to gratify their other appetites, including sexual desire. Could this be why the males have developed weapons of offense and defense? Could they be in pursuit of conquest rather than sex? Sex, of course, is one of the rewards of the conquest strategy, but not its primary objective. Trivers fails to consider this possibility.

As the male of the species takes little or no interest in what happens to his offspring, except occasionally to devour them, the maximization of the number of his progeny does not appear to be high on his list of desires. It is curious that in his application of the neo-Darwinian hypothesis, Trivers (1985: 357–58) does not appear to recognize that the high positive correlation between longevity and the size of harems in red deer can be explained by the competing materialist model—as a function of the success in capturing and defending scarce natural resources.

Sociobiologists have particular problems in explaining the sexual behavior of individuals in the minority of species in which males do invest significantly in the raising of their offspring. Human males join a select and strange company of species—sea spiders, butterflies, birds, and wolves—in which males invest in their offspring. Curiously this is not something we share with our closest relatives, the great apes, nor with the monkeys closest to them. And, we are told, where there is bonding of a sort between males and females in baboons or chimpanzees, it involves "trading, in effect, grooming and protection from others for increased sexual access" (Trivers 1985: 239). Surely this can be explained more persuasively by the materialist hypothesis than the reproductive-success hypothesis. Unperturbed, Trivers attempts to explain the contrast between the parental investment of human males and great ape males in the following curious way: "human male parental investment is an example of convergent evolution." Is it not more likely to be the result of a self-interested response to different economic circumstances?

The third issue is the "choice" of sexual partners. Sociobiology runs into acute difficulties over the matter of male selection in species where males invest in their offspring. The main case studies concern bird behavior, and the most interesting of these are where females "choose" males on the basis of the quality of their nests and their physical attractiveness. Where male birds build nests in order to attract the opposite sex, as in the case of weaver birds, the females carefully inspect the nest and either accept or reject its maker. If rejected, the male bird demolishes the offending nest and builds another. Sociobiologists assert that each female is "selected" to "choose" a male who is most likely to invest in her offspring. Trivers (1985: 252), for example, claims that "females at the time of pairing may be able to choose males who will later strongly invest." But exactly what is the relationship between the nest-building skills of males and the ability of females to maximize the survival opportunities of their offspring. (Or to put it in Dawkinsese: How are the genes in the yet to be born offspring able to manipulate its future parents and their artifacts in order to maximize their future influence? Surely this is just a small logical step in extending the extended phenotype!) It appears very tenuous. Trivers (1985: 254) appears to recognize the problem when he says that "even when females choose according to the ability to invest, they may also choose according to genetic criteria." The materialist interpretation of specialization and exchange of services appears more convincing.

A similar problem exists in the case of female birds, such as finches, that "seek out" attractive males. Trivers (1985: 259) expresses some uncertainty when trying to explain why this might be so: "Our best guess is that physical attractiveness usually correlates in nature with superior genes. Just how this is true in any given case is usually a mystery." It is a problem compounded by the evidence that female birds with an attractive mate are willing to shoulder a higher proportion of the workload in raising their young and to turn a blind eye to his sexual involvement with other females. Trivers's (1985: 260) explanation is particularly unconvincing: "If attractiveness indicates genetic quality, then each individual will want to match higher investment to higher genetic quality." This is a rather muddled *economic* incantation!

Far more convincing is the cogent materialist hypothesis that female finches seek out attractive males who excite them sexually and who appear to be able to offer protection and material comfort (a stronger companion, nest, and food) in return for sexual favors. (As we have seen, Trivers is prepared to use this type of argument in the case of primates.) It is not surprising that, when free to choose, female finches invariably select healthy, attractive males who are likely to possess the qualities they require to satisfy both their desire to survive and their other appetites. In part III I argue that male selection is central to the success of their dynamic strategy.

Nor is it surprising that females with unusually attractive mates take on a greater share of family work. These females need to offer greater inducements to persuade their mates to stay when they are much in demand elsewhere. In other words they need to compete for male attention just as in other cases it is recognized that males compete for female attention. What is involved is an active trade in the goods and services required to satisfy the primary needs and dynamic strategies of individual animals. It is absurd, as the Dawkinsesque example above suggests, to claim that this evidence has anything to do with investment in the genetic quality of a female's future offspring. Genetic quality of offspring is just a fortuitous outcome of the health (and hence attractiveness), vigor, and other physical and instinctive attributes required for individuals to survive the intense competition for scarce resources by pursuing their dynamic strategies. While the evidence of animal behavior directly supports the common-sense materialist hypothesis, it cannot be used to demonstrate the fanciful neo-

Darwinian idea that genes—even of unborn offspring!—are attempting to maximize their presence in the gene pool. The sociobiological view of the world is not only metaphysical; it reads like science fiction.

Finally, the issue of differential gender mortality provides a direct confrontation between the reproductive and materialist hypotheses. It is well-known that the mortality rate in humans is higher for males than females from conception to old age. While the male/female "population" ratio is at least as high as 120:100 three months after conception, by the time of birth it has fallen to 106:100. Thereafter the annual death rate is substantially higher for males than for females, so that by seventy years of age the male/female ratio is 84:100. This phenomenon is also common in other animal species. Why? The answer is not directly related to our chromosomes but is the outcome of male hormones: castrated male mammals have the same lifespan as female mammals.

This simple fact has not deterred sociobiologists from insisting that differences in gender mortality can be explained in terms of the genetically based reproductive-success hypothesis. According to Trivers:

> For species with negligible male parental investment the key predisposing factor is the high potential reproductive success of a male. Imagine that an animal is trading mortality rate prior to adulthood for an increase in its reproductive success at adulthood, assuming survival. Such an animal will be selected to suffer a one-half reduction in survival if this is associated with more than double the reproductive success in adulthood. . . . In effect males pursue a high-risk–high-gain strategy, and the high potential gain selects for any traits that give the gene, even at a cost of high mortality, as long as this mortality is not so great that it cancels the gain. (Trivers 1985: 311–12)

But, he admits, there are no data to test this very strange hypothesis. Can there be any doubt that sociobiology inhabits the realm of metaphysics? And of fantasy: would you knowingly trade years of life for a greater survival rate for your genes? Here is further evidence of the absurdity of neo-Darwinism.

A more sensible and testable hypothesis is that of materialism presented in part III. Owing to the nature of their hormones, males (and their sexual cells) exhibit much greater vigor and aggression than do females. This leads both to higher ratios of males to females at conception and to higher death rates for males—because of more reckless behavior, physical conflict, stress, and suicide—throughout life (Snooks 1997: 106–7). Curiously, even sociobiologists admit that higher male death rates are a direct response not to their chromosomes but rather to hormones. Clearly the average individual who has any choice in the matter is not going to trade years of life for reproductive success. It would violate the very force that has driven life for the past three thousand five hundred myrs—the strategic desire to survive and prosper. In human society there is abundant evidence that, whenever there is a direct choice between life (or even the quality of life) and children, families take steps to limit the number of children through contraception, abortion, and/or infanticide (Snooks 1999: ch. 13). The same, as we have seen, is clearly the case with animal life.[1]

The Model of "Choice"

In sociobiology there is no effective model of choice. At least not in the same sense as is normally employed in the social sciences, or the everyday world, where individuals operating within the usual environmental constraints are able to choose between a variety of alternatives. "Choice" for the sociobiologist is something that individuals make because they are "selected" to do so. While a number of examples of this view of choice have already been given, one more will focus our attention on this choiceless form of "choice." Trivers (1985: 353) asks us to "imagine a gene in a female leading her to choose a male whose genes improve her daughters' chance of surviving by 10%." (No doubt many will find it extremely difficult to image such a strange event.) What we are being told is that animals in the sociobiological model are genetically programmed to behave the way they do. They have no real choice. Sociobiology, in applying the theory of neo-Darwinism, has constructed a world of robotic inhabitants.

When attempting to reach a lay audience, Richard Dawkins in *The Selfish Gene* deals with the relationship between genetic structure and animal behavior in a more explicit way than most sociobiologists. The following extract illustrates my point:

> animal behaviour, altruistic or selfish, is under the control of genes in only an indirect, but still very powerful, sense. By dictating the way survival machines [the individual animals] and their nervous systems are built, genes exert ultimate power over behaviour. But the moment-to-moment decisions about what to do next are taken by the nervous system. Genes are the primary policy-makers; brains are the executives. (Dawkins 1989: 60)

Of course, in the world of sociobiology, genes not only determine the policy for animal behavior but they construct the brains that implement this policy. Edward Wilson (1998), as we shall see in the next chapter, makes this particularly clear. Hence, both behavioral policy and "moment-to-moment decisions" are under the control of genes. In Wilson's terminology, animals are held by genes on a tight "genetic leash." This leaves little room for individual choice.

Dawkins illustrates the sociobiological concept of animal behavior using the analogy—a "deceptive guide" according to Darwin—of the digital computer programmed to play chess. As the different combinations of chess moves are virtually endless, the chess-playing software establishes a set of general rules and tactics, together with a feedback instruction to increase the weighting given to successful moves. Once programmed in this way it is up to the computer to "select" the most appropriate moves. While the control of the programmer is "indirect," it is also powerful because the computer has no choice but to play the game according to the guidelines dictated via the software. It is inconceivable that the computer could, in an emotional outburst, *override its instructions* and play recklessly and suicidally like some of its human opponents. Hence, an organism of the chess-playing type has no real choice; it must follow the dictates of its genes—of its programmer.

But we should let Dawkins speak for himself. "Like the chess programmer," he explains, "the genes have to "instruct" their survival machines not in specifics, but in the general strategies and tricks of the living trade"; and he continues, "every decision that a survival machine takes is a gamble, and it is the business of genes to program brains in advance so that on average they take decisions that pay off" (Dawkins 1989: 55). In other words, while genes do not directly manipulate every animal action, they build and program the nervous systems that do. Like the programmed computer, animals can respond only to immediate challenges within predetermined guidelines. They are unable, therefore, to respond to any type of situation that has not been foreseen by the "immortal" genes. But, as is well documented, animals are able to adapt to radically changed and totally unforeseen circumstances, as in the case of the blood-drinking finches of the Galapagos or in the case of modern humans in our totally unprecedented and "artificial" civilizations. As we shall see, the latter is of particular concern to Edward Wilson (1998). It is argued in part III that, although genes are employed in constructing the brains of animal species, this is a response by individuals to the **strategic pursuit**. I call this the process of **strategic selection**.

This genetic determinism becomes even more evident when Dawkins moves from metaphor to the real world. In dispensing with the computer analogy, Dawkins tells us:

> But now we must come down to earth and remember that evolution in fact occurs step-by-step, through the differential survival of genes in the gene pool. Therefore, in order for a behaviour pattern—altruistic or selfish—to evolve, it is necessary that a gene 'for' that behaviour should survive in the gene pool more successfully than a rival gene or allele 'for' some different behaviour. (Dawkins 1989: 60)

Although Dawkins warns against a simplistic, literal interpretation of the idea that there is a gene for altruism or a gene for selfishness, in the final analysis animal behavior in the sociobiological model is genetically determined.

On the whole, the leading sociobiologists are just as committed to genetic determinism as is Dawkins. They argue that there is a strong genetic link between genes and behavior in animals. Edward Wilson (1998) argues, as we shall see, that genes are responsible not only for constructing the brains of animals but for framing the "epigenetic rules" that shape behavior. This is the rationale behind his claim that genes hold their host animals on a tight "genetic leash." Robert Trivers also makes strong claims for the genetic determination of behavior:

> genes can affect the transmission of neural impulses and can, in principle, have minute and specific effects on behavior. We are still largely ignorant of how most of these effects come about, but evidence from breeding experiments in animals leaves no doubt that many behavioral traits have a genetic basis. (Trivers 1985: 95)

Throughout his book *Social Evolution*, Trivers operates from the unqualified assumption that "genes control development of traits" (Trivers 1985: 107). For those not convinced by this argument about the existence of genetic determinism in the theorizing of the sociobiologists, consider the implication of its absence—the sociobiologists would have no role to play.

As always with the sociobiological model, there is a problem with evidence. Dawkins, who admits that evidence for the alleged link between genes and behavior is scarce, appears to pin his hopes on a study of bees by W. C. Rothenbuhler (1964). This interesting study investigates the differential susceptibility of bees to "foul brood" disease, which attacks bee grubs in their wax cells. There is a "hygienic" strain of bees that uncap the wax cells and throw out the infected grubs and an "unhygienic" strain that does not and, as a result, is decimated. In his famous study, Rothenbuhler crossed a queen of one strain with a drone (male) of another in order to observe the behavior of the workers (daughters). The cross produced only unhygienic hives. He then "backcrossed" the first generation hybrids with a pure hygienic strain. The daughter hives fell into one of three groups displaying behavior that was either hygienic, unhygienic, or intermediate (where workers uncapped the wax cells but did not throw the grubs out). Guessing that there might be two separate genes involved, one for uncapping and one for discarding, Rothenbuhler intervened and uncapped cells in the hives of the unhygienic group and found that the workers then discarded the infected grubs.

At first sight this does seem like persuasive evidence that animal behavior is determined by the presence or absence of a particular gene, no matter how complex the connection. But there are at least two problems, both mentioned by Dawkins in the spirit of disinterested discourse, which are much more serious than he admits. First, in the endnotes to the second edition of *The Selfish Gene* (Dawkins 1989) he reports that Rothenbuhler's results were not unambiguously clear—there was one case that in theory should have been unhygienic but in fact was hygienic. This suggests that other influences can, at least on some occasions, override the genetic. Second, and more important, Dawkins suggests that the "uncapping" bees may have a taste for infected wax (and equally the throwers-out may have a distaste for the smell of infected grubs). Yet, Dawkins (1989: 62) feels able to argue that "even if this is how the gene works, it is still truly a gene 'for uncapping' provided that, other things being equal, bees possessing the gene end up by uncapping, and bees not possessing the gene do not uncap." To the contrary, what the evidence suggests is that bees, like all other animals, possess genes that influence their tastes or preferences, not genes determining their behavior. This involves a fundamental distinction that will be explored in part III.

While not compiled for the purpose, some of the case studies of animal behavior assembled by sociobiologists inadvertently suggest how a model of choice in animal "society" could operate. When discussing a detailed case study, Trivers tells us that

> the lek-breeding system of the black grouse allows females to choose males whose sons are expected to be successful at attracting the daughters of other females. Females can easily choose in this manner by (1) *watching* male–male

interactions, (2) *watching* male–female interactions, and (3) *watching* the skill of each male's courtship. (Trivers 1985: 352; my emphasis)

For the sociobiologist this cannot be a model of choice because all "choice" in this discipline is programmed by the genetic structure of individuals. Females are "selected" to respond to these signals in order to maximize the survival of their genes. It would be a model of choice only if individual animals—with only their desires, tastes, and personal characteristics influenced by their genes—were able to respond freely (subject to environmental constraints) to these signals. Such a model is developed in part III.

What is interesting about black grouse—or indeed of other animal species, such as chaffinches, starlings, baboons, monkeys, and schools of small fish (Wilson 1975: 51–52)—is that individuals gather information in order to make what appear to be real choices. What is *particularly* interesting is that the type of information they gather focuses on those individuals who are successful and popular and on why this is so. This information, which is gathered by individuals *watching* the actions and interactions of others, concerns what mate is likely to help them survive and prosper, what new food sources can be safely eaten, and what systems of communication are useful (Davis 1973). What this amounts to is the *imitation* of the behavior of those who are successful in the pursuit of survival and prosperity. In other words, animals appear to be watching their successful peers in order to collect **imitative information**.

If a member of the opposite sex is clearly popular with one's own sex, then he or she is worth pursuing. If other members of one's group are happily feeding on a new food source without suffering any ill effects, then it makes sense to follow their lead. This imitative process, which I call **strategic imitation**, saves time, economizes on limited brain power, and helps to avoid disastrous mistakes. Following the (successful) leader comes naturally to animals (including humans), because the outstanding characteristic shared by all animals is mimicry. Imitation would appear to be the basis of real choice in the animal world in which intelligence is the scarcest of all resources. Hence, a nongenetically determined form of choice does not need to be an intellectual exercise.

What is even more interesting is that the human world is no different. As shown in *The Dynamic Society* (Snooks 1996: 212–13) and *The Ephemeral Civilization* (Snooks 1997: 25–51), human decisionmaking is also only a limited intellectual activity. In a manner similar to animal species we make decisions by gathering information not about the benefits and costs of material investments (and certainly not about "reproductive success") but about which individuals and projects are successful and why. Although we are able to make complex intellectual calculations, on average we find this difficult and costly. Such calculations are left to a very small proportion of our species who pursue academic interests. The rest of us prefer to make choices that economize on the scarcest resource in nature—intelligence. This is explored in part III.

Social Organization

Social organization in the animal world is, according to the sociobiologists, genetically determined through natural selection. This is a logical, if not a realistic, extension of their genetic explanation of animal behavior. They argue that institutions (informal rules and conventions) and organizations (family and wider social groups) emerge from the genetic basis of kinship and from "reciprocal altruism." Trivers (1985: 84), for example, claims that "kinship and reciprocity . . . provide bases on which selection can mold larger cooperative units"; and Maynard Smith and Szathmary (1995: 258) claim that the "three processes—kin-selected altruism, enforcement and mutual benefit . . . can lead to the evolution of cooperation." We need to examine what sociobiologists have to say about these key issues.

Kinship

By "kinship" sociobiologists mean not the socioeconomic bonds between members of the extended family but rather their "genetic relatedness." This concept of kinship, suggested by John Maynard Smith (1958) and pioneered by W. D. Hamilton (1964), provided the foundations for sociobiology. While it has long been known that an individual is genetically related in varying degrees to all members of his or her extended family, Hamilton argued that the social relationships of individuals depend on the degree of "genetic relatedness" between them. Further, he argued that the probability of this genetic relationship can be measured by the "generation distance."

The concept of generation distance has been effectively illustrated by Richard Dawkins (1989: 92) in the following way. In the case of identical twins who share the same genes, the genetic relatedness is equal to unity (1). For siblings it is, on average, $^1/_2$ because half the genes possessed by one sister will be found in her sibling. It is also $^1/_2$ between parent and child, and $^1/_4$ for grandparents, grandchildren, aunts/uncles, and nieces/nephews. And, completing the family scale, it is $^1/_8$ for first cousins, $^1/_{32}$ for second cousins, and $^1/_{128}$ for third cousins. By the time we get to third cousins, Dawkins tells us, "we are getting down near the baseline probability that a particular gene possessed by A will be shared by any random individual taken from the population."

With perfect information and a copy of this index of genetic relatedness, individual organisms can maximize the spread of their genes! They do this, say the sociobiologists, by assisting other individuals according to the degree of the genetic relationship between them. This curious alleged behavior is called "altruism." For example, an identical twin would be treated as oneself; one would be indifferent as between parents, children, and siblings; and all of these would be more important than grandparents, grandchildren, aunts/uncles, and nieces/nephews, all of whom would be treated equally.

Quite clearly no one acts like this in reality. Recognizing this they introduce a qualification that takes into account the average remaining lifespan of individuals in any category. An individual's children, we are told, will receive more care and support than his or her parents, because they have longer to live. Presumably if the child has an incurable illness it will receive less support. This is clearly not the case in human society, as care is lavished on terminally ill children. In addition, this qualification is awkward for the sociobiologist because it is economic rather than genetic in nature owing to its implicit recognition of the variation in *material* rates of return on investment in time and effort.

An obvious problem with this genetic approach to family relationships is that the mate of any individual making these complex calculations should be no more important than any other member of the opposite sex selected at random from the population. In some species, including our own, genetically unrelated sexual mates or strategic associates (to be explained in chapter 12) are usually more important to us than our closest genetic kin. The same is true of adopted (genetically unrelated) children. This is a fatal flaw in kinship theory and it lies at the heart of sociobiology. And there are other problems. It is demonstrably incorrect to argue that one would treat an identical twin as oneself (or even twice as favorably as one's child) or treat an ordinary sibling as one's child (even if both were the same age). The underlying problem is that in reality decisions are made by organisms that have preferences that cut across those alleged to be exhibited by genes.

In order to resolve some of the less important of these difficulties, Dawkins introduces a further index—the "index of certainty"—arguing, not very convincingly, that

> although the parent/child relationship is no closer genetically than the brother/sister relationship, its certainty is greater. It is normally possible to be much more certain who your children are than who your brothers are. And you can be more certain still who you yourself are! (Dawkins 1989: 105)

We are told that this uncertainty about close genetic relationships leads to "selfish" rather than "altruistic" individual behavior, even when the degree of genetic relatedness is the same.

This argument is not only unconvincing, it is also unnecessarily, indeed unworkably, complex. To recapitulate: Dawkins advocates Hamilton's "genetic relatedness index," which cannot be usefully applied until it is qualified by another, and this time an *economic*, measuring stick—the expected average length of remaining life. But even this adjustment is not sufficient. It is necessary to combine the mortality-qualified genetic relatedness index with a further index of certainty. In other words, the genetic-relatedness hypothesis, which is based on perfect information, can only be regarded as operational in a world in which information is imperfect! But even then it cannot explain the care that individuals in some species devote to their sexual/strategic partners or to adopted children. Hence it would be a hopeless practical guide for those wanting to know how to treat those around them, whether it is the individual or the gene that is in charge of the process.

But these crippling problems are merely swept aside by the sociologists. Trivers, for example, states that we expect mechanisms of choice to evolve that reflect differential degrees of relatedness. Each individual will seem to value the reproductive success of others, compared to its own, according to the *r*'s [degrees of relatedness] that connect them. (Trivers 1985: 109)

In other words, individuals will act toward each other as if they know what the underlying genetic relatedness is because they are "selected" for that purpose. Trivers tells us that

it is not proximity that is causing the kinship effects we observe in nature; on the contrary, kinship [genetic relatedness] causes animals to be near each other, presumably in order to enjoy an increase in altruism and a decrease in selfishness. (Trivers 1985: 116)

Nevertheless, Trivers does feel compelled to discuss some possible "mechanisms of kin recognition" in various species such as sweat bees, Beldingo ground squirrels, monkeys, and birds. These studies, he claims, suggest that kin recognition is probably based upon smell and learning through early association. But, as Trivers frankly admits, there is no firm evidence for a genetic basis for these relationships. Indeed it is more likely, as I argue in part III, that family relationships are based upon economic bonds developed at an early stage of life.

On the other hand, Maynard Smith and Szathmary (1995: 259–60) claim that Hamilton's formulation has been widely misunderstood. They assert that it is a "fallacy" to believe that "for kin selection to operate, the actor must be able to calculate its relatedness to the recipient . . . no-one has to know what *r* is, except the biologist trying to test Hamilton's idea." Their argument is mathematical rather than empirical and, as such, is unconvincing to realists. Even they do not seem convinced because they express concern about whether individuals in social groups can recognize their kin: "This capacity is of obvious importance in ensuring cooperation between [in the case before them] nest-mates" (Maynard Smith and Szathmary 1995: 268).

Whatever the outcome of this disagreement, the kinship model leaves unresolved a number of important puzzles in the evidence. These include the positive relationship between "mild aggression" and the intensity of personal relationships, the willingness of females to "adopt" unrelated children, and the devotion of mating couples in some species and homosexual partners in our own.

First the problem of mild aggression. In reporting the behavior of monkeys, Trivers is puzzled by the observation that

mild aggression is much more frequently directed toward unrelated individuals than toward relatives, but when the aggression is directed toward relatives, closer degree of relatedness does not reduce frequency of threat or attack. Quite the contrary, there is a steady, though non-significant, increase in mild aggression with increasing degrees of relatedness . . . close association permits more altruism but also throws individuals into closer competition, thus engendering more opportunities for selfishness. (Trivers 1985: 119)

This is a major problem for kinship theory, which predicts greater "altruism" rather than greater "selfishness" as the "generation distance" diminishes. Because Trivers is unable to explain the tension and/or violence in close family relationships using the kinship model, he is forced to introduce an economic argument—competition for scarce resources. This explanation is a confusion of genetic "altruism" and economic "selfishness."

Compare this with what I wrote in *Portrait of the Family*—*before* I had read any sociobiology—regarding my materialist "concentric spheres" model of human behavior (discussed in chapter 14 below):

> There is always tension between the centre [the self] and the periphery [other individuals and groups] no matter how short the **economic distance** may be, because all personal relationships are built up by the central individual during his or her lifetime in order to maximize his or her utility . . . the degree of tension appears to be inversely related to the economic distance, with most conflict and violence occurring between people who are closely associated with each other. (Snooks 1994: 50)

Economics, in contrast to genetics, can explain this aspect of personal relationships.

Second, Trivers acknowledges the problems of the kin-selection model in explaining the observation that females in some species—he nominates Japanese monkeys, and we could add human beings—"adopt" unrelated infants and care for them as if they were their own. Trivers discusses the incidence of "alloparenting" in Japanese monkeys as follows:

> Kurland [1977] began his work assuming that alloparenting was altruistic. He expected it to be directed preferentially toward close relatives. Sometimes it is, as when older siblings babysit younger siblings . . . but in general, Kurland found, alloparenting was common by females who had not yet given birth, and in 122 of 140 cases was directed at *unrelated* individuals. (Trivers 1985: 120; emphasis in original)

How does the sociobiologist respond to these most inconvenient observations? By simply but inconsistently arguing that those species contradicting kinship theory are acting selfishly. In Trivers's (1985: 120) words: "by taking kinship theory as true, we are able to infer that in Japanese macaques, alloparenting is usually a selfish act that prepares the young female to be a better mother when her chance comes." Once again we are presented with an economic argument—maximization of individual utility—when the genetic-relatedness model fails to explain reality. And this occurs regularly.

This continual attempt by sociobiologists to tidy up loose ends actually undermines the credibility of the genetic model. It is much like Charles Darwin employing the concept of sexual selection whenever natural selection failed to explain the real world. The construction of flying buttresses, particularly when they are borrowed from the competing materialist theory, is a sure sign that the structure is about to collapse. As demonstrated in chapter 14, my materialist dynamic-strategy model provides a complete explanation of social organization in the animal and human worlds.

Reciprocal Altruism

Kinship theory, therefore, is seen as the cornerstone of the sociobiological explanation of social organization in the animal and human worlds. Trivers (1985: 135) insists that "kinship is critical to distinguishing the various levels of organization." Further, he argues that

> there is a kinship structure in every social group, and this kinship structure selects for biased exchanges in which individuals tend naturally to favor those to whom they are related by higher degrees of relatedness. Thus there is a fundamental change as we pass the level of the individual and go on to higher levels of organization: degrees of relatedness fall below 1 [less important than the self] and social conflict is expected. (Trivers 1985: 135)

But this does not take us very far. There are many higher levels of social organization that do not come into the alleged sphere of influence of kinship theory—apart from the suggestion that greater social conflict can be expected (greater than family rape and murder?).

The attempt to explain higher levels of social organization has led some sociobiologists to develop group-selection theories (Wynne-Edwards 1962), by which individuals evolve to devote themselves to the higher good of the group or species. This widespread, genetically determined altruism would provide a basis for explaining wider forms of social organizations. Unfortunately for its supporters, Dawkins (1989: 7–10) and Trivers (1985: 67–86), among others, claim that this "fallacy" is based on an inadequate knowledge of evolutionary theory. It would take, we are told, only one selfish rebel determined to exploit the altruism of the rest of the group to provide the genetic basis for his progeny to eventually overwhelm the entire group. To fill the void, sociobiologists like Trivers have employed the neo-Darwinian concept of "reciprocal altruism."

The concept of "reciprocal altruism" was developed by Robert Trivers (1971) to extend the neo-Darwinian version of natural selection into a broader range of social relationships beyond the family. It was an attempt both to build upon Hamilton's work on kin selection and to fill the void left by the demolition of group selection. Trivers argues that, although two strangers do not possess a close genetic relationship, if an altruistic act by one of them is followed by a reciprocal act of altruism by the other, both will have gained by substituting a lower risk of death for a higher one. To illustrate this idea he argues that if the risk of dying on the part of a drowning man is one in two, and that on the part of the rescuer is one in twenty, and if they both attempt to save each other when one is in danger of drowning, then each individual will have traded a one-half chance of dying for about a one-tenth chance. And if the entire society adopts reciprocal altruism, there will be an increase in its genetic fitness. Natural selection, Trivers argues, will "favor" a situation in which reciprocal action emerges as a permanent response, operating through an emotional system involving friendship, moralistic aggression, gratitude, sympathy, guilt, and a sense of justice. This is an unwitting attempt to solve Darwin's unresolved problem in *Descent* (1871), one hundred years before, of how to explain the emergence of "social virtues" in successful human societies.

But even among sociobiologists there is disagreement about the existence of reciprocal altruism. In response to the claim by Edward Wilson (1975: 120) that while human society abounds with this type of reciprocal behavior the animal world is largely devoid of it, Trivers (1985: ch. 5) attempts to detail examples in nature. These include vampire bats sharing blood with neighbors unable to find any; birds that respond less aggressively to near neighbors than to strangers, provided they do not violate the territory of the "owners"; baboons that support their neighbors in conflicts with more distant individuals, provided there is reciprocation; chimpanzees that develop alliances within small social groups based on mutual support; and dolphins and whales that appear to provide assistance to others (he neglects to mention the pack rape by young male dolphins of available females). As I argue in part III, while these examples are intriguing, they merely indicate the tendency of animals to *cooperate* with one another whenever that is in the best *material* interests of the individuals concerned.

The sociobiological attempt to explain social relationships in the animal and human worlds encounters a number of major difficulties. In the first place it is entirely misleading to employ the term "reciprocal altruism" to refer, as Trivers does, to social relationships based on mutual gain. "Altruism" is not the same as "cooperation." Any attempt by individuals to maximize their own utility by cooperating with each other cannot under any circumstances be regarded as altruism, even reciprocal altruism. Why? Because, if their expectations are fulfilled, all individuals will gain more materially through cooperative action than if they battled on alone. There are very many circumstances in which individuals can only maximize their utility by cooperating with other maximizing individuals. This cooperative action, which does not preclude competition between coworkers, is motivated and sustained by material self-interest, not by altruistic self-sacrifice. Despite this, sociobiologists have used the word "altruism" for mutual self-interest because it arises from kinship theory—a fantasy they are determined to impose on an unaccommodating reality.

By employing the term "reciprocal altruism" to embrace cooperative behavior aimed at maximizing the individual utilization of all parties, the sociobiologists have been responsible for creating considerable confusion in the minds of those not party to the specialized debate over the genetic basis for family and social relationships. In this debate the word "altruism" should be replaced with the word "cooperation." Failure to do so has clouded the more general issue of whether individual action is motivated and sustained by self-interest or by genuine altruism (the attempt to help others even at the expense of oneself). Those who reject the reality of self-interest have used this specialized debate in sociobiology and, more recently, in economics, to support the more general proposition that mankind is driven by genuinely altruistic impulses. It amounts to a confusion of ends and means (Snooks 1994: 42–46). An individual can maximize his or her material advantage through either individual or cooperative action depending on the prevailing circumstances.

It is interesting to realize that the sociobiological approach is very similar to economic game theory based not on genetic inheritance but on material self-interest. Indeed, this sociobiological approach owes much to the game-theory interests of John Maynard Smith (1972; 1974; 1976), which were stimulated by this particularly sterile branch of neoclassical economics.[2] Trivers (1985: 389–92) also adopts this approach and attempts to place "reciprocal altruism" in a game-theoretic framework. Here we see a direct connection between the deductive absurdities of neoclassical economics and neo-Darwinian sociobiology.

Our second major criticism of the "reciprocal altruism" concept is that it is not at all persuasive. How is the genetic model ever to become operational? As Edward Wilson (1975: 120) has said: "Granted a mechanism for sustaining reciprocal altruism [a "selected" emotional system], we are still left with the theoretical problem of how the evolution of behavior gets started." Like economic game-theorists, Trivers emphasizes the importance of repeated exchanges between individuals. But to explain those repeated cooperative interactions we need a materialist theory, because those interactions are required to provide *individuals* (not genes) with the information they require to make economic decisions. And if we can construct a persuasive economic theory why do we need a genetic one with all its limitations and metaphysical overtones?

Finally there are a host of limitations facing the genetic theory of social organization. First, apart from use of the word "altruism," there is little connection between the kinship and the reciprocal-altruism models. Sociobiology has no encompassing theory. But they are not alone in this, as Charles Darwin faced the same problem. There is just no general dynamic theory of evolution. Second, it is hard to see that there has been enough time—only ten thousand years—for natural selection to operate on cooperative behavior in human civilization. Third, more consistent and broad-ranging alternative, materialist models can be developed to explain social institutions in both the animal and human worlds. In particular, human culture is so varied and complex that these simple genetic models cannot hope to do them justice. The "genetic view of social life" (Trivers 1985: 138) is severely circumscribed. As we have come to expect, sociobiology has completely failed the ultimate test of being able to explain the ecosociopolitical structure and dynamics of human society.

Conclusions

Sociobiologists employ a deductive rather than an empirical approach to animal behavior and social evolution. Their case studies constitute an attempt to *illustrate* their neo-Darwinian theory. They claim that this theory has been adopted from Darwin. What they do not admit to is that they have completely transformed Darwin's theory of natural selection. They have stripped the exogenous driving force—the life-and-death struggle for scarce resources—from Darwin's theory and have replaced it with a passive genetic sorting device called "repro-

ductive success." Darwinian fitness for the "war of nature" has also been redefined: "fitness refers to reproductive success, or the production of surviving offspring" (Trivers 1985: 69). And "reproductive success" is code for the way the
selfish gene maximizes its presence in the gene pool. Sociobiologists have, in
other words, converted Darwin's materialist model into a sociological model.

Sociobiologists have transformed Darwin's model in another way. They
have converted it from a dynamic to a comparative-static model. This is why
they have been able to dispense with the exogenous driving force in Darwin's
theory. Indeed, they have little practical interest in dynamics—the exception is
the coevolution model of Wilson and his followers examined in the next chapter—because the data they employ are cross-sectional rather than temporal.
They are compelled to observe and explain the behavior of animals at a given
point in time—their own era. Their discipline does not, indeed cannot, have a
historical focus. The only historical data in evolutionary circles are those of the
paleontologists and their subjects have long since ceased to exhibit any notice-
able behavior. Essentially their work involves testing the predictions of their
neo-Darwinian models—such as "kin selection" and "reciprocal altruism"—using contemporary observations. In the process they distort our perception of reality.

Yet even this limited, static exercise throws up critical difficulties for the
sociobiology model. Much of the data they employ challenge the central assumption that individuals in all species attempt to maximize their reproductive
success. First, many of the case studies presented by sociobiologists suggest that
the forces underlying social relationships in the animal world are more direct
and immediate than those required for the maximization of an individual's genes
in the gene pool. One need only recall the examples of baboons and chimpanzees, in which males trade grooming and protection services for sexual favors,
or of territorial mammals and birds that trade food and protection for sex. Most
of these males show no interest in the survival of their offspring: a strange way
to maximize reproductive success. Second, the role of physical attractiveness is
another stumbling block. Trivers (1985: 260), for example, is unable to answer a
fundamental question that arises in his work: Why should an individual be willing to increase the level of its parental investment when its mate is attractive?
Why? The answer is economic rather than genetic. Third, similar problems are
encountered with core issues of why reproduction largely occurs sexually rather
than asexually (which maximizes the transfer of genes from parent to child), and
why there is a high correlation between male longevity and the size of the families (as in deer). The problems cannot be satisfactorily resolved while it is assumed that individuals are "selected" to maximize the survival of their genes.
Finally, sociobiological models of kinship are unable to explain either the attention that individuals devote to genetically unrelated spouses and adopted children, the occurrence of homosexuality (no gene transfer), or the increase in
"mild" aggression (including rape and murder) as the generation distance declines.

Sociobiology, therefore, is just not able to persuasively explain the real-
world relationships in animal families, let alone in social groupings beyond this

primary level. The reason is that genetic kinship theory, which is a product of arid deductive reasoning based on fanciful assumptions, is not operational. It is not borne out by the evidence from the real world. What we need is an entirely new theory of animal behavior based on a close observation of reality. This is provided in part III.

Chapter 8

The Image-Makers and Evangelists II: Sociobiology and Human Society

The ultimate test of any dynamic theory is, as has been maintained throughout this book, to explain and predict the changing fortunes of human society. While most Darwinists have been reluctant to expose themselves to this test ever since Charles Darwin failed it so badly, some sociobiologists seem eager to do so. The boldest among them even claim that the social sciences and humanities are the final chapters to be written in the great book of sociobiology. It is their self-appointed task to construct the biological foundations for those mathematical models that will finally reveal the true nature of human society and its future. These true believers assert that sociobiology will form the basis for the unity of all knowledge, reaching back to chemistry and physics, and forward to the social sciences, humanities, ethics, and religion. The leading evangelist of this cause is Edward Wilson, one of the founders of sociobiology.

What *Is* Human Nature?

This is the central question addressed in Wilson's attempt to provide a sociobiological explanation of human culture. He begins by rejecting the way human nature has been treated by social scientists in general and economists in particular. Wilson tells us that

> in economics and in the remainder of the social sciences as well, the translation from individual to aggregate behavior is the key analytic problem. Yet in these disciplines the exact nature and sources of individual behavior are rarely considered. Instead, the knowledge used by the modelers is that of *folk psychology*, based mostly on common perception and unaided intuition, and *folk psychology* has already been pushed way past its limit. (Wilson 1998: 202; my emphasis)

Human nature, he insists, should be the subject of scientific biological and psychological study. In fact he regards "evolutionary psychology" as part of sociobiology.

In answer to the rhetorical question "What is human nature?" Wilson tells us:

> It is not the genes, which prescribe it, or culture, its ultimate product. . . . It is the epigenetic rules, the hereditary regularities of mental development that bias cultural evolution in one direction as opposed to another, thus connecting the genes to culture. (Wilson 1998: 164)

In other words, human nature is merely the outcome of human genes operating at one remove in a social context. And the human brain is the link between genes and culture—a link made possible by natural selection. In Wilson's words:

> Thousands of genes prescribe the brain, the sensory system, and all other physiological processes that interact with the physical and social environment to produce the holistic properties of mind and culture. Through natural selection, the environment ultimately selects which genes will do the prescribing. (Wilson 1998: 137)

This is why Wilson believes that biology is the linchpin of human knowledge.

According to Wilson (1998: 165), "the search for human nature can be viewed as the archeology of the epigenetic rules." But what are epigenetic rules? They are the genetically determined constraints that operate on the "anatomy, physiology, cognition, and behavior of organisms." And, of course, they are the outcome of the neo-Darwinian version of natural selection:

> During hundreds of millennia of Paleolithic history, the genes prescribing certain human epigenetic rules increased and spread at the expense of others through the species by means of natural selection. By that laborious process human nature assembled itself. (Wilson 1998: 165–66)

As usual no evidence is given, nor could it be given for this assertion. It is metaphysical in nature.

Epigenetic rules are said to operate at two distinct levels. The "primary" epigenetic rules are "the automatic processes that extend from the filtering and coding of stimuli in the sense organs all the way to perception of the stimuli by the brain. The entire sequence is influenced by previous experience only to a minor degree, if at all." And the "secondary" epigenetic rules are "regularities in the integration of large amounts of information. Drawing from selected fragments of perception, memory, and emotional coloring, secondary epigenetic rules lead the mind to predisposed decisions through the choice of certain memes and overt responses over others" (Wilson 1998: 151). Wilson regards this division as subjective, and suggests that there will also be intermediate levels for these genetically influenced rules.

It should be quite clear by now that Wilson—and other sociobiologists interested in coevolution—regards variations in behavior as well as in physical characteristics as genetically determined to a large degree. It is difficult to discover from reading his work, however, to what degree exactly. Clearly it must be to a high degree otherwise the sociobiological position could not be distinguished from that of the existing social sciences that Wilson wants to transform. And he is determined that the social sciences will be totally rewritten from a sociobiological perspective. In the mid-1970s Wilson proclaimed somewhat triumphantly that

> It may not be too much to say that sociology and the other social sciences, as well as the humanities, are the last branches of biology waiting to be included in the Modern Synthesis [or, neo-Darwinism]. One of the functions of sociobiology, then, is to reformulate the foundation of the social sciences in a way that draws these subjects into the Modern Synthesis. (Wilson 1975: 4)

Some twenty-five years later the social sciences are still *waiting* to be embraced. The great sociobiological vision has yet to be realized. Accordingly, by the late 1990s Wilson's tone was more conciliatory. Now he is offering not an unconditional takeover but a junior partnership to social scientists. Yet he still insists, a little more shrilly, that human culture will only be fully understood through a biological study of the human brain and of human nature.

Wilson's belief in the central role played by genetics in human behavior is conveyed rather aptly by his provocative metaphor of the "genetic leash." Employed at least as early as the late 1970s in *On Human Nature* (1978), the "genetic leash" was still part of his vocabulary as recently as the late 1990s in *Consilience* (1998). In responding to a rhetorical question on that earlier occasion, Wilson proclaimed:

> Can the cultural evolution of higher ethical values gain a direction and momentum of its own and completely replace genetic evolution? I think not. *The genes hold culture on a leash.* The leash is very long, but inevitably values will be constrained in accordance with their effects on the human gene pool. (Wilson 1978: 167; my emphasis)

The human "dog," therefore, is under the control of its "master," the gene. This is, of course, merely an alternative—although a far more provocative and offensive—metaphor to Richard Dawkins's "selfish gene." There have always been some evolutionary geneticists, such as Richard Lewontin, who find these metaphors offensive and who are highly critical of the sociobiological focus on genes rather than organisms. Lewontin has even joined forces with paleontologists such as Stephen Gould (Gould and Lewontin 1979) in attacking the neo-Darwinist position, while the geneticist Steve Jones (1999: xxxvi) regards the various attempts "to apply Darwinism to civilization" as "more or less infantile."

Despite his more conciliatory tone in *Consilience*, Wilson continues to employ the "genetic leash" metaphor. When discussing the parallel "evolution" of genes and culture throughout mankind's prehistory, Wilson asks: "How tight was the genetic leash? That is the key question, and it is possible to give only a partial answer." And that answer is:

> In general the epigenetic rules are strong enough to be visibly constraining. They have left an indelible stamp on the behavior of people in even the most sophisticated societies. (Wilson 1998: 158)

Also, when explaining that different scholars have different views on how tight the "leash" is, Wilson writes:

> Nurturists think that culture is held on a very long genetic leash, if held at all, so that the cultures of different societies can diverge from one another indefinitely. Hereditarians believe the leash is short, causing cultures to evolve major features in common. (Wilson 1998: 143)

There are some leading geneticists—indeed, those at the frontier of biotechnological research—who do not accept Wilson's claim. Ian Wilmut, who with Keith Campbell pioneered cloning (or nuclear transfer) technology in mammals ("Dolly" the sheep was cloned from an adult body cell), when explaining why even identical twins are different, argues:

> But of course—and biologists are at least as aware of this as anybody—genes do not *determine* in tight detail how a creature turns out. In general, the genome merely sets broad limits on the possibilities. Genes, in short, merely propose possibilities. . . . It is the environment that shapes the final outcome. (Wilmut, Campbell, and Tudge 2000: 302–3)

And on the role of genes in forming the brain and influencing its processes, Wilmut says:

> We can see the physical realities of this [the above-quoted] principle within the brain. Thoughts and memories seem to be contained within and framed by the brain cells—neurones—of which there are billions. But what matters most is not the number nor the distribution of the cells, but the connections between them: called the synapses. Two individuals may possess similar neurones, but the neurones in the two individuals become wired up to each other in different ways, and so form different neural 'circuits'. Thus two brains with similar cells are nevertheless qualitatively different. . . . Do the genes determine the circuitry in fine detail? . . . It seems rather that the genes lay down the broad structure, but the structure then works itself out in ways that are beyond their fine control. . . . *Once the brain is formed it does its own thing.* (Wilmut, Campbell, and Tudge 2000: 303; my emphasis)

He concludes that "all in all, then, the genes lay down the ground rules—but in the end our upbringing and experience make us what we are" (Wilmut, Campbell, and Tudge 2000: 303). Hence, scientists at the forefront of biotechnology are convinced that there is no tight genetic leash on animal behavior.

Despite this, Wilson and his fellow sociobiologists—who claim that they have something distinctive, even revolutionary, to say about the dynamics of human society—regard the genetic leash as tight and binding. If the "leash" is loose, or worse, has been slipped, they have no theory at all about human society. Their refusal to say how tight is "tight" could be seen as a response to the unpopularity of the idea of biological determinism among the humanists they are attempting to persuade to take sociobiology seriously.

Yet, having argued a case for the "genetic leash," Wilson is forced to admit that there is very little hard evidence of any connection between genes and epigenetic rules. This, he insists, is not because the relationship does not exist but because coevolution is "an infant field of study." Hardly a convincing argument! As shown in part I, Charles Darwin argued that the lack of evidence for the slow but gradual evolution and replacement of species that he expected to find in the fossil record was due to the primitive state of paleontology in the mid-nineteenth century. More detailed fossil evidence proved Darwin wrong. As is widely appreciated, the lack of opposing evidence is not proof of a hypothesis.

The evidence that Wilson does provide—dyslexia and irrational aggressive behavior—hardly inspires confidence in his genetic-leash hypothesis. Despite this he asserts that "variation in virtually every aspect of human behavior is heritable to some degree, and thus in some manner influenced by differences in genes among people" (Wilson 1998: 154). For the sociobiologist, no matter what the nature of the evidence, human behavior is "in some manner" genetically determined. Neo-Darwinism is ultimately a matter of faith and Wilson is one of its chief and more extreme evangelists.

There is, however, a fatal flaw in Wilson's method. As shown in part III, it is just not possible to build dynamic theories about human society using the genetic building blocks of human nature. Even assuming that human nature can be understood through its genetic building blocks, this merely provides us with a driving force, not a dynamic mechanism. That has always been the geneticists' dilemma and was the reason for their rapprochement with Darwinian dynamics in the 1930s. Yet even this is lost on the neo-Darwinists who, if they believe in a driving force at all, locate it in the "selfish gene" (although even Dawkins admits that genes do not "strive").

In part III the driving force in my dynamic-strategy model is the **materialist organism**—which in human society becomes **materialist man**—who struggles in a competitive environment to survive and prosper. "Desires" are genetically determined, but behavior is a response to **strategic demand** through a process of strategic imitation. Hence, while a study of the genetic building blocks of the brain can only be beneficial to our understanding of the physical structure of mankind, the nature of human nature can only be explored by its systematic observation in historical space and time, rather than in the sociobiological laboratory. To dismiss such observation as "folk psychology" is to miss the boat entirely.

The Myth of Coevolution

It is alleged by sociobiologists that coevolution is a dynamic process in which genes and culture interact under the shaping influence of natural selection. As this is a rather shadowy process it is best to let the evangelist of coevolution explain it himself. In a chapter of *Consilience*, entitled "From Genes to Culture," Wilson tells us:

> Culture is created by the communal mind, and each mind in turn is the product of the genetically structured human brain. Genes and culture are therefore inseverably linked. But the linkage is flexible, to a degree still mostly unmeasured. The linkage is also tortuous: Genes prescribe epigenetic rules, which are the neural pathways and regularities in cognitive development by which the individual mind assembles itself. The mind grows from birth to death by absorbing parts of the existing culture available to it, with selections guided through epigenetic rules inherited by the individual brain. (Wilson 1998: 127)

Hence, not only is the human brain genetically determined at birth but the subsequent growth of the mind is subject to the shaping influence of epigenetic rules that are "prescribed" by one's genes. And all of this is the outcome of natural selection as genes attempt to maximize their influence in the gene pool. Wilson further explains that

> The nature of the genetic leash and the role of culture can now be better understood, as follows. Certain cultural norms also survive and reproduce better than competing norms, causing culture to evolve in a track parallel to and usually much faster than genetic evolution. The quicker the pace of cultural evolution, the looser the connection between genes and culture, although the connection is never completely broken. (Wilson 1998: 128)

Darwin, therefore, is never far away. As Wilson (1998: 128-29) puts it: "Gene-culture coevolution is a special extension of the more general process of evolution by natural selection," which he regards as an "impersonal force." Only those genes that "confer higher survival and reproductive success on the organisms bearing them, through the prescribed traits of anatomy, physiology, and behavior increase in the population from one generation to the next." He concludes that

> To genetic evolution, putting the matter as concisely as possible, natural selection has added the parallel track of cultural evolution, and the two forms of evolution are somehow linked. We are trapped, we sometimes think, for ultimate good or evil, not just by our genes but also by our culture. (Wilson 1998: 130)

But, of course, for the sociobiologist even our culture is shaped by our genes.

It merely begs the question: How can this imprecise model of coevolution—which has genetic and cultural evolutionary selection "somehow linked"—explain the structure and dynamics of human society? There is no escaping this ultimate and most demanding test. As Wilson acknowledges in *Consilience*, to explain human society is far more difficult than to explain either the physical or biological worlds.

To answer this question we need to consider what Wilson means by "culture." It is, as we shall see, a very simple concept that does little to improve our understanding of the complexities or dynamics of human civilization. Belatedly and briefly we are told (Wilson 1998: 166) that, "strictly defined," culture is the "complex socially learned behavior" of humans, which, of course, is constrained by strong genetic influences. Wilson's only other comment on the nature of human "culture" concerns what he calls the "cultural unit." He tells us:

> We [Lumsden and Wilson 1981] recommended that the unit of culture—now called meme—be the same as the node of semantic memory and its correlates in brain activity. The level of the node, whether concept (the simplest recognizable unit), proposition, or schema, determines the complexity of the idea, behavior, or artifact that it helps to sustain in the culture at large. (Wilson 1998: 136)

This definition of "meme" is very different to that by Richard Dawkins, for whom the cultural "replicator" is largely independent of genetic constraint. In *The Extended Phenotype* Dawkins notes the disapproval of other sociobiologists of this degree of freedom from the "genetic leash."

In effect, Wilson equates the idea of culture with the way our genetically engineered brains view the world. Human ideas, behavior, and artifacts are somehow extruded from our brains to form the complexities of our sophisticated civilizations. In his own words:

> The epigenetic rules of human nature bias innovation, learning, and choice. They are gravitational centers that pull the development of the mind in certain directions and away from others. Arriving at the centers, artists, composers, and writers over the centuries have built archetypes, the themes most predictably expressed in original works of art. (Wilson 1998: 223)

He further believes that a biological study of these archetypes will place the "flawed" psychoanalysis of Freud and Jung back on the right track. Wilson (1998: 74) claims that "mysticism and science meet in dreams."

Coevolution is a rather naive supply-side "world view." It cannot be called a model because we are not told how we get from genetically shaped products of the human brain to the sophisticated and varied ecosociopolitical structures of human civilization. Exactly how does "coevolution" explain, for example, the rise and fall of ancient Greece, the rise and fall of Greek democracy, the sudden and radical transformation of Japan's ecosociopolitical structure following the Meiji restoration, the rise and fall of the U.S.S.R., or the recent economic and political crisis and recovery of Southeast Asia? The fact is that it cannot. Nor will it ever form the basis of such a model owing to the flawed concept of natural selection. Both Darwinism and neo-Darwinism have failed the ultimate test.

We should, however, permit Wilson to put the coevolution concept through its paces. He argues that before the emergence of *Homo sapiens* about one hundred and fifty thousand years ago "genetic and cultural evolution were closely coupled," so that primitive social behavior and social structures operated largely under the influence of genes. But with the "advent of Neolithic societies, and especially the rise of civilizations, cultural evolution sprinted ahead at a pace that left genetic evolution standing still by comparison" (Wilson 1998: 157–58). Surely this is an account of an "elastic" genetic leash! To claim any credibility for the coevolution approach, Wilson has to demonstrate that in these circumstances natural selection transfers its attention from genetic to cultural change, so that societies with a "cultural" edge become more widespread and dominant. But even if this could be done, it would not explain the fluctuating fortunes of individual societies or civilizations.

Wilson's coevolution concept raises more questions and problems than answers and solutions about the dynamics of human society. A few of the more problematic issues are raised here. First, if the "genetic leash" were tightly held prior to and during the emergence of modern man, how were we suddenly able to slip the leash and construct the first civilizations? What accounts for this timing? And why did it occur in the Old World first? Coevolution cannot answer any of these fundamental questions. In reality, the historical data suggest that the "leash" did not constrain humanity at all, so that we were able to build our civilizations at the very first opportunity (Snooks 1996; 1997). What I am suggesting is that ecosociopolitical change is driven by forces on the demand rather than the supply side. When and where the strategic demand arose, mankind was able to respond without genetic constraint.

Second, if the "genetic leash" was so easily slipped, or indeed if it never existed, how can sociobiology explain the dynamics of human society? If culture bolted some ten thousand years ago, as Wilson readily acknowledges, how can it be argued that genes or epigenes have played a "constraining"—let alone a shaping—role in the affairs of man? Could the realization of this difficulty be why sociobiology's announcement in the 1970s of an imminent takeover of the social sciences and humanities has, more than twenty years later, been diluted into an offer of a junior partnership? Of course, if genes do not drive human behavior, sociobiologists have no explanation of the dynamics of culture at all. A shaky basis for any projected takeover/partnership.

In part III I present a dynamic model that has been constructed from the systematic observation of individuals, families, and societies in historical space and time. I argue, contrary to the reductionist thesis underlying Wilson's coevolution model, that **existential historicism** (Snooks 1998a: 185–92) is the only way to understand and model the dynamics of human society, or indeed of life itself.

The dynamic-strategy model, unlike the coevolution concept, can explain the causes and timing of the Neolithic (agricultural) Revolution in both the Old and New Worlds. It was the outcome of a technological paradigm shift (see figure 13.3) made necessary and possible by the eventual exhaustion of the paleolithic (hunting/gathering) technological paradigm that had spread around the accessible world through the family-multiplication strategy (procreation and migration) pursued in order to bring additional natural resources under family/clan control. This was achieved about ten thousand six hundred years ago in the Old World and seven thousand years ago in the New World.

The neolithic technological paradigm shift generated an entirely new demand for urban institutions, which the societies of man were able to respond to effectively and rapidly without any sign of effective supply-side (including genetic) restraint. The nature, timing, and rate of this response depended on the type of dynamic strategy (conquest or commerce) pursued by a society within the context of this new technological paradigm. My demand-side model—the dynamic-strategy theory—can explain what the supply-side coevolution concept has failed to account for. Rather than being a "leash" on this well-documented response to strategic demand, our genetic inheritance provided a flexible platform for the urban takeoff. It is this physical and intellectual flexibility, which was fashioned not by epigenetic rules but by millions of years of experience with the **strategic pursuit**, that accounts for the remarkable success of our species.

In contrast, Wilson puzzles over how natural selection, which he admits cannot anticipate future needs, prepared humans for civilization before it existed. "That is," he tells us, "the great mystery of human evolution: how to account for calculus and Mozart" (Wilson 1998: 48). It is a mystery easily penetrated by dynamic-strategy theory. As life is a strategic pursuit rather than an outcome of natural selection, the human mind did not need to be programmed through genetic anticipation of civilization to generate civilized ideas such as calculus and classical music. If there is a strategic demand for these services, the human mind, shaped by the strategic pursuit, will respond in the most effective way it can—strategically. Innovation is a response to strategic demand which changes as the prevailing dynamic strategy unfolds. It is not extruded from the mind at the command of the selfish gene.

The dynamic-strategy model can also account for—to take an earlier example of the failure of coevolution—the rise and fall of ancient Greece, the rise and fall of Greek democracy, the radical transformation of Japanese ecosociopolitical institutions since the Meiji restoration, the rise and fall of the U.S.S.R., and the emergence and resolution of the recent Southeast Asian crisis, or indeed any other issue of societal dynamics, no matter how big or small (Snooks 1996; 1997; 1999; 2000). In essence all these developments are the outcome of a response to the systematic exploitation, exhaustion, and replacement of the four dynamic strategies of society (and of life).

Owing to its importance to the "unity of knowledge" argument in Wilson's *Consilience*, something needs to be said here about the scientific revolution in Europe since the sixteenth century. It is curious that Wilson takes a rather unscientific approach to the scientific revolution. But then this is part of a pattern. Scientists, both natural and social, who are rigorous in their own disciplines invariably take a very casual approach to history. They tend to employ it as a source of highly selected facts in order to justify their more general views about society or life. They indulge, as Wilson in a different guise might say, in "folk history."

In attempting to explain the rise of modern science, Wilson (1998: 48) asserts, without any evidence, that "three preconditions, three strokes of luck in the evolutionary arena, led to the scientific revolution." His nominations, apparently at random, for these "three strokes of luck" are human curiosity, an inborn facility for abstraction, and the wonders of mathematics. Unfortunately this short shopping list of history explains nothing: there is no driving force and no dynamic mechanism that could possibly account for the timing, location, and nature of the scientific revolution. Why in Europe and why there only since the sixteenth century? Why not in China in the Song period; and why not earlier, say in ancient Greece? The only hint of any answers to these questions given by Wilson are that it did not occur in China for "historical and religious" (whatever that means) reasons, and that it did occur in Europe because of the reductionist approach of its scientists (is that genetically determined?). Nothing is said about timing. In other words, coevolution has absolutely nothing to tell us about the rise of modern science.

In contrast, the dynamic-strategy model can explain the emergence of the scientific revolution in Europe rather than China and from the sixteenth century rather than earlier or later, all in terms of the dynamic strategies employed, the stages reached in the unfolding of these strategies, and what happened after these strategies were exhausted (Snooks 1996: chs. 9, 10, and 11; 1997: chs. 10 and 12; 1998a: ch. 9). In a nutshell, the fundamental difference between Europe and China from 1500 to 1750 was that the former pursued the commerce strategy which generated an entirely different strategic demand to the family-multiplication strategy of the latter. The unfolding European commerce strategy (which had replaced an exhausted conquest strategy) from about 1500 generated a growing strategic demand for new scientific ideas required for more sophisticated forms of global transport, communication, engineering, and urban production as the commerce metropolises of Europe grew in size, wealth, and power. In contrast, China's family-multiplication strategy had no need for scientific ideas (except for military defense), only the means to access underutilized natural resources in the south and west. It is merely a myth that China came close to generating a scientific/industrial revolution in the Song period.

Why did the scientific revolution pass through a number of surges and retreats? While this question has Wilson stumped, the answer is quite straightforward. It has to do with the exploitation, exhaustion, and replacement of dynamic strategies. The commerce strategy of Europe had exhausted itself by the mid-eighteenth century, which in turn terminated the demand for new scientific ideas. As the old commerce empires languished and flirted with a new conquest strategy, the enlightenment foundered—much to Wilson's puzzlement—and left a generation of intellectuals to their own fancies. These fancies, no longer disciplined by strategic demand, turned to the mysteries of romanticism, a reality that so exasperates Wilson in *Consilience*. Only when the industrial technological strategy took hold in Europe in the mid-nineteenth century (as early as 1780 to 1830 in pioneering Britain) did a new source of strategic demand for scientific ideas emerge. It is this strategic demand that has driven science ever since. Because it was not inevitable that the technological strategy replace the exhausted commerce strategy—in the premodern world the exhaustion of commerce had always led to a replacement with conquest—there was no guarantee that the deflating enlightenment would ever be reinflated. Equally there will be no guarantees in this respect when the industrial technological paradigm exhausts itself later in the twenty-first century (Snooks 1996: ch. 13).

Finally, it should also be mentioned that the dynamic-strategy model can explain the so-called "archetypes" that Wilson insists have emerged in the mind of man under the control of genes and epigenetic rules. Sociobiology, he believes, can strengthen the work of the great psychoanalysts Sigmund Freud and Carl Jung. Wilson tells us:

> These fragments [images and episodes in the unconscious mind] may correspond in a loose way to Freud's instinctual drives and to the archetypes of Jungian psychoanalysis. Both theories can perhaps be made more concrete and verifiable by neurobiology. (Wilson 1998: 78)

As I have shown in detail elsewhere (Snooks 1997), the recurrence of images and stories about mythic characters is not a result of the "collective unconscious" as suggested by Freud and Jung, nor of "epigenesis" as asserted by Wilson, but rather they are a response to strategic demand. In oral, visual, and written tradition the heroes, giants, and gods emerged from a need to justify and reinforce the conquest strategy; the earth mothers, great migrations, serpents, and holy men are a response to the family-multiplication strategy; the sea monsters, serpents, and other demons of the desert are generated by the commerce strategy; and the apocalypse is a response to the violent collapse of a dynamic strategy, particularly of conquest. The recurrence of these images through time and space is the outcome of the reemergence of the same dynamic strategies. Dreams and art, therefore, are shaped not by the collective unconscious or epigenetic rules but by the strategic pursuit. Quite clearly the study of the biological basis of the brain, while important in its own right, is no substitute for existential historicism.

Sociobiology's Vision of the Future

The demonstrable fact that the coevolution approach cannot explain the past or present has not deterred Edward Wilson from making predictions about the future. These predictions concern the future of both the human species and its society. Not surprisingly in neither case are these predictions based on formal sociobiological models.

A Bleak Future for Human Society?

In *Consilience* Wilson is concerned about the impact of our species on the natural environment. The only problem is that existing sociobiological models are unequal to the task. Rather than modeling the dynamics of human society and its impact on the natural environment, he adopts a naive neo-Malthusian approach. There is a certain irony here: having stripped the "doctrine of Malthus" from Darwin's theory of natural selection, some sociobiologists are now employing it in a nonbiological way for predictive purposes! Instead of employing a realist dynamic model to predict future outcomes, Wilson warns us about population growth in the (unanalyzed) event that it runs out of control. There is nothing here to distinguish his treatment from that of any other neo-Malthusian—any other "folk futurist."

This unscientific exercise is hardly a good advertisement for the sociobiological theory that we are told is to form the basis for the unity of knowledge. While Wilson quite justifiably criticizes the orthodox economists' approach to this issue, his own method is no more enlightening. We are merely told that "the wall toward which humanity is evidently rushing is a shortage not of minerals and energy, but of food and water," and we are subjected once more to the tired old homily about the lily pad which, in doubling in size each day, goes from half covering the pond on the penultimate day to completely covering it on the next. The well-worn implication being that the Malthusian crisis will descend upon us before we can take its implications into our decisionmaking.

No attempt is made to explain why this neo-Malthusian argument is any more correct today than when Malthus first advanced it almost two hundred years ago. Instead of providing scientific analysis, prediction, and policy, we are exhorted to gamble on his guess that the worst possible scenario of the folk futurist will eventuate. Is this what Wilson means when he says that sociobiology will transform the social sciences?

To support his vision of doom, Wilson refers with approval to the compatible conclusions of *selected* social scientists who also fail to employ a dynamic model when examining the fluctuating fortunes of human society. We are told that

> Archaeologists and historians strive to find reasons for the collapse of civilizations. They tick off drought, soil exhaustion, overpopulation, and warfare—singly or in some permutation. *Their analyses are persuasive.* Ecologists add another perspective, with this explanation: The populations reached the local carrying capacity, where further growth could no longer be sustained with the technology available. (Wilson 1998: 287; my emphasis)

What he means is that the *conclusions*, not the *analyses*, are persuasive because that is what he wants—has been "selected"—to hear. It is quite clear from Wilson's discussion of the social sciences in *Consilience* that he does not regard their method as scientific, but he is willing to accept their conclusions if they agree with his own. Wilson cannot have it both ways. In the end all he has demonstrated is the poverty of sociobiology as a method for investigating the dynamics of human society. It is curious that Wilson has overlooked a recent attempt from within the social sciences to resolve these issues of population, economy, and environment by developing a realist dynamic model (Snooks 1996) rather than resorting to what I call the great shopping list of history—the random nomination of all those issues that are believed to have some influence on the question under study—to which Wilson gives his approval. This is discussed in part III, where I show that there is no Malthusian crisis, only a strategic crisis: we exhaust dynamic strategies, not natural resources.

The Future of Human Nature

Wilson is also concerned with the present and future role of natural selection, particularly in relation to genetic engineering. In the final chapter of *Consilience* he begins by asking: "Is natural selection still operating to drive evolution? Is it forcing our anatomy and behavior to change in some particular direction in response to survival and reproduction?" (Wilson 1998: 271–72). In answering this question he says that "the big story in recent human evolution is not directional change, not natural selection at all, but homogenization through immigration and interbreeding." This is a remarkable admission for a Darwinist. It suggests not only that natural selection is no longer an active force in the modern world, but that it has been impotent since the emergence of *Homo sapiens* one hundred and fifty thousand years ago. This conclusion has some interesting implications that are usually conveniently ignored by neo-Darwinists. Most importantly it implies that modern man has been able to survive and prosper by abandoning genetic change in favor of technological change. And it begs the question: How is this possible if Darwin was correct about the importance of natural selection?

Whatever the present role of natural selection, its future will be insignificant, Wilson tells us, because "the rules under which evolution can occur are about to change dramatically and fundamentally." Owing to the

> advance of genetics and molecular biology underway, hereditary change will soon depend less on natural selection than on social choice. Possessing exact knowledge of its own genes, collective humanity in a few decades can, if it wishes, select a new direction in its evolution and move there quickly. (Wilson 1998: 273)

He calls this radical new process "volitional evolution." Hence, "humanity will be positioned godlike to take control of its own ultimate fate. It can, if it chooses, alter not just the anatomy and intelligence of the species but also the emotions and creative drive that compose the very core of human nature" (Wilson 1998: 274).

The implications arising from direct human control over its own genetic structure are huge not only for our very humanness but also for the status of the theory of natural selection. I will confine my remarks to the latter. The first and most obvious point is Wilson's realization and admission that natural selection is not a universally applicable theory and that in the future the selfish gene will no longer be in control of the human species—the "genetic leash" will be severed forever. The organism will have finally triumphed over the gene. In other words, this pioneering sociobiologist is admitting that natural selection does not constitute a universal law of life. At best it is only relevant to life on Earth for the four billion years prior to the emergence of modern mankind, and will not be relevant to the remaining three to four billion years that we have left on this aging planet before an expanding sun forces us to move to the outer reaches of our solar system, nor to the eternity that might lie ahead of us elsewhere in the universe after our solar system has totally self-destructed.

But we might ask, if "social choice" is to replace natural selection in the future, how confident can anyone be about its status in the past? Similarly, if the selfish gene can be dethroned so easily—it is only ten thousand six hundred years since the first urban societies emerged and merely four centuries (about sixteen generations) since the first stirrings of the scientific revolution—how confident can we be that it ever held sway in the past? Both questions become even more significant in face of the further admission that natural selection has been dormant since the emergence of modern man.

What this means is that the neo-Darwinists are without a dynamic model to predict the entire future of the human species and its society or even to explain their past over the last one hundred and fifty thousand years, let alone life in general over the three thousand five hundred million years (myrs) before that. How then can sociobiology possibly become the linchpin of human knowledge and how will its practitioners be able to explain the process of genetic engineering (part of the technological strategy) in the future?

Clearly we need a universal dynamic model that can explain not only the deep past of life on Earth but also the more recent past of human civilization, and of course all future developments. The dynamic-strategy model, presented in part III, suggests that the future of hereditary change will not be a radical break with the past, as claimed by Wilson and his colleagues. My theory suggests hereditary change has *always* been influenced by "social choice"—what I call **strategic selection**—through the strategic pursuit in which genetic change is just one of four possible dynamic strategies that mankind (and all other species) has (have) always employed. *The future will involve a difference of degree, not of kind, because in the future "social choice" will operate on genetic change, not for the first time in history, just more directly, more precisely, and more rapidly.*

Mankind's influence over genetic change has passed through three main stages during the past two to three million years. During the first stage—prior to one hundred and fifty thousand years BP—"social choice" over genetic change was indirect, imprecise, and very slow acting. Genetic change from two million years to one hundred and fifty thousand years BP was a human-influenced response—through strategic selection—to the demands generated by our pursuit of the family-multiplication strategy. As will be shown in chapter 12 below, genetic change enabled mankind to adopt new food sources that were essential to the expansion of its geographical range.

The second stage, from 150,000 BP to the present, saw technological change—at first as a facilitating device for other dynamic strategies and then (from the late eighteenth century) as a dynamic strategy in its own right—displace genetic change entirely. Once the brain size of mankind had, with the emergence of *Homo sapiens*, passed the necessary threshold, technological rather than genetic "ideas" were much more easily and precisely controlled and they generated vastly quicker results. Because of this, genetic change fell into "disuse" for one hundred and fifty thousand years.

With the twenty-first century we are entering a third stage in which genetic change, after a lapse of one hundred and fifty thousand years, is being reintroduced into life's strategic pursuit as part of the ongoing dynamic strategy of technological change. And it is being reintroduced with infinitely more directness, precision, and speed. What is happening for the first time in history is not that we can influence genetic change through "social choice," as Wilson and many others claim, but that we can have a more *direct* influence. This is a matter of degree rather than of kind, which has been brought about by the adoption of a more effective dynamic strategy—of technological change. Like the broader technological strategy since the late eighteenth century, the substrategy of genetic change promises to be a *continuous* agency for the dynamics of both human society and life. For the first time in history human society and life will become one.

As the dynamic-strategy theory can explain all three stages encompassing the entire past, present, and future history of mankind (and of life), it can claim far greater generality than Darwin's theory of natural selection. Even if natural selection was not fatally flawed, it could not claim to constitute a universal law of life. Such a claim, however, can be made on behalf of the dynamic-strategy theory. If natural selection was not fatally flawed it would only be a special case of my more general and universal theory. But natural selection and strategic pursuit are totally incompatible theories. The rise of the dynamic-strategy theory necessarily involves the fall of Darwinism in all its forms.

Conclusions

What has sociobiology got to offer thirty-five years after its initial manifesto? Its great strength is a focus on the behavior of real animals rather than on the theoretical fantasies of neo-Darwinism. No doubt much of the detailed curiosity-driven research on insects, birds, mammals, and so on will stand the test of time. Its great weakness, however, is that much of this empirical material has been forced into a neo-Darwinian straitjacket. While this mainly constitutes lipservice to the Darwinian icon of natural selection, the more ideological exercises will have distorted the empirical results. Much of this research will need to be reworked within the context of the dynamic-strategy theory.

Where sociobiology fails completely is in its claim to be able to explain the dynamics of both life and human society. An attempt has been made here to show that the core kin-selection theory is unable to explain animal behavior, and that the theory of coevolution is incapable of accounting for the dynamics of human society. Their basic problem is the flawed neo-Darwinian paradigm of the "selfish gene," which is alleged to hold humanity on a tight "genetic leash." It is a fantasy that not only defies common sense but cannot be empirically supported. Worse still, it cannot explain human dynamics. The supply-side coevolution view has human culture being extruded from the genetically constructed and controlled human brain. Like all supply-side models, coevolution is unable to explain the past or predict the future of human society. Further, it is grotesque in its antihumanist stance.

To bring this critique to an end I wish to reflect on the striking contrast in styles between the leading advocates of neo-Darwinism and sociobiology. Richard Dawkins, one of the leading publicists of neo-Darwinism, is crystal clear about his intentions. He wishes to take the neo-Darwinian argument to its logical conclusion. In doing so he unwittingly exposes the absurdity of neo-Darwinism and of sociobiology. He appears less interested in persuasion than in exposition. Wilson on the other hand is far more guarded about what a belief in neo-Darwinism actually means. He wants to persuade outsiders, particularly in the social sciences and humanities, to accept and adopt the "modern synthesis" as it applies to the animal and human worlds. Although he accepts the logical conclusion of neo-Darwinism that genes are attempting to maximize their presence in the gene pool, he realizes that such talk is anathema to most humanists. Possibly for this reason he now (Wilson 1998) focuses his attention (in contrast to Dawkins) on the organism rather than the gene, and talks rather obscurely about the extent to which genes exercise control over individuals, their behavior, and their societies.

When discussing the role of "consilience researchers," for example, Wilson tells us:

> Their grasp of cultural change is enhanced by insights from evolutionary biology, which interpret the species-wide properties of human behavior as products of genetic evolution. *They are careful how they express that idea*—avoiding the assumption that genes prescribe behavior in a simple one-on-one manner. Instead, the analysts use a more sophisticated formula that conveys the same meaning more accurately: *Behavior is guided by epigenetic rules*. (Wilson 1998: 193; first emphasis mine; second, Wilson's)

Wilson, who is now more careful not to expose the logical implications of neo-Darwinism, must find the enthusiastic exposition of Dawkins an embarrassment. His one major slip, however, was, in a flush of youthful enthusiasm of his own, to introduce the term "genetic leash." But do not be misled by the recent more moderate tone. Wilson and all other sociobiologists subscribe to the same worldview expounded so brilliantly by Dawkins: individuals are the survival machines constructed by genes and they are held in check by a tight genetic leash. It is a profoundly antihumanist view, which is the product of science fiction rather than science reality.

While close cooperation between the natural and social sciences is an excellent idea—and one I have been pursuing since my book *The Dynamic Society* (Snooks 1996)—it is unlikely to occur along the lines advocated by Edward Wilson in *Consilience* (1998). Certainly we will all gain from a better understanding of the genetic basis of the human brain. But this will not open the door to a complete understanding of human nature, nor will it provide a satisfactory analysis of the structure and dynamics of human society. Reductionism may increase our knowledge of some static issues but definitely not of dynamic systems. The fact that the past generation of sociobiologists has been unable to provide a workable model of human society despite Wilson's confident claims in the 1970s is adequate proof of this conclusion.

The only way to understand human nature is through the systematic observation of human beings in history, and the only way to construct realist models of the dynamics of human society is by exploring the causal regularities that *underlie* the dynamic patterns of society throughout time and space. This is the basis for the new analysis in part III to which we now turn.

Part III

Rising from the Ruins

When the dust had settled it became clear even to the high priests that their work, together with the master's original chapel, had been destroyed. . . . Totally demoralized, the religious community disintegrated and quickly left the region forever.

With their departure the townspeople resumed removing the fossil-bearing stones and within a relatively short time a towering new edifice rose within the town's center. They called it La Pedrera. . . . Those who organized the construction of the new secular structure told a different and persuasive story of life based on careful observation of what had gone before. In this way they were also able to see what was to come. The great cathedral of metaphysics had collapsed but a new realist structure dedicated to humanism had risen from the ruins.

Prologue

Chapter 9

A New Story of Life

The story of life has been told ever since early humans gathered around their campfires at night. Constantly changing, this story has only recently gained any precision, because only recently have the storytellers been able to assemble the necessary fossil evidence. Yet even now that we have a better understanding of what has happened, we have little idea of what made it so. Why? Because our explanations are based on flights of fancy rather than on the growing body of fossil evidence. We either cling to badly flawed Darwinian explanations and reject the evidence, or we embrace the evidence and offer fanciful explanations about exploding mountains, crashing asteroids, biblical-like floods, and even regular visitations from "Death Stars." The inhabitants of ancient Sumer, mankind's first great civilization, would have felt perfectly comfortable with these stories of life.

Any persuasive theory of life must be based on a sound foundation of evidence. We need to sketch the changing patterns of life—or what I call **timescapes**—before beginning the process of theory construction. It is the timescapes that need to be explained. Among the Darwinists only the paleontologists, through their detailed study of the fossil evidence, have taken the historical patterns of life seriously. Their tragedy is that, instead of developing a new dynamic model to explain these patterns, they have attempted to force them into a Darwinian straitjacket. The outcome is complete confusion. They could have succeeded only had they rejected Darwin's theory of natural selection and begun anew. While sympathetic to their historicist approach I cannot accept their *analytical* explanations—particularly the statistical pseudo-explanations of the biometricians (Sepkoski 1978; 1979; 1984) which possess the same kind of technical aridity as econometrics in the social sciences.[1]

In this chapter I present a series of timescapes outlining the fluctuating fortunes of life over the past 3,500 million years (myrs) and provide an initial historicist explanation of these macrobiological developments.[2] This forms the foundation upon which my formal dynamic-strategy theory of life and human civilization is constructed in the rest of the book. The objectives in this chapter are to show what any dynamic theory of life must be able to explain, and why all forms of Darwinism are not up to the task.

The Timescape of Life—The Past 3,500 Million Years

The fossil evidence suggests that life on Earth has experienced four distinct long-term phases of development. These comprise the 2,000 myrs dominated by the single-celled bacteria known as blue-green algae (until about 900 myrs BP—before the present); the 600 myrs following the decline of bacteria, dominated by primitive plant and animal (including reptiles) life in the seas and on the land; the 180 myrs following the mass extinction of these primitive life forms at about 245 myrs BP, which was dominated by dinosaurs on the land and in the air; and finally the 60 myrs following recovery from the extinction of the dinosaurs, which has been dominated by birds, mammals and, finally, mankind.

The Diversity of Life

The best quantitative measure available to substantiate these main phases of life is the number of families of marine and land animals (metazoa) as shown in figure 9.1. These timescapes are really pictures of the fluctuating diversity of life over the past 600 myrs (known as the Phanerozoic eon), a vast expanse of time several times greater than even Darwin was willing to postulate in the mid-nineteenth century. Little diversity existed before that time owing to the dominance of blue-green algae. While this general pattern of diversity has been known since John Phillips first sketched its outlines in 1860, John Sepkoski did so with greater precision in the late 1970s and early 1980s (Cowen 2000: 80). When examining this figure it should be realized that, as the recovery rate of fossils increases the nearer we approach the present, it exaggerates the expansion of families over time. Also the apparent downturn in land families in the early and late Jurassic is largely a figment of the poor fossil record (Benton 1985: 813). Finally we should not confuse fluctuations in families with changes in biomass, because there is a trade-off between size and quantity of individuals on the one hand and diversity on the other.

What these timescapes show is a succession of surges and retreats in the diversity of life. The initial radiation of marine and land animals between 570 and 245 myrs BP (the Paleozoic era) was terminated by a major extinction resulting in a sharp decline in the number of marine (by 52 percent) and land (by 49 percent) families; the renewed expansion between 245 myrs and 65 myrs BP (the Mesozoic era) was brought to an end by widespread extinction of marine (11 percent) and land (14 percent) families that was much worse in terms of species; and the very rapid increase in diversity since 65 myrs BP (the Cenozoic era) has recently been eroded by the activities of mankind.

Figure 9.1 Family number for marine and land organisms—the past 800 myrs: (a) marine organisms

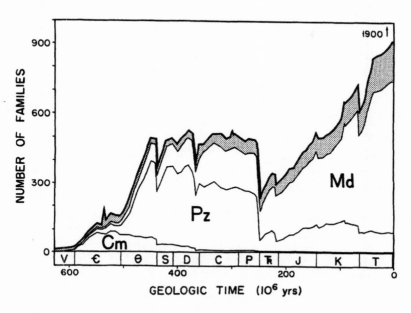

"The upper curve shows the total number of fossil families. . . . The fossil diversity of . . . 'poorly preserved' families (which largely reflects range extension between various exceptional fossil deposits and the Recent) is indicated by the stippled field in the figure. The two curves below the stippled field divide the diversity of heavily skeletonized families into three fields, representing the three 'evolutionary faunas' that dominate total diversity during successive intervals of the Phanerozoic. . . . The fields are labeled 'Cm' for Cambrian fauna, 'Pz' for Paleozoic fauna, and 'Md' for Mesozoic–Cenozoic, or Modern, fauna. . . . Data on fossil families are from Sepkoski (1982)."
Source: Sepkoski 1984: 249 (reproduced with permission).

It is important to realize that the timescapes presented in figure 9.1 are a reflection of genetic diversity rather than total biological output. The general pattern is one of exponential growth in genetic diversity in the early history of any dynasty, followed by stagnation, or even slow decline, and later by collapse and extinction (Sepkoski 1978; 1979; 1984). This should not be interpreted as indicating that total biological activity by any dynasty followed the same pattern. Quite the contrary, dynasty population continues to increase rapidly after genetic diversity stagnates and reaches a peak just prior to collapse and extinction. The pattern of genetic diversity reflected in figure 9.1 is, it will be argued, the outcome of the pursuit of different dynamic strategies. The exponential growth of diversity is an outcome of the dynamic strategy of genetic change, while the long phase of stagnation in diversity (wrongly identified by Eldredge and Gould [1972] and Sepkoski as a state of "equilibrium"), is an outcome of the pursuit of the dynamic strategy of family multiplication (procreation and migration).

Figure 9.1 Family number for marine and land organisms—the past 800 myrs: (b) land organisms

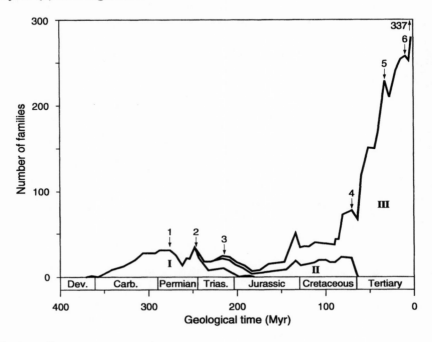

I   = reptiles
II   = dinosaurs
III = mammals, birds, etc.
Source: Benton 1985: 811 (reproduced with permission from *Nature*, Macmillan Magazines Ltd).

The timescapes in figure 9.1 also show that these major radiations and extinctions comprise a number of overlapping developments in both marine and land fauna. Each new radiation of families and species was launched only after an earlier radiation was on the decline or had suffered major extinctions. Made by a number of paleontologists, this important observation will be returned to later in the chapter. What it means is that speciation—the formation of new species—has only taken place once competition has been eliminated. Why is this important? Because Darwin's theory of natural selection is irrevocably based on the assumption that evolution is the outcome of intense competition for scarce resources. The evidence suggests the very opposite: that speciation occurs only during periods of minimal competition and of resource abundance. Hence, natural selection is unable to explain the major phases of genetic change and speciation during the past 3,500 myrs—a major limitation.

Any survey of the fluctuations of life, no matter how brief, would be incomplete without mention of the world's flora. As can be seen from figure 9.2, there have been three main stages in the development of Earth's land flora involving three very different plant types. As in the case of the fauna, with which a close association was formed, each new radiation occurred only after the decline of the previously dominant one.

Figure 9.2 The rise and fall of major land plants over the past 410 myrs

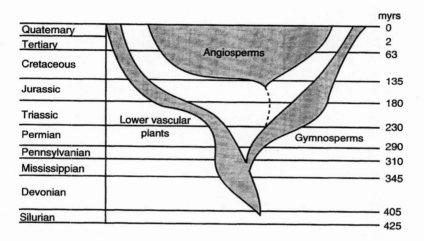

The changing flows are roughly proportional to the number of genera of plants in each group.
Source: Based on Newell 1978: 183.

The first stage of plant development on the land concerned the lower vascular plants—such as ferns, lycopods (club mosses), and horsetails—which expanded rapidly from about 405 myrs BP to reach a peak at about 290 myrs BP. This coincided with the invasion of the land by animals such as insects, amphibians, and reptiles. The second stage involved the gymnosperms (plants with naked seeds)—such as conifers, cycads, and ginkgos—which diversified from about 310 myrs BP to reach a peak at about 180 myrs BP. This was the period associated with reptiles, protomammals (therapsids), and protodinosaurs (thecodonts). Thereafter their number of genera declined and, like the lower vascular plants, they have continued to the present by playing a more modest role. The last stage involved the great expansion of angiosperms (flowering plants), beginning at about 135 myrs BP and diversifying rapidly to the present. This incomparable radiation began during the age of the dinosaurs—with some (Bakker 1986: ch. 9) arguing that it was made possible by changes in their feeding activities—and has reached its greatest expression during the era of birds and mammals.

The Great Waves of Life

Using what paleontologists have learned about the fluctuating diversity of the Earth's flora and fauna, it is possible to identify the longest swings in biological development. I have elsewhere called these the **great waves of life** (Snooks 1996: ch. 4). To describe the great waves we need to identify the main turning points in life, when one major form or dynasty of life gave way to another. These include the shift from prokaryotic life (blue-green algae) to eukaryotic life (plant and animal organisms) at about 900 myrs BP; the shift from ectothermic life (cold-blooded reptiles) to endothermic life (warm-blooded protomammals) at about 245 myrs BP; and the replacement of dinosaurs with mammals at about 65 myrs BP. Other setbacks such as the widespread extinctions around 435 myrs, 370 myrs, and 215 myrs BP constitute major fluctuations within these great waves, just as minor fluctuations can be recognized within the major fluctuations, and so on—waves within waves. Something like a dynamic version of a Russian doll.

The reasons for identifying the great-waves timescape are twofold. First, to provide a convenient visual structure for organizing the detailed story of life that is assembled later in this chapter. And second, to identify the macrobiological pattern that must be explained by any dynamic model of life.

Owing to our modern fossil-dating techniques, it is possible to provide a reasonably accurate timescale (horizontal axis) for figure 9.3. The vertical axis, however, presents a greater problem. What I wish to achieve with this timescape is a sense of the expansion in the quantity of life on Earth over vast periods of time—a sort of impossibly gigantic population census adjusted for size and complexity. Something equivalent to the fluctuating timescape of gross domestic product (GDP) for a modern nation. The idea is to provide a macrobiological view of life on Earth—in effect, *the changing outcome of life's use of the Earth's resources*—rather than the usual micro view of the rise and fall of individual species. Accordingly the vertical axis should be thought of as a measure of total biomass. Of course it can only be represented impressionistically because, although we have reasonable evidence on the number of families, we know little about the number of species and even less about the number of individuals. This is not a major limitation, however, because I am interested in the length rather than the amplitude of these great waves.

The dynamic timescape presented in figure 9.3 shows a series of four great waves of biological activity. The first great wave is about 2,000 myrs in duration, the second about 600 myrs, the third about 180 myrs, and the fourth, which is by no means complete, about 60 myrs. It is interesting that the declining duration of each wave in figure 9.3 is a mathematical constant—each wave is approximately one-third as long as the one that preceded it (Snooks 1996: 79–80). Also we know enough to say that the amplitude of these great waves is increasing, probably in a similar exponential way. There can be little doubt, therefore, that the momentum of life on Earth is accelerating, owing to the increasing access to natural resources by its life forms, made possible initially by genetic change (the **genetic option**) and more recently by technological change (the **technology option**).

Figure 9.3 The great waves of life—the past 3,000 myrs

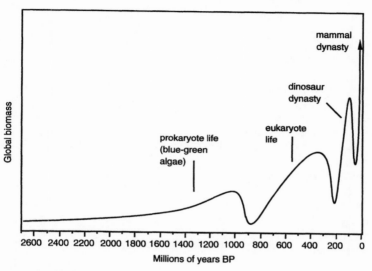

Source: Snooks 1996: ch. 4..

Either this systematic relationship between the great waves is purely coincidental or it is powerful evidence that the great waves of biological activity are largely endogenously determined, with the resource-accessing "knowledge" (either genetic or technological) of each wave of life feeding into its successor, rather than being exogenously and randomly driven by natural catastrophes or fortuitous genetic occurrences. What are the implications of this discovery? It appears to constitute an important law of life (Law #10, examined in chapter 15): that any dynamic life system (including human societies) involving an accumulating stock of "ideas" (either genetic or technological), where the output of "ideas" in one wave becomes an input of the next wave, will grow exponentially. What is more difficult to explain is why this overall exponential expansion occurs through a series of waves—involving expansion and collapse—of increasing intensity.

Interpreting the Great Waves of Life

To understand the great waves of life over the past 3,500 myrs it is necessary to explore the nature of their upward surging, their stagnation as they appear to defy gravity, their cascading collapse and, eventually, their rejuvenation. This requires the use of a theoretical framework that must be based on a close examination of the fluctuating fortunes of both life and human society. I have called it the dynamic-strategy theory (Snooks 1996; 1997; 1998a; 1998b; 1999). At this stage my inductive theory is merely used to tell a new story about life on Earth, and only in subsequent chapters is the theory set up formally and explained in detail. A brief sketch here will help the storytelling.

The story I tell about the great waves of life focuses on the determination of life forms to survive and prosper by devising **dynamic strategies** to exploit available natural resources as sources of energy and shelter. These dynamic strategies include genetic/technological change, family multiplication (procreation and migration), commerce (or symbiosis), and conquest. While each phase in the **strategic pursuit** is dominated by one dynamic strategy, it will be supported by one or more of the other strategies. For example, the conquest strategy of the dinosaurs (or Rome) was supported by the genetic (or technological) development of offensive and defensive weapons. We can now run through a typical sequence in the rise and fall of the great waves of life. The why and how will be examined in later chapters.

Following the widespread extinction of an earlier dynasty of life—the collapse of a great wave—surviving organisms are suddenly faced with an abundance of unutilized natural resources which they are able to exploit through genetic change. This, of course, is a "non-Darwinian world" in which natural selection is impotent. Once an extensive range of new species—what I call **genetic styles**—has been developed by the surviving organisms, unutilized resources are exploited through the pursuit of the more economical family-multiplication strategy. It is this shift from the genetic to the family-multiplication strategy by organisms that accounts for the genetic profile, at both the species and dynasty levels, that has become known, misleadingly, as "punctuated equilibria." It was first discovered in 1859 by Hugh Falconer, a contemporary of Darwin, and rediscovered by Gould and Eldredge (1972). With the exhaustion of its genetic style a species overpopulates its ecosystem, becomes subject to internal strife ("civil war") and/or an aggressive takeover (conquest strategy) by other species, and finally goes extinct. New species may emerge to exploit any unutilized natural resources.

In addition to exhausting individual genetic styles, organisms in the new dynasty eventually exhaust the entire **genetic paradigm**. This paradigmatic exhaustion is reflected in the "depressed origination rate" that some paleontologists (Benton 1985) have noticed in the data on families of land animals. What this means is that a point is reached when the dominant life form is unable to extract any further resources from the accessible world. At this time global populations press heavily on a fixed—for the time being—supply of natural resources creating intense competition between individuals and species. Here at last is the "Darwinian world."

But rather than pursuing genetic change (speciation) the herbivores battle with each other for scarce resources and begin to destroy the physical environment by overgrazing, while the carnivores not only overexploit their stressed prey but also turn on each other. This is the conquest strategy, which may generate **strategic demand** for support from genetic change in the form of add-on "technology" involving offensive and defensive weapons. The end result of this endogenous dynamic process—in which the "origination rate" falls and the "extinction rate" rises, possibly only marginally—is collapse and extinction. While extinction of an entire dynasty is the logical outcome of this endogenous process, it may be hastened either by a **strategic struggle** with individuals in an aggressive new dynasty or by a series of physical catastrophes.[3] It is important to realize that physical catastrophes cannot generate the systematic rise and fall of the great waves of life. Also it is significant that natural selection is impotent even in a "Darwinian world" of intense competition, which leads to conquest and extinction rather than genetic change and speciation.

Once oppression by the old dominant forms of life—the old **strategists**—has been removed in this way, previously minor forms—the new strategists—take center stage and, through the pursuit of the genetic strategy, exploit their monopoly position to flourish and prosper. Once again this is a "non-Darwinian world" of systematic genetic change and speciation in response to an *abundant* supply of natural resources and *minimal* competition. In these circumstances the constraints on the new strategists are released and they now have the time and opportunity to develop a range of genetic styles to gain access to resources from which they were previously excluded. New speciation takes the form either of the generation of superior species (in terms of their resource-accessing abilities) or of replacement by species with similar abilities to the former inhabitants.

The emergence of a range of superior species is the outcome of a genetic revolution—what I call a **genetic paradigm shift**—while a replacement with similar species is the outcome of the exhaustion of the entire genetic option. In the latter case it is the beginning of the **eternal recurrence**—or the **great wheel of life**—in which dynasties rise and fall without improving their access to the Earth's resources (see figure 13.4). The only way to break out of this eternal recurrence is through the replacement of the genetic option with the technology option. As we shall see, while the dynasty of dinosaurs was unable to achieve this and went extinct (as had the protomammals before them), a branch of the dynasty of mammals succeeded, if only just, before they also nearly collapsed and fell into oblivion.

This conceptual outline has left many theoretical questions unanswered, such as how can life forms pursue a genetic strategy, or indeed any other dynamic strategy; what are the dynamic mechanisms driving these great waves; and what are the underlying laws of life implied by this dynamic-strategic system? These issues can only be dealt with in the chapters that follow. Here I merely wish to marshal the evidence within my great-waves framework and to present the dynamic-strategy story of life.

The First Forms of Life on Earth

Sometime before 3,850 myrs BP protocells emerged through the concentration of organic chemicals, possibly by evaporation, freezing, or one of a number of other suggested causes (Nutman et al. 2000: 3052; Cowen 2000: 9). These protocells grew by absorbing simple organic substances from seawater, and they divided as they grew. As this occurred in a passive and accidental way in the beginning, these cells cannot be regarded as being "alive." While the dividing line between living and nonliving protocells is rather unclear, the conventional wisdom tells us that "a protocell could have been called living as soon as protein formation, nucleic acid formation, and cell division became integrated and reliable" (Cowen 2000: 12).

This is, however, a rather mechanical and passive definition of life. It reflects a widely held view in biology, indeed in the natural sciences generally, that organisms merely respond to a changing environment in a passive way. Most explanations of the history of life treat exogenous physical conditions as the driving force. Paleontologists do not even consider the possibility that the driving force might reside *within* the life forms themselves. No one to my knowledge has successfully attempted to develop an endogenous dynamic theory of life. It would require a very different vision to that held by the Darwinists.

I want to suggest a more realistic definition of life that focuses on the origin of the driving force rather than on the mechanism of replication. The central thesis of the dynamic-strategy theory is that life contains a dynamic that drives it to grow and prosper. This dynamic, which can be observed operating throughout the history of life, had its origin in the emergence of biochemical processes in simple cells. In order to generate and sustain life, these simple cells had to be able to maintain an internal metabolic process. This involved the capture and fermentation of organic molecules (fuel) to produce the energy needed to sustain life. Once an individual metabolic system is established—there is still no agreement about how this occurred—the simple cell generates a **metabolic demand** (or hunger) for fuel to feed this chemical process. If this demand for fuel is not met, the cell begins to starve and its structure begins to break down. Starvation leads the cell to frantic efforts to find a new source of fuel. This need to meet metabolic demand was the beginning of what I call **strategic desire**. And the different ways that organisms have attempted to meet this need I have called the dynamic strategies of life.

Life emerges, therefore, when a simple cell is able to develop a workable metabolic system. Without it life cannot be sustained. Replication, the core of the neo-Darwinian definition, is a secondary matter, because it is not essential in sustaining the life of the initial simple cell. In the beginning of life on Earth, new cells emerged without the ability to replicate. It was a situation that probably existed for a considerable period of time before the trick of systematic replication was learnt. In fact, replication—and the development of the tools of replication such as RNA, DNA, and protein formation—was a response to strategic desire. It was one—family multiplication—of the four dynamic strategies by which organisms have attempted to meet metabolic demand. Strategic desire, therefore, emerged before—long before—the need or ability to reproduce.

There is considerable controversy about how the first living cells obtained the energy they needed to meet their metabolic demand. The conventional wisdom

is that the first forms of life were cells that absorbed fuel for their simple metabolic systems directly from seawater. This fuel, which is still employed by organisms today, was adenosine triphosphate (ATP), composed of three phosphates. By detaching one of these phosphates from ATP a living cell is able to
create metabolic energy, which provides the fuel of life. Recent research, however, suggests that before the use of ATP, living cells probably absorbed py-
rothermic energy from hot springs.

Whatever the first source of fuel, we know that supplies of freely available
ATP were relatively quickly exhausted. This meant that existing cells had to
develop a more complex fermentation process, known as heterotrophy, in order
to extract energy from captured organic molecules and to convert adenosine
diphosphate (ADP), consisting of two phosphates, into ATP. In effect living
cells had to employ genetic change as a strategic instrument in the pursuit of the
protostrategy of conquest. By chemically breaking down organic molecules
captured in and absorbed from seawater, these first dynamic strategists were
able to effectively fuel their internal energy systems. This fermentation process
is illustrated in figure 9.4.

Figure 9.4 Heterotrophy—the first dynamic strategy

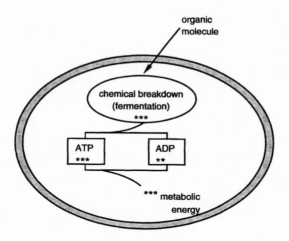

Source: Based on Cowen 1990: 25.

This, then, is what the history of life is all about: devising more effective
ways of obtaining fuel—what we call food—to fire up the internal metabolic
system in order to survive and prosper. In this case the heterotroph pursued the
protostrategy of conquest, which involved the capture and consumption of organic molecules. Only later did it pursue the protostrategy of family multiplication, which involves replication and migration. Right from the very beginning
life was devising dynamic strategies to achieve its objectives. Clearly, intelligence was not required to join in the strategic pursuit.

The Prokaryotic Revolution

This revolution occurred some time before 3,500 myrs BP when the heterotrophic life form had exhausted the supply of freely floating organic molecules. It was created by what has been called "the world's first energy crisis" (Cowen 1990: 30). The response to this crisis by heterotrophs was the development of a metabolic system based on photosynthesis, which involved the substitution of a **genetic strategy** for the earlier conquest and family multiplication protostrategies. This prokaryotic transformation constituted a genetic revolution, or a genetic paradigm shift, because it greatly increased the potential development of existing life forms by opening up their access to the Earth's natural resources.

The new autotrophic cells had constructed a metabolic system based on photosynthesis. As is well-known, this process employs sunlight through the agency of chlorophylls together with carbon dioxide and hydrogen to fuel the ATP/ADP process shown in figure 9.5. These autotrophic cells were also able to build up and store energy-rich molecules such as glucose that could be used to feed a fermentation process when sunlight was not available. The waste product from this revolutionary metabolic system was either sulphur or oxygen depending on the form in which hydrogen was employed.

The first and least efficient autotrophic process was employed by sulphur (or green and purple) bacteria, which broke down hydrogen sulphide (H_2S) releasing sulphur as a waste product. More efficient (by a factor of six) is the process employed by cyanobacteria (or blue-green algae) which break down water (H_2O) to gain access to hydrogen and release oxygen as a waste product. Water not only fuels a more efficient chemical process, it is one of the Earth's most abundant substances. But there is one unfortunate side effect: oxygen is highly toxic to life and must be handled with extreme care. It created acute problems for blue-green algae when the oxygen levels in the seas rose to significant levels. This was the first case of a dominant life form on Earth polluting its own environment—the "oxygen holocaust" (Duve 1996: 44)—and it led to the transformation of biological systems to cope with it. Once this transformation had occurred the rising levels of oxygen in the oceans and the atmosphere actually enabled the further expansion and development of other life forms.

Following the Prokaryotic Revolution, blue-green algae spread steadily in colonies in the shallow waters around continents and islands and probably in a thin floating layer across the ocean surface through the pursuit of the **family-multiplication strategy**. Family multiplication usually follows genetic transformation as it enables organisms in the new species to exploit their capacity to access natural resources. The colonial approach became the most effective and enduring way to survive. Through simple cell division blue-green algae built new colonies called stromatolites consisting of layers of bacterial mat and sand cemented together with protective layers of slime secretions. These colonies also harbored uninvited guests that flourished by employing a **commerce** (symbiosis) **strategy**. The top layer of the stromatolites consisted of blue-green algae, which operated as the powerhouse of the colony through its use of photosynthesis to produce organic matter that could be exploited by other life forms. The middle layer comprised green and purple bacteria, which used a less efficient form of photosynthesis and produced sulphur rather than oxygen as a waste product. And the bottom layer consisted of heterotrophs, which lived off the organic matter generated by the higher layers of life. Together they produced a colony that gave protection to them all, particularly from the worst effects of ultraviolet light and oxygen.

Figure 9.5 Autotrophy—the first genetic revolution

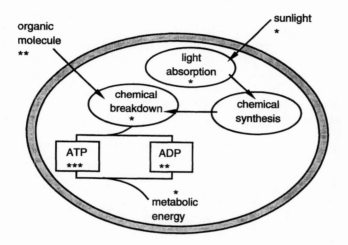

Source: Based on Cowen 1990: 28.

Fossil evidence for stromatolites can be traced back 3,500 myrs in the Warrawoona series in the northwest of Western Australia, and their living counterparts can be seen today in nearby Shark Bay. By 3,100 myrs BP the evidence shows that stromatolites had diversified into two different styles and by 2,800 myrs BP they had invaded salty lakes as well as coastal regions. Thereafter the expansion of this life form in these environments through the family-multiplication strategy was rapid, continuing for a further 2,000 myrs. During this later phase, at least, blue-green algae was the dominant form of life on Earth.

This is not to imply that the first great wave of life developed smoothly or without interruption. By generating oxygen as a waste product, the stromatolites polluted their own environment in two ways, both of which threatened their survival and caused a major interruption to their expansion. First, as oxygen is toxic to all forms of life, special methods have to be developed to cope with its presence. After initial setbacks before 2,500 myrs BP blue-green algae finally developed an antidote—an enzyme called superoxide dismutase—which saved them from the fate experienced by competitors that were less innovative. They also invented ways of employing oxygen in a new process called respiration, whereby the by-products of fermentation (lactic acid) are oxidized down to carbon dioxide (CO_2) and water, thereby releasing up to eighteen times more energy per sugar molecule than by simple fermentation. Once these "add-on" pieces of genetic technology were developed, the early reverses were left far behind, at least until overexpansion of blue-green algae took place around 1,000 myrs ago. This dominant life form, unlike all those to follow, had the time to develop new genetic technologies because they had no serious competitors until after they had resolved this problem and the rapid growth of oxygen levels enabled the emergence of dangerous new life forms (such as snails and other grazing animals).

The second problem facing blue-green algae owing to the increasing levels of oxygen was the growing iron deficiency in the oceans. In the early years of this life form's development the oceans contained a high content of dissolved iron, which was essential for its success. The iron deficiency arose because the rising levels of oxygen in the sea water caused the dissolved iron to precipitate, laying down vast banded iron formations on the sea floor between 2,500 myrs and 1,800 myrs ago (Cowen 2000: 33–35). Today these rich iron deposits are to be found in the Hamersley Range in northwest Western Australia, Krivoi Rog in Russia, Labrador and Minnesota in North America, and Brazil in South America. Hence, a by-product of the Prokaryotic Revolution provided, some 2,000 myrs later, a major input for the Industrial Revolution. But this gift for a very distant ancestor threatened the very existence of blue-green algae.

Survival and renewal of expansion for blue-green algae depended on their ability to invent a further piece of add-on technology. Remarkably they were able to develop molecular "cages"—complex molecules called siderophores—that, triggered by iron levels falling below a critical level, can pass through cell membranes in pursuit of iron atoms. Having captured its prey the siderophore reenters the parent cell through a special valve. This remarkable exercise is repeated until the cell's iron level rises to the required level. A set of genes is responsible for switching this process on and off. As explained in chapter 12 below, this is the role of what I have called the **strategic gene**, a role which was later played by the **strategic cerebrum**.

This genetic response occurred not in just one bacterium, but in several hundred bacteria, all generating different types of molecular "cages." In fact these bacteria employed their siderophores to compete with one another. Some produced "Trojan horses" that were close copies of the siderophores of their competitors which, when accepted by their neighbors as their own, poisoned their hapless victims, while others produced iron-absorbing siderophores to starve their competitors of iron. It is just possible that this ingenious solution to the general problem of iron deficiency arose from bacteria pursuing the conquest strategy against their neighbors—it arose, in other words, as a weapon rather than as an antipollution innovation. But it could be employed for both purposes.

While these various forms of add-on genetic technology could, and did, extend the reign of blue-green algae it could not enable this life form to remain forever dominant. It only enabled them to reach their full potential. A potential met by pursuing the family-multiplication strategy—a cloning process—until blue-green algae had fully exploited all available ecosystems throughout the accessible world. The role of this add-on technology, which was a response to strategic demand, was to facilitate the dominant dynamic strategies of family multiplication (replication and migration) and of conquest.

By about 1,000 myrs BP blue-green algae appear to have completely exhausted the prokaryotic paradigm. Further pursuit of the family-multiplication strategy led to overcrowding, intense competition, and starvation due to pressure on nutrition. To survive, bacteria substituted the conquest strategy for the family-multiplication strategy, employing earlier developed weapons (siderophores) to devastating effect. Also in these overcrowded conditions it probably became increasingly difficult for bacteria of various types to cope with their own waste products of oxygen and sulphur. The outcome for this dominant life form was slow but irreversible decline, which released resources for the newly emerging eukaryotic life forms. Renewed expansion was impossible because blue-green algae had reached the ceiling imposed by their genetic paradigm, and even recovery from crises of increasing intensity was hampered by competition from new life forms (grazing animals) that were able to take full advantage of the growing oxygen levels and the associated ozone layer (providing necessary protection for organisms living outside protective colonies) that blue-green algae had created. It is not surprising that from 550 myrs BP stromatolite fossils are rare. This was the world's first strategic struggle between the old strategists (in this case the prokaryotes) and the new strategists (the eukaryotes). The subsequent decline of blue-green algae was the first great collapse, if not extinction (they live on only in small protected enclaves), of life on Earth. Hence, while self-contamination can be overcome by modest genetic change—which I have called add-on technology—paradigm exhaustion can only be resolved by a new biological revolution or paradigm shift.

Before we pass on to the second genetic paradigm shift, we should briefly consider the long-running success of blue-green algae. There is, as we shall see, a critical problem here for the neo-Darwinist. Blue-green algae were remarkably successful in their pursuit of the universal objective of survival and prosperity, dominating the Earth for over 50 percent of the history of life. Despite this unparalleled achievement, which involved the solution of major problems resulting from their pollution of the environment, it is generally acknowledged that blue-green algae was not held on a tight "genetic leash." We are told that

> the DNA content of prokaryotes is small, and they have only one copy of it. There is little room to store the complex "IF . . . THEN . . ." commands in the genetic program that would turn on one gene as opposed to another. Therefore, *genetic regulation is not well developed in prokaryotes.* (Cowen 2000: 38; my emphasis)

Yet without well-developed genetic regulation, bacteria were able to successfully—very, very successfully—pursue a number of dynamic strategies including genetic change, family multiplication, commerce, and conquest. These are the universal dynamic strategies that even the most sophisticated human societies still employ. And we have yet to demonstrate that as a species we can solve our own pollution problems as effectively, or last as long. So far we have been the dominant species (if we adopt a generous span of 1.6 myrs) for only 0.08 percent of the time that blue-green algae held sway. Further, if this life form with its limited genetic control could successfully pursue the full range of dynamic strategies, how is it that more complex life forms need to be held on a tight genetic leash as claimed by the sociobiologists?

The Eukaryotic Revolution

The second genetic paradigm shift, or the Eukaryotic Revolution, was under way by 900 myrs BP. What was the genetic basis of this revolution? It was the emergence of a cell—the eukaryote cell—that could go beyond the construction of simple colonies of other identical cells and build complex, integrated biological structures or organisms. The first stage of this revolution was confined to cold-blooded, or ectothermic, life forms, because the emergence of warm-blooded, or endothermic, life forms some 550 myrs later constituted a revolution of its own.

Eukaryote organisms were far more mobile than their prokaryote forebears and, as a result, were capable of exploiting natural resources that lay beyond the reach of blue-green algae. This is the hallmark of a paradigm shift, whether genetic or technological, that it enables a radical increase in potential access to natural resources. Eukaryote organisms were able to make this advance by gaining greater control over genetic change owing to the "invention" of sexual reproduction. While reproduction through cloning produces offspring with the same DNA as its sole parent, sexual reproduction, by combining the DNA of two individuals (half each) enables parents to exercise some discretion over the characteristics of their offspring through the choice of a mate. This is discussed more fully in chapter 12.

When and how did the eukaryote cell emerge? Until recently the earliest known eukaryote fossil was dated at 1,400 myrs old, some 400 myrs before prokaryotic life went into decline (Cowen 2000: 41). The latest discovery in northwestern Australia suggests, however, that this might have to be pushed back to 2,700 myrs (*Science* 1999: 981). Quite clearly eukaryotic life was not competitive with bacteria at this early stage. It would seem that, like mammals in the mesozoic period (245 myrs to 65 myrs BP), this new life form only emerged fully once the earlier dominant form had exhausted its paradigm and released nutrients for the use of other life forms.

It is thought by some that the eukaryote cell first emerged through symbiosis. This is the so-called endosymbiotic hypothesis, which suggests that the components of the more complex eukaryote cell—including mitochondria, plastids, and flagella—were originally free-living bacteria that became so closely associated through their symbiotic relationship that they eventually formed an integrated cell. The implication of this hypothesis is that the eukaryotic transformation was merely a matter of chance association—an accident. While it may explain how the eukaryote cell emerged, it does not explain why. I want to suggest that it was an outcome of the commerce strategy employed by a number of prokaryote cells to overcome regional exhaustion of the more usual family-multiplication strategy. It was not until the entire prokaryotic paradigm was exhausted some 1,700 myrs later that this new form of life seized its chance and diversified rapidly.

What were the fundamental differences between prokaryote and eukaryote cells that made it possible to develop more complex forms of life, even if not more complex dynamic strategies? First, in a prokaryote cell the DNA floated freely, whereas in a eukaryote cell the DNA, which was much more abundant (by a factor of one thousand), was contained within a nucleus. This provided the more abundant DNA with the protection required to construct complex organisms. Second, unlike the prokaryote cell it was subdivided into organelles that specialized in different functions. This enabled a "division of labor" that generated greater efficiency of energy use. Third, the eukaryote cell was much larger, with a diameter and a volume greater by a factor of ten and a thousand respectively. Fourth, in contrast to the prokaryote cell it was able to expand its membrane to engulf other cells. Finally, and most significantly, while prokaryote cells reproduced by dividing themselves to create clones, most eukaryote cells reproduced sexually by combining their own DNA with that of another cell to create unique offspring.

It is thought that sexual reproduction emerged in eukaryote cells at least as early as 1,400 myrs ago. I intend to argue that the significance of this development resides in the greater control over genetic change that it conferred on individual organisms. Through sexual reproduction an organism was able to manipulate the genetic structure of its offspring in two ways. First, it was able to influence the genetic makeup of its offspring by a judicious choice of sexual partner. **Selective sexual reproduction** is based on those perceived characteristics in other individuals that will improve the prospect of survival and prosperity of the selecting individual. If this perception turns out to be correct, those characteristics will also be passed on to the individual's offspring. As I show in chapter 12, mate choice is an important technique for implementing all four dynamic strategies, not just that of genetic change.

Second, sexual reproduction increases the probability of genetic mutation. With each act of conception, life goes back to the very beginning. The DNA of both parents is split, reproduced, and recombined in an entirely new way. This process increases the frequency not only of copying errors—of mutation—but also of favorable mutations that can be exploited by the organisms involved. When required, the favorable mutations can be amplified through appropriate mate selection. Why sex? Because it is the way individuals can gain greater control over the genetic strategy.

Early Life in the Seas and on the Land

The Eukaryotic Revolution made possible the emergence of a remarkable variety of new and increasingly complex life forms. It provided the genetic technology for the second great wave of life that surged upwards, stagnated, and then collapsed between about 900 myrs and 245 myrs BP. It also enabled individual organisms to seek survival and prosperity through the application of the usual quartet of dynamic strategies. In the process, the eukaryotic paradigm was exploited until it was finally exhausted in the sense that no further access to natural resources could be attained with this "technology."

In the beginning the main dynamic strategy was genetic change. Eukaryotic life radiated out from simple beginnings to form increasingly complex and specialized species of plants and animals in the seas and on the land. What is most unsettling for the Darwinist is that all this remarkable genetic innovation and speciation occurred in the absence of serious competition and in the presence of an abundance of natural resources. What is more, this revolution occurred only once the earlier paradigm had been exhausted and the dominant bacterial form of life had collapsed. Of course, this decline was hastened, but definitely not caused, by the grazing activity of some eukaryotic life forms.

The Eukaryotic Revolution, therefore, was a response not to scarcity, as the Darwinian model predicts, but to abundance. And the genetic strategy was pursued by eukaryote organisms to gain access to this abundance. They were like monopolists in pursuit of supernormal profits: the prospect of large returns and adequate time provide the incentive and the "funds" to innovate, whereas "normal (or subnormal) profits" under intense competition and short-run time horizons are an incentive for conflict (the conquest strategy).

Once a new genetic technique for gaining access to natural resources had been developed, and a new species had been created, this technique—or genetic style—was exploited through the more efficient and quick-acting dynamic strategy of family multiplication. New species continued to emerge, however, until all possible genetic styles in the paradigm had been adopted. Once a new genetic style had been perfected by individuals in the new species, no further directional genetic change occurred. Eventually that style was exhausted and the species was extinguished. Sometimes add-on technology would be developed in response to the strategic demand generated by other dominant dynamic strategies. This was particularly so in the case of weapons of offense and defense required to facilitate the conquest strategy that emerged during the intense competition resulting from the exhaustion of a genetic style or genetic paradigm. Even the threat of final extinction for an entire dynasty—*extreme* Darwinian scarcity—was not sufficient to generate major genetic change as the master had predicted. This was because conquest was regarded as a more effective dynamic strategy when time and returns were of the essence.

Between 900 myrs and 245 myrs ago a remarkable variety of plant and animal life emerged in the seas and on the land. As the details are well-known only the main developments will be sketched here. One of the earliest major life forms to exploit the eukaryotic paradigm was the protozoan (plankton), which appeared at about 800 myrs BP. Plankton emerged to gain access to food sources in the below-tide shallows around continents and islands—waters deeper than stromatolites could colonize. Then from about 700 myrs BP the metazoans—sponges, worms, and cnidarians—appeared in order to exploit food sources on the sea floors. And from about 600 myrs BP organisms with hard bodies (trilobites) and, later, skeletons (fish) developed to harvest the oceanic depths. During the following 100 myrs, as can be seen from figure 9.1, there was a "spectacular burst of innovation" during which most existing body plans emerged (Cowen 2000: 61). By this time the once dominant stromatolites had become "quite rare" (Cowen 1990:85).

Paleontologists appear quite puzzled about the sudden appearance of organisms with hard bodies and skeletons. They are puzzled because Darwin's theory of natural selection is unable to explain these genetic innovations at a time of resource abundance. Paleontologists are forced, therefore, to seek non-Darwinian explanations while maintaining the fiction, to themselves as well as their readers, that their ideas are consistent with the concept of natural selection. The usual ad hoc suggestions—or flying buttresses—are that these innovations were the outcome of higher levels of oxygen in seawater that enabled the production of hard body parts for the first time through biochemical reactions, or of the need for defense against predators.

Neither argument is very persuasive as far as the development of skeletons is concerned. The dynamic-strategy argument, on the other hand, is more systematic and part of a larger explanation. Skeletons were needed by organisms to develop strong muscles in order to propel themselves more effectively through the seas or along the sea floors in search of new supplies of food. In other words, like all major genetic innovations, the development of skeletons was necessary to explore strategic opportunities in the constant pursuit of survival and prosperity. It was, therefore, a response to strategic demand—inputs required to exploit strategic opportunities—and was not driven by the abundance of resources itself. But, of course, it did exploit changing supply conditions. The dynamic-strategy theory is a general demand-side theory rather than a series of ad hoc and partial supply-side hypotheses—or flying buttresses—suggested somewhat hopefully by the paleontologists.

The emergence of vertebrates with functioning jaws was essential to the great invasion of the land. It was begun about 500–400 myrs ago probably by millipedes and then by primitive fish (rhipidistians and lungfish) that gradually transformed their lower fins into limbs and certain internal organs into lungs and became amphibians. While it is more efficient to extract oxygen from air than from water, this is not the reason that these animals invaded the land. Once again it was the internally driven strategic pursuit of living organisms for access to more abundant resources both for food (there had been a prior explosion of land invertebrates, such as grubs, worms, and insects) and for shelter. Nor was it a result of Malthusian pressures, because, as the paleontologists tell us, "only one line of rhipidistians took this evolutionary path," and that "*aberrant* rhipidistians (or evolving amphibians) spent more and more time at and near the water's edge, sunning and basking, while *normal* rhipidistians remained creatures of open water" (Cowen 2000: 134, 135; my emphasis). Clearly those that remained were not all immediately extinguished.

Nevertheless these "aberrant" individuals were responding to the changing conditions in the seas and on the land as the second great wave gathered momentum. They were responding in an intuitive way to the material benefits and costs in their actual and potential environments. Between 500 myrs and 400 myrs BP (see figure 9.1a) there was an exponential increase in the number of marine families—and, of course, in their species and populations—which would have increased the energy costs of access to resources. At the same time there was an explosion of plant and invertebrate life along coastal regions, thereby increasing the potential benefit for those "aberrant" vertebrates that could also make the transition from water to land. The important point, however, is that the exploitive response to these changing conditions—the driving force—was made by innovative individuals that some followed and others ignored.

Life possesses a vitality that results in organisms actively seeking out an improvement in their living conditions. They are not just the playthings of external forces. In terms of my dynamic-strategy theory, the "aberrant" individuals are the strategic pioneers who "pursue" the genetic strategy to explore economic opportunities that consist of unused or underused resources for both food and shelter. This was not a matter of Darwinian survival of the fittest—because the "normal" individuals continued their old fishy way of life—but rather of the materialist pursuit of "supernormal profits" that accrue to those who can enter into the competitively restricted world of the innovating monopolist. In this context the Darwinian mantra sounds particularly hollow.

The Age of Reptiles
While the "aberrant" rhipidistians, or pioneering land strategists, had turned themselves into amphibians by at least 368 myrs BP, it was not until some of these had been transformed into reptiles by 350 myrs BP—a further attempt to improve access to natural resources—that the land invasion gathered momentum (see figure 9.1b). This was the age of the reptiles and was the last phase in the exploitation of the eukaryotic paradigm by ectothermic life forms. In this period, from 350 myrs to 245 myrs BP, reptiles extended their dynasty to all those ecosystems—employed all those genetic styles—that were available to cold-blooded animals.

The reptile dynasty included species ranging from small herbivores such as *Trimerorhachis* (3 kg) to large herbivores such as *Ophiacodon* (150 kg), *Eryops* (175 kg), and *Diadectes* (250 kg), to predators such as *Sphenacodon* (60 kg) and the finned *Dimetrodon* (150 kg). Fins were employed by the latter to more effectively extract heat from a low sun at the extremities of the day and season.

So successful were the reptiles in gaining access to natural resources through these various genetic styles that by about 240 myrs BP they had exhausted the Eukaryotic/ectothermic Revolution. In other words this genetic paradigm could not be employed to extract any further resources because it had, through the family-multiplication strategy, been taken around the accessible world. Any further resource access by life forms would require a genetic paradigm shift. As for the reptiles, the exhaustion of their paradigm meant the stagnation of their dynasty and, ultimately, its extinction.

When a genetic paradigm approaches exhaustion, as reflected in a declining origination rate and a growing extinction rate, dynastic extinction normally follows. This is an outcome of the adoption of the conquest strategy which offers a short-term solution in these intensely competitive circumstances to the strongest individuals and species. In the case of the reptiles, collapse occurred more suddenly than it might have done, harried as they were by a newly emergent and superior life form—the therapsids or protomammals. This is another example of the strategic struggle between the old and new dynamic strategists.

It is important to realize that the dynamics of life necessarily involves a process of rise and fall as genetic styles/paradigms are successively exploited and exhausted through the pursuit of a series of dynamic strategies. Physical catastrophes play only a marginal and nonsystematic role in this dynamic process. While they might add their weight to a downturn that was going to happen anyway as a result of strategic exhaustion, they could never be responsible for driving a dynamic process of biological change. Hence, if the major volcanic eruptions in Siberia, roughly at the time of the collapse of the reptile dynasty, did have a short-term impact on life as some claim (Campbell et al. 1992), it was merely to reinforce the inevitable outcome of endogenous strategic forces.

The Endothermic Revolution

The primitive reptiles lived and died at a pace that we would regard as unbearably slow and ponderous. They were slow-growing, slow-breeding, slow-moving, and slow-witted creatures possessing a slow metabolism. Their cold-blooded (ectothermic) biological system meant they were unable to colonize the Earth's colder regions toward the poles or on the higher slopes of low latitude regions. They were, in other words, unable to exploit the natural resources of all parts of the planet as warm-blooded animals can. Clearly there was still scope in the old genetic option for further development.

Cold-bloodedness, however, was not a disadvantage for as long as reptiles were the dominant form of life. But it would be an entirely new ball game if they ever had to face a competitor that could generate body heat internally. Such competitors would be unstoppable, because this new breed would be able to live and die at a furious pace. They would be fast-growing, fast-breeding, fast-moving, fast-digesting, fast-thinking creatures with a high metabolism. And as they could maintain a constant body temperature it was even possible that they might eventually be able to develop larger brains. They would also speciate more rapidly, would exhaust their genetic styles more quickly, and their carnivores would have the advantage of an early start and a late finish to their day and would not slow down during winter. They would also have greater speed and stamina. They would, in other words, be a super race of animals. As far as the reptiles were concerned it would be a complete mismatch. Yet this endothermic lifestyle has its costs as well as its benefits. It requires a higher input of food and it leads to a higher turnover of species and families.

The Protomammals
The Endothermic Revolution began with the emergence of the therapsids, or protomammals, at the end of the Permian period at about 245 myrs BP (see figure 9.6). The first protomammals were "slender-limbed, wolf-sized predators,"

but they diversified rapidly by pursuing the genetic strategy and quickly sweeping aside the remaining cold-blooded reptiles that were struggling to cope with the collapse of their old paradigm. Within a few million years of the collapse of the age of reptiles—a period when there was no serious competition for resources—the protomammals "had taken over all the carnivorous roles—large, medium, and small—nearly all the herbivorous roles, and produced dozens of small insect-eating species as well" (Bakker 1986: 410). This was the most rapid genetic diversification the world had yet seen, and it was an outcome of the Endothermic Revolution—of warm-bloodedness.

Figure 9.6 Predator–prey systems of land vertebrates—the past 300 myrs

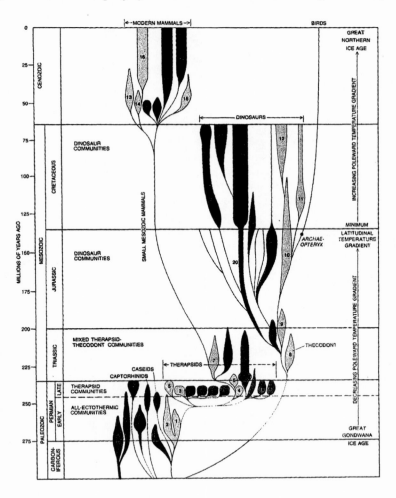

Source: Bakker 1978: 135 (reproduced with permission).

Not only was their speed of emergence unprecedented, but the protomammals diversified to a greater degree—developed more genetic styles—than the old reptile dynasty. In the first few million years there were four families and eight to ten species of predators—including the anteosaurs (500 kg) and the gorgons (200 kg)—and five families with twenty or so species of herbivores—including the struthiocephalids (1,000 kg) and *Titanosuchus* (1,500 kg). But this faster pace of life, during which these genetic styles were exploited more rapidly, led to a high turnover of species and families. Protomammals, unlike the reptiles, actually generated three waves of rapid speciation and extinction between 250 myrs and their final disappearance by 200 myrs BP.

The dynamic-strategy model can explain how these waves of speciation and extinction occurred. The protomammals emerged only once the older paradigm had been exhausted and the reptilian species, which were going extinct owing to overpopulation of their ecosystems, were not being replaced by new species. In order to survive and prosper, the early protomammals adopted the conquest strategy to eliminate the remaining large reptiles in this great Permian strategic struggle. Owing to their greater speed, stamina, and logistical ability, the early protomammals quickly drove the remaining large reptiles to extinction.

Once this competition for resources had been eliminated—once they had created a non-Darwinian world—the protomammals, with time and resources on their hands, were able to diversify biologically by employing the dynamic strategy of genetic change to exploit food supplies in both warm and cold regions. Having created the optimum number of genetic systems, which gave access to all types of habitat, the protomammals switched to the family-multiplication strategy to exploit them to the full all around the globe. In this way they asserted their world dominance.

Once a species had spread around the world and exhausted its genetic style, the family-multiplication strategy led to overpopulation, degradation of their ecosystem, and to the adoption of an insidious form of the conquest strategy—civil war. Such conflict left a species vulnerable to a takeover by existing or new species. When this genetic-style exhaustion was more general—when it took place in groups of species—the more usual conquest strategy was pursued, endangering entire orders and classes. This appears to have occurred on three occasions—at about 230 myrs, 225 myrs, and finally 210 myrs BP—and may have occurred again but for the intervention of an opportunistic interloper, the archosaurs or "protodinosaurs."

It is interesting that once again intense competition—the Darwinian scenario—led not to genetic change but to conquest and extinction. When resources were scarce, life forms had no time to transform themselves biologically. In turn the resulting widespread extinction created a non-Darwinian world of minimal competition and abundance, which enabled diversification through genetic change. Once again natural selection is unable to explain the rise and fall of dynasties, families, or even species. There can be no doubt by now that Darwinism is as dead as these long extinct species.

The Age of the Dinosaurs

The age of the dinosaurs was built on the ruins of the first stage of the Endo-thermic Revolution. The protomammals had emerged rapidly and powerfully, and overwhelmed the remaining large reptiles, had flourished for a brief season (50 myrs!), had passed through a number of coordinated genetic styles, had embarked on a war with the upstart archosaurs, and had disappeared forever once they had exhausted their genetic paradigm. With one exception. The descendents of the carnivorous cynodont, which regressed in size from an 80 kg animal to a very small shrewlike nocturnal mammal about 10 cm long and 25 g in weight, departed from the therapsids before their demise to spend the following 160 myrs as very minor players on the margins of the main game (Cowen 2000: 250–51). To the archosaurs and the later dinosaurs the defeat of the protomam-mals appeared all but complete. This is one of the wonderful ironies of life.

The war between the protomammals and the archosaurs has been referred to as a "titanic ecological battle" (Bakker 1986: 416). In fact it was the greatest and most destructive strategic struggle the world had ever witnessed, beginning at about 225 myrs ago during the Triassic period. Gradually the archo-saurs—whose top predators included the erythrosuchid (300 kg) and coelo-physid (200 kg) and the giant "crimson" (as its fossilized bones were stained red by the soil) crocodile that was terrestrial—advanced and the protomammals re-treated, until by the late Triassic (about 215 myrs BP) the surviving protomam-mals had been reduced to a group of small to medium-small predators and her-bivores.

Because the archosaurs won this great strategic struggle, it is usually con-cluded that the "protodinosaurs" had a biological edge over the protomammals. When viewed in terms of dynamic-strategy theory it is not clear that this is cor-rect. The protomammals were unfortunate to encounter their determined adver-sary at a critical stage in their own development. They had, as we have seen, already passed through two earlier stages of expansion and extinction and ap-peared—as reflected in a declining "origination rate" (Benton 1985)—to be facing the exhaustion of their entire dynastic paradigm. In this overpopulated and weakened state they were an easy target for the upstart archosaurs. But even so, the archosaurs could not have won this world war if they had not also shared the benefits of warm-bloodedness (Ricqles 1974; Bakker 1986: 421–22).

With the demise of the protomammals, and the creation of a non-Darwinian world of minimal competition and of resource abundance, the triumphant new strategists—the archosaurs—quickly filled all accessible ecosystems with predators by the pursuit of the genetic strategy. By the late Triassic they had also branched out into heavily armed herbivores (aetosaurs). But, having won their great war against the protomammals and flourished briefly, they disappeared completely from the scene by about 210 myrs BP. Indeed, as can be seen from figure 9.1b, by about 180 myrs the number of families of land animals was at an all-time low—lower than that achieved during the time of the most celebrated extinctions of 245 myrs and 65 myrs BP. Most paleontologists are at a complete loss to explain why.

The dynamic-strategy model, however, has the likely answer. Like all other endothermic dynasties, the archosaurs rapidly generated an optimum number of genetic styles through the genetic strategy and then even more rapidly exhausted them through the family-multiplication strategy, whereupon the resulting over-population and intense competition led to conquest and extinction. This process passed, as it did for the earlier protomammals, through a number of cycles until the entire dynastic paradigm for archosaurs had been exhausted. In the end a great world war broke out between carnivores and their prey, resulting in a deterioration of the environment and the extinction of the archosaurs. The extent of the devastation accounts for the slowness with which the surviving true dinosaurs—which had branched off from the archosaurs in the mid-Triassic (about 225 myrs BP)—were able to diversify and expand. Indeed in the early Jurassic the fate of the dinosaurs appears to have hung in the balance.

In the early Jurassic (200 myrs ago) the dinosaurs had the world before them. All their competitors—the protomammals and the archosaurs—were extinct and they were faced with an abundance of natural resources. After a shaky start they took up this non-Darwinian challenge, as all other dynasties did before and after that time, by adopting the dynamic strategy of genetic change. They had all the time and potential resources they needed to do so.

According to the most perceptive authority on dinosaurs, Robert Bakker (1986: 422), "a horde of new species" took advantage of these "ecological opportunities." Hence, while "in the Late Triassic times the dinosaurs had been a minority group . . . in the Jurassic every single land predator and herbivore role was filled by their newly evolving species." He also argues that the speed with which this was achieved was only possible because dinosaurs were warm-blooded creatures. When first promulgated by Bakker this idea caused quite a stir because, in the main, his colleagues regarded dinosaurs as cold-blooded reptiles. But in fact they, together with the archosaurs and the protomammals (widely regarded as warm-blooded), were products of the Endothermic Revolution. It would be a remarkable backward step if the evidence had shown dinosaurs to be cold-blooded, because they and the archosaurs would have been no match for the warm-blooded protomammals.

Bakker provides convincing independent evidence for his argument about the endothermic nature of dinosaurs. This consists of data on growth rings in bones and teeth, their predator/prey ratio, the upright stance of dinosaurs, and the contrasting geographical distribution of reptiles and dinosaurs. First, the less pronounced growth rings in the bones and teeth of dinosaurs in comparison with reptiles suggests that the former experienced more continuous growth throughout the year, something that is characteristic of warm-blooded creatures (Ricqles 1974; Bakker 1986: ch. 16). Second, it is typical of warm-blooded animal species, which require a higher intake of food than cold-blooded animals for any given body weight, that they achieve a lower predator/prey ratio. The fossil evidence also confirms this for dinosaurs: predator/prey ratios for dinosaurs and fossil mammals (1 to 5 percent) were much lower than those for protomammals

(10 to 12 percent) or early Permian finback reptiles (25 percent) (Bakker 1986: 381).

Third, as warm-blooded animals could achieve higher sustained speeds than cold-blooded animals, we can expect their hip, shoulder, limb bones, and joints to differ. We are not disappointed by the evidence. Reptiles with their shallow-socketed hipbones, and shoulder sockets that faced sideways (like modern lizards), were low-set creatures only able to waddle relatively slowly over long distances. By contrast, dinosaurs with their deep-socketed hipbones, shoulder sockets that faced down and to the back (like modern rhinos), and more massive limbs that would have permitted greater muscle power were high-set creatures able to run at great speeds over relatively long distances (Bakker 1986: ch. 10; Carrier 1987). It has been suggested, for example, that *Tyrannosaurus* (2,000 kg), the size of a small elephant, could, when necessary, sprint at 75 kilometers per hour—a terrifying prospect! Finally, while ancient reptile bones are only found in areas that had warm climates at the time, dinosaur bones can be found in regions that had climates far too cold for reptiles (Bakker 1978: 128). Taken together, this wide-ranging evidence should leave us in little doubt that dinosaurs were indeed warm-blooded—despite the continued support by a few paleontologists for cold-bloodedness (Ruben et al. 1999)—and that they had participated fully in the Endothermic Revolution.

Since writing these paragraphs, further and more direct evidence (Fisher et al. 2000) has emerged that confirms Bakker's hypothesis about the warm-blooded nature of dinosaurs. A fossilized heart belonging to a *Thescelosaurus*—a four-meter, two-legged herbivore found in South Dakota in 1993 and acquired by the North Carolina Museum of Natural Sciences in Raleigh—was subjected to medical imaging techniques. The resulting three-dimensional images show that the heart was "highly advanced," being more like that of a mammal or bird than a reptile. These techniques revealed that the organ was a double-pump heart with two upper and lower chambers (ventricles) and a single aorta that carries blood to other parts of the body. Modern reptiles have two aortas and a simple heart structure that allows some mixing between oxygen-rich blood from the lungs and oxygen-poor blood from the body, which reduces the oxygen delivered by blood to the body tissues. Reptiles, therefore, have a lower metabolism than mammals, birds and, on this direct evidence, also dinosaurs. This is surely final confirmation of Robert Bakker's brilliant intuition and research. Dinosaurs were certainly warm-blooded.

The dinosaurs were able to take advantage of the strategic opportunities opened up to them once the devastating effects of the world war between the archosaurs had subsided. As is always the case in a non-Darwinian world of resource abundance, the dinosaurs energetically pursued the dynamic strategy of genetic change. The outcome was the generation of a diverse range of dinosaur genetic styles. The carnivores included, amongst others, the short-lived coelophysids (200 kg), the mighty allosaurs (3,000 kg), the small but vicious deinonychids (70 kg) and, later (about 100 myrs ago), the frightening *Tyrannosaurus* (2,000 kg). The more numerous herbivores included the early prosauropods

(1,500 kg), the diplocids (17,000 kg), the camarasaurs (25,000 kg), the stegosaurs (3,000 kg), the brachiosaurs (40,000 kg), the ornithopods (2,000 kg), the
ankylosaurs (2,000 kg), and the ceratopsians (2,000 kg). Of the herbivores the
largest had gone extinct by the early Cretaceous (about 120 myrs ago) leaving
their smaller, armed relations to struggle for survival against the fearsome tyrannosaurs. In addition to these land animals we should note the Pterosauria or flying dinosaurs that emerged in the Triassic period and, of course, *Archaeopteryx*
(from the separate Theropoda line) that possessed feathers and is thought to have
given rise to modern birds. Hence by the end of the Jurassic (145 myrs ago) the
dinosaurs had effectively generated a sufficient number of genetic styles to fully
exploit all ecosystems available to land and air-bound creatures in both warm
and cold climates. No other dynasty had, until that time, been able to spread so
widely throughout the world or to exploit so effectively the Earth's natural resources. It is interesting that the dinosaurs had no interest in returning to the
seas, possibly because reptiles (sea and swan lizards) in the oceans did not suffer
the same disadvantages of those on the land.

As with the protomammals, the dinosaurs generated a number of waves of
species diversification and extinction. The species that emerged in the early Jurassic—particularly the early predators and the gigantic, long-necked herbivores—were extinct by the early Cretaceous. Paleontologists appear puzzled
about the causes of these waves of extinction: Bakker (1986: 193; my emphasis), for example, refers to it as "the *mysterious* hand of worldwide disaster."
Yet, as we have seen, they can be explained quite simply by the dynamic-
strategy theory. These earlier versions were replaced by the heavily armored,
low-grazing herbivores and by the king of the carnivores, *Tyrannosaurus rex*.

Robert Bakker (1986: ch. 9), in his usual innovative fashion, has argued that
this changing of the guard had a profound effect upon the nature of the Earth's
vegetation. The replacement of high-browsing, long-necked brontosaurs by low-
grazing, beaked herbivores conferred an advantage on plants that could grow
quickly to heights above grazing level and that could distribute their seeds
widely onto overgrazed areas to produce fast-growing plants that could extend
out of reach before the low-grazers returned. The early angiosperms met both
these conditions and thereby gained a competitive advantage over the longer-
established gymnosperms that included conifers, cycads, and tree ferns (see figure 9.2). Once they gained a foothold, the angiosperms employed the familiar
dynamic strategies of genetic change, family multiplication, commerce, and
even conquest (by invading the territories of other plants and plundering their
sunlight, water, and other nutrients) to populate the Earth with a vast variety of
species and individuals. Like animals, plants diversified in the absence of intense competition.

As with every preexisting dynasty the dinosaurs substituted the family-
multiplication strategy for the genetic strategy once a genetic style had been
successfully created. In doing so members of that species spread gradually
around the accessible world. Only once the genetic style had been exhausted and
overpopulation had occurred did individuals switch from family multiplication

to conquest. An interesting feature of the dinosaurs' conquest strategy was the demand it generated for weapons of offense and defense. While there were insufficient time and resources to employ genetic change for speciation it could, and was, employed to develop "add-on" technology. Add-on technology is easier, quicker, and cheaper than a strategy required for complete genetic transformation. What we have here, therefore, is genetic change designed to support the zero-sum game of conquest, rather than an independent strategy designed to give direct and revolutionary new access to natural resources.

The conquest strategy of the dinosaurs generated a strategic demand for either additional armor or more effective defensive weapons in herbivores, or sharper/larger teeth and claws or more powerful jaws in carnivores. There were a number of stages in the development of this "military technology." In the conquest strategy of the Jurassic, herbivores such as the stegosaurs developed warclub tails with spikes about a meter in length, together with sharp defensive triangular-shaped plates along their backbones, while those species without armor developed slashing claws and powerfully long lashing tails. The carnivores, such as the allosaurs, were as large as elephants, were fast and agile, had large powerful jaws, and possessed slashing claws on their hind legs.

In the period of Cretaceous conquest the herbivores were also able to deal out as good as they had to take. The nodosaurs were heavily armored with large shoulder spikes, the ankylosaurs were built like small tanks with war-club tails, the pachycephalosaurs had giant bony head domes that were used like flying projectiles, and the *Triceratops* were built like giant battering rams with monstrous heads and meter-long horns. With all this offensive and defensive armor, the top carnivore had to be a frightening creature. And it was. *Tyrannosaurus* stood six meters tall, was as heavy as a small elephant, yet was fast and nimble with huge, powerful jaws and, a major advantage, stereoscopic vision. Both sides in this conflict were equipped for Armageddon—a world war that ended the era of the dinosaurs.

The role of genetic change in the conquest strategy of the dinosaurs was just the same as the role of technological change in the Roman conquest strategy. It was a response to military demand. Conquest is a strategy to gain control over resources, not through new methods of gathering, hunting, or production, but through plunder that accrues to more effective techniques of waging war. It is fascinating how closely the offensive and defensive weapons developed genetically by the dinosaurs paralleled those developed technologically by the Roman (and other ancient) conquerors: tanklike structures, battering rams, flying projectiles, armored and armed foot soldiers, and armored cavalry with lances.

This brings us to the relatively sudden extinction of all dinosaurs—with the sole exception of the line that developed into modern birds—around 65 myrs ago. Within the space of probably five hundred thousand years the mighty dynasty of dinosaurs that had flourished for more than 135 myrs was brought to a dramatic end that has fascinated the expert and layperson alike.

As we have come to expect, natural scientists argue that large-scale extinctions are the result of exogenous physical events. It is curious that not one of them, even among those concerned with plants and animals, is prepared to grant living organisms the dignity and credit of being responsible for their own rise and fall. Either a training in the natural sciences brainwashes otherwise intelligent people in this respect, or it attracts those people who prefer to view the dynamics of life in physical and mechanical terms. Perhaps it is a combination of both. Anyway, the list of external physical causes of extinction is rather long and there is little agreement on the subject. Everyone has his/her own pet hypothesis.

As Robert Bakker has said of his own colleagues:

> I keep a file of published "solutions" [to the problem of dinosaur extinction]. Among its contents, it is suggested the dinosaurs died out "because the weather got too hot;" "because the weather got too cold;" "because the weather got too dry;" "because the weather became too hot in the summer and too cold in the winter;" "because the land became too hilly;" "because new kinds of plants evolved which poisoned all the dinosaurs;" "because new kinds of insects evolved which spread deadly diseases;" "because new kinds of mammals evolved which competed for food;" "because new kinds of mammals ate the dinosaur eggs;" "because a giant meteor smashed into the earth;" "because a supernova exploded near the earth;" "because cosmic rays bombarded the earth;" or "because massive volcanoes exploded all round the earth." (Bakker 1986: 425)

We could also add: because a Death Star returns regularly every 26 myrs (see Raup and Sepkoski 1984; 1986).

Yet what has Bakker, after a long career devoted to reconstructing the lives of the dinosaurs in an interesting and exciting way, got to offer that differs from this natural scientists' wishlist? As it turns out, not a great deal. Bakker's (1986: 427) conclusion that "some shift in the habitat must have doomed the dinosaurs" is a truly great disappointment. Like all his colleagues in the natural sciences, Bakker fails to see, despite marshalling all the evidence that can be used as the basis for a radically new and realistic interpretation, that the answer lies within rather than without the dinosaurs themselves. These scientists all suffer from a form of myopia called Darwinism. They are all too busy building buttresses around the crumbling cathedral of Darwinism to see the obvious.

Bakker (1986: 431) conveniently summarizes the chief characteristics of the impact of major extinctions. They
- kill on land and sea at the same time;
- strike hardest at large, fast-evolving families on land;
- hit small land animals less hard;
- leave large cold-blooded animals untouched;
- do not strike at freshwater swimmers—most of these creatures are cold blooded;
- strike plant-eaters more severely than plants.

He also investigates the changing structure of families and species in the lead-up to all major extinctions, and finds that the universal pattern is one of initial faunal "evenness" (or diversity) followed by a decline in evenness ending in collapse (Bakker 1986: 436–38). This collapse took thousands to millions of years.

While all this is good historicism it leads to the wrong conclusions because it is forced into a Darwinian straitjacket.

But at least these facts do eliminate the popular catastrophe stories about asteroid impacts (Alvarez et al. 1980) and massive volcanic eruptions (Rampino and Stothers 1988), because the timing is not right (would take only years or decades to deliver a death blow), and because too many species on the land and in the water survived. In any case there have been other catastrophes in the past and they have had no long-term impact on life; and most earlier extinctions were not associated with catastrophes. Of course, the suggestion that an unidentified "Death Star" returns every 26 myrs (Raup and Sepkoski 1984; 1986) to precipitate a deadly shower of comets (that conveniently leave little trace) is just out of this world! Some natural scientists clearly prefer to invoke metaphysical ideas rather than to empirically examine the dynamics of life forms on Earth.

Like Bakker, most paleontologists reject the catastrophe stories of the few, and opt for the idea of extinctions being induced by climatic change brought about by the physical dynamics of the Earth expressed through continental drift. In this vision, catastrophes are, at most, only the final blow to a system operating under stress from climatic change. As Bakker, more than any other, is responsible for changing our perceptions of the lives of the dinosaurs, his hypothesis is outlined in more detail.

Bakker argues that, owing to a cooling of the planet around 65 myrs ago, the oceans fell, the shallow seas drained off the continents, separate continents were connected by land bridges or island chains, and the weakening of mountain-building forces reduced barriers to migration. This sequence of events led both to the extinction of shallow sea animals and to an exchange of land animals between the emerging modern continents, which in turn led to an outbreak of disease and to takeovers by certain species at the expense of others.

In the light of the richness of the information that Bakker has uncovered about dinosaurs, this extinction explanation is extremely disappointing, and not all that different from those on his ridicule list. And despite his closing rhetoric (Bakker 1986: 444)—"The grand rhythm of extinction and reflowering of species on land and in the sea must surely go to the earth's own pulse and its natural biogeographical consequences"—his hypothesis cannot be regarded as a general theory. Like all other natural scientists, Bakker overlooks the grand *internal* rhythms of life. A close study of life—like his own study of the dinosaurs—shows that the physical environment merely establishes the rules by which the game of life must be played, and it is up to the players—the various life forms—as to if and how this game is to be played. This interaction between the physical rules of the game and the players is explored in detail in *The Dynamic Society* (Snooks 1996: ch. 2).

In the end the explanation provided by Bakker—of the changing rules of the game of life—is no different in kind to those provided by other natural scientists about asteroids, volcanoes, and climatic change. The point is that none of these events is able to shape the *systematic* rise and fall of species or of dynasties. This rise and fall is, as argued throughout this book, the outcome of an endogenous dynamic in life. The ball is in the players' hands. It is this internal dynamic that explains the success of the dinosaurs as well as their ultimate extinction. What I am arguing is that the dinosaurs would have risen and fallen in the absence of any one or all of these external happenings, even if they actually happened. At the very most they might have acted, either individually or together, as a marginal influence, a final straw in the extinction of the dinosaurs. But they will never account for the dynamics of life of which the age of the dinosaurs was an interesting and important part.

More specifically there are a number of critical problems with Bakker's argument. First, he has misinterpreted his own hard-won evidence. He has done this in a general way by failing to recognize the endogenous dynamic model underlying the rise and fall of the dinosaurs. Of course he is not alone in this. It is true of his entire profession. At the more detailed level Bakker has misread the implications of the expansion and contraction of faunal diversity—"what ecologists call equability" and what he calls "evenness" (Bakker 1986: 436). This is the situation in which one species comes to dominate a dynasty. He claims that the diversity of the dinosaur dynasty declined—a "decay in evenness"—a few million years before the catastrophes are thought to have happened. It was due, he argues, to the "badly weakened" state of the dinosaur dynasty owing to a change of habitat forced by climatic change (Bakker 1986: 438, 437).

While I agree that the decline of diversity reflected the fact that the end for the dinosaurs was near, in reality it had an entirely different source. As the endothermic paradigm was progressively exhausted, the conquest strategy was adopted to play out a final zero-sum game (that became a negative-sum game). It was world war that reduced diversity and heralded the end. There had been a predictable shift of dynamic strategies from genetic change (increase in diversity), to family multiplication (status quo), to conquest (reduction in diversity). While Bakker's important evidence supports the dynamic-strategy theory, he and other paleontologists have misread it owing to their remarkable faith in Darwinism.

Second, Bakker's arguments about why cold-blooded species survived and *some* warm-blooded species disappeared as the world's climate changed do not hold water. He claims that as the cold-blooded species could not move so easily and quickly they were not exposed to the same disease and competition. This is a weak argument. In an era of mass migration not even those who stay at home (if they did; certainly they had migrated in the past) will be isolated from the tourists. Also we are told that warm-bloods tend to incubate diseases more readily than cold-bloods because they maintain an accommodating steady temperature. Even if this is true, why were mammals and birds, all warm-blooded creatures, immune? There are just too many holes in Bakker's argument.

Third, the dinosaur migrations were not unique. Mass migration is an integral part of the family-multiplication strategy pursued by every dominant life form over the past 3,500 myrs. And as such it is associated with the phase of status quo and not of decline in diversity. The latter, as we have seen, is an outcome of the final phase of conquest. Take the case of the great diaspora of early man from Africa. New diseases may have been encountered but this did not cause complete extinctions and did not prevent the colonization of the world. Even in the case of the Old World's invasion of the New World (early sixteenth century) and of Australasia (late eighteenth century), while indigenous populations were decimated by the diseases (small pox, venereal disease, measles, influenza) carried by Europeans, they recovered robustly. Disease can cause a severe yet temporary setback but it does not eliminate its host species entirely because that would be counterproductive to its own cause.

The dynamic-strategy theory provides an entirely new explanation for the fossil evidence, without resort to convenient catastrophes. It does so by focusing on the internal dynamics of life forms. We have already seen how individual dinosaur species or groups of species exhausted their genetic styles after about 5 to 6 myrs through the rapid pursuit—made possible by warm-bloodedness—of the family-multiplication strategy; and then how, through population pressure, the herbivores degraded their environment and terminally weakened their ability to survive, and how the carnivores overkilled their prey and suffered accordingly.

Once the entire endothermic paradigm, of which the dinosaurs were the third great dynasty, approached exhaustion, unsustainable pressure was placed on available natural resources, and there was an increasing degradation of the global environment, a loss of ecological balance, and a widespread adoption of the conquest strategy. This led to a "world war" between the various species of the dinosaur dynasty. It was a struggle to the death. Not just of individuals but of entire species, families and, ultimately, of the dinosaur dynasty itself. This world war raged with varying degrees of regional intensity for hundreds of thousands, even millions, of years. And at its conclusion the global environment was devastated and not a single dinosaur was left standing. Only the ancestors of modern birds that had sought refuge in flight survived into the age of the mammals.

In the process of this world war the military technology of the dinosaurs, both herbivores and carnivores, was brought to deadly perfection. This was genetic change in support of the dynamic strategy of conquest—a response, in other words, to strategic demand. As food sources for the armored herbivores diminished under intense population pressure, they turned upon each other. One can imagine the clash between the giant *Triceratops* (5,000 kg) as they charged at each other with their huge heads (more than two meters in length) lowered and their meter-long double horns deployed to kill the intruder on their diminishing territory. No Darwinian nonsense here about the evolution of male weapons through "sexual selection" for access to the females. The prime *imperative* was survival, not sex. Also we can visualize the king of the carnivores, *Tyrannosaurus*—heavy as a small elephant but as agile as a prizefighter—desperate to bring down its heavily armored prey with its massively powerful jaws. And as these prey declined in numbers to critical levels, *Tyrannosaurus* began to turn on its own kind as they battled over degraded territories.

Victory, no matter how glorious, was ultimately hollow as the great dynasty of dinosaurs continued its long and inevitable journey into eternal night. Much as Rome was to do, and for the same reasons, between AD 180 and 476. In the process much of the world's vegetation at one time or another was destroyed, some (such as the cycadeoids) permanently. Plants recovered fully only after the warring dinosaurs had fought each other to extinction.

Mammals and protobirds escaped the worst effects of the dinosaur wars. Like other small societies (such as Switzerland during the twentieth century) they were "neutral." Mammals lived in logs and burrows and fed on insects and worms that remained largely unaffected by the carnage. Any difficulties experienced were probably balanced by the declining attentions paid to them by their traditional dinosaur predators such as the turkey-sized *Stenonychosaurus*. Protobirds could use flight to escape the combatants and to seek protection in the larger trees or higher crags and to locate diminishing food supplies. Small reptiles were able to find refuge in logs, under rocks, or in the ground, while some of the larger reptiles retreated to the depths of inland waters or even put out to sea. While the survival of these animals is an embarrassment for the catastrophe and the climate-and-habitat-change theorists, it can be easily and consistently explained using dynamic-strategy theory.

Nor does the apparent coincidence of widespread extinctions in the seas as well as on the land present the **stratologist** with a problem. It should be realized that marine animals—particularly the long-bodied sea lizards and the long-necked swan lizards that we know experienced waves of expansion and extinction (Bakker 1986: 429–30)—went through a similar dynamic process to land animals. That extinctions in the seas and on the land coincided at some point in a period that extended over millions of years should not surprise us. The endogenously generated decline would have been exacerbated for shallow-sea creatures that were unable to move into deeper oceanic waters—particularly plankton, sponges, and shellfish, but not the sea lizards—by the new "ice-house" period that emerged from about 68 myrs BP with its growing polar caps, falling ocean levels, and diminishing continental shelves. Yet while there may have been a marginal role for physical events for certain types of shallow-sea and continental-shelf creatures, they played no significant role in the fluctuating fortunes of ocean-going and land-based animals that could move as physical conditions changed. Even Darwin (*Origin*: ch. 11) recognized this. Life, it must be concluded, is driven by an internal dynamic rhythm.

The Age of the Mammals

The age of the mammals is the fourth and most recent cycle of biological expansion to take place within the endothermic paradigm. As we have seen the first three—those of the protomammals, the archosaurs, and the dinosaurs—flourished for a season, exhausted their versions of this genetic paradigm, collapsed and disappeared. Each cycle was more complete and impressive than its predecessor. Each cycle involved greater genetic fine-tuning of the same endothermic paradigm.

This process of biological baton-passing is what I call the **great wheel of life**, whereby a dynasty emerges to exploit a genetic paradigm, exhausts their version of it, collapses and is followed in the same way by a new dynasty operating within the *same* paradigm (figure 13.4). The great wheel of life rotates without gaining traction, without being able to generate a new paradigm shift to replace the old exhausted paradigm. Without gaining more intensive access to the Earth's natural resources. This eternal recurrence, of dynasty replacing dynasty without generating further significant progress, is a sure sign that the **genetic option** which began with life some 3,850 myrs ago had finally been exhausted—that it was no longer possible to gain further access to natural resources through genetic change alone. When the age of the mammals dawned, following the collapse and extinction of the dinosaurs, this new endothermic dynasty promised to go the same way as the rest.

If life had relied only on the genetic option it would never have escaped the eternal recurrence. New dynasties would have emerged as old ones disappeared and would have gone through the same strategic process with only slight physiological variations. And the great wheel of life would have continued turning in space. Forever.

But toward the end of the age of the mammals something remarkable happened. Although there was no key innovation behind the radiation of the mammals, they carried with them, completely by accident, the potential for the greatest innovation of them all—intelligence. But even this did not lead to a new genetic paradigm shift as in the past, whereby the innovating species (in this case mankind) generates a large number of genetic styles. There is only one extant species of man. Rather the emergence of intelligence enabled the most intellectually advanced line of the mammals to replace the genetic option with the **technology option**, which was to spawn a series of **technological paradigm shifts** and, within each of these, a series of **technological styles**.

Clearly the importance of the technology option was that it enabled one line of mammals to break free from the great wheel of life and to escape the fate of all previous dynasties of endotherms—the fate of extinction. If the dynasty of mammals had not discovered the technology option, there can be no doubt they would have collapsed and gone extinct once they had reexhausted the endothermic paradigm. This would have provided some other species with its chance of escape, even if it was a very, very slim chance.

The dynamic-strategy theory raises the equally fascinating question of why the mammals succeeded where the dinosaurs had failed. Some paleontologists claim that although dinosaurs had only small brains—the largest in relation to body size, comparable with that of modern birds, was possessed by the turkey-sized *Stenonychosaurus*—a large-brain species would have "evolved" had their dynasty survived longer. Dale Russell (1969) has speculated that with time *Stenonychosaurus* might have evolved into a one-hundred-pound (45 kg) biped with an enlarged forehead, scaly skin, clawed hands capable of dexterous manipulation, and an intelligence equivalent to our own.

The only problem is that Russell, and any other author prone to this type of fanciful speculation, has missed the central point in the dynamics of life. The dinosaurs lacked not time but inclination. To develop superintelligence it would have been necessary for the dinosaurs to begin in the manner they wanted to end, by focusing on brain development rather than muscle power—on finesse rather than force. It is because the mammals favored finesse that they succeeded in breaking out of the eternal recurrence in about 65 myrs, whereas the dinosaurs, who favored force, failed to do so in twice that amount of time. Indeed, they were no closer to an intellectual revolution at the end of the Cretaceous than they had been at the beginning of the Jurassic. The dinosaurs, therefore, had exhausted their version of the endothermic paradigm without getting close to transcending it. "Dinosaurman" was never an option.

It was as if the dynasty had been wrongly advised by some sort of dinosaur policy maker who, having anticipated Darwin's theory of natural selection, advised his leaders that as genetic transcendence would arise automatically from population pressure and resource scarcity, it would be counterproductive to invest in the development of large brains—intelligence would arise from the mechanical operation of natural selection. Far-fetched? In effect this is what Dale Russell and his ilk are telling us. But the dinosaur version of Darwin misled his leaders because, as we now know, genetic transcendence arises from abundance rather than scarcity. Scarcity gives rise to the conquest strategy, which leads only to collapse and extinction. How lucky we are that our leaders do not take their advice from Darwinists![4]

From the very beginning the mammals relied to a far greater degree than their predecessors on intelligence in the pursuit of their dynamic strategies. They depended on finesse rather than force. While mammalian strategies were the same as those pursued by other dynasties, they were undertaken more effectively because of this emphasis on finesse. The point is that finesse requires larger brains. But why did the mammals favor finesse?

To answer this question we need to go back to the time of the dinosaurs. During their long exile throughout the age of the dinosaurs, the shrewlike mammals had to live by their wits as their main predator appears to have been the relatively large-brained—literally bird-brained (not a term of abuse in the dinosaur age)—*Stenonychosaurus* (Russell 1969). The interaction between these two adversaries probably led to the increase in brain size of both. But it was the smaller mammalian creature that was forced to specialize entirely in finesse in the pursuit of its dynamic strategies. The larger *Stenonychosaurus* only needed to use finesse when force failed, which was not often enough.

Once *Stenonychosaurus* was extinct the mammals had no intellectual peer, and once the entire dynasty of dinosaurs had passed into oblivion the mammals could come out of hiding and use their intelligence to more effectively facilitate the usual sequence of dynamic strategies—genetic change, family multiplication, commerce and, when necessary, conquest. With their greater intellectual potential the expanding mammals were able to head off a short-lived challenge from large flightless birds and savage crocodiles, both survivors of the dinosaur age. But it remained to be seen whether mammals could employ this intelligence to discover the technology option before their version of the endothermic paradigm was exhausted. If not they would be extinguished and the great wheel of life would continue to turn slowly, silently, and unobserved.

The role of the Endothermic Revolution was critical to the eventual success of the mammals in this unwitting race between brain size and paradigmatic extinction. The fabric of advanced brains, as is well-known, is sensitive to even modest changes in temperature. While it functions well at 37°C, a drop of just

5°C of body heat causes higher cerebral activity to become erratic, while an increase in temperature by the same amount will damage brain tissue. Constant body temperature made possible by the Endothermic Revolution, therefore, was a prerequisite for the development of large and sophisticated brains. Yet while the precise control of body temperature was a necessary condition for the emergence of intelligence it was not a sufficient condition. And it was a passive rather than an active force. The prime role in this, as in all dynamic issues, is played by strategic demand generated by an unfolding dynamic strategy.

Animals with small brains can pursue successful dynamic strategies, but animals with large brains can pursue them more flexibly, creatively, and effectively. Primitive mammals had, during the late Cretaceous, finally outwitted their dinosaur predators. And it would appear that this lesson was not lost on their Cenozoic or modern descendents. The greatest prizes in this new era went to those who had the most powerful brains. Strength, size, ferocity, and defensive and offensive weapons were all important, but intelligence was even more so in the strategic pursuit of the mammals. It was the age of finesse rather than force.

Intelligence was a response to strategic demand generated by those individuals who were pioneering the exploitation of strategic opportunities opened up by the demise of the dinosaurs. And as new species emerged from the pursuit of the genetic strategy they took with them a greater intellectual capability than the dinosaurs had ever bothered to do. Most of this brain development occurred in the centers of higher learning—in the cerebral lobes—which were responsible for improving the effectiveness of the strategic pursuit.

Enlargement of the brain appears to have occurred at speciation, that initial phase in which individuals pursued the genetic strategy in order to gain favorable access to natural resources. And the size of the mammalian brain increased from species to succeeding species. Hence, the modern jaguar has a brain twice the size of the saber-toothed cats of some 30 myrs ago; and, among the hominids, modern humans have a brain twice the size of that of *Homo habilis* about 2.4 myrs ago. The reason for this steady increase in brain size is that success in the strategic pursuit by a mammalian species depended not on the sheer size, strength, or effectiveness of its weapons, but on the ability to outwit one's opponent and to pursue one's dynamic strategy with finesse. Sheer, crushing force was being replaced by degrees with a more subtle and skillful execution of the strategic pursuit. But, of course, when finesse failed, force was always employed.

This emphasis on finesse rather than force can be seen in the characteristics of the first wave of mammals that replaced the dinosaurs during the early Cenozoic. The ferocity of the predators, which included the creodonts (10–200 kg), the miacids (20 kg), and the mesonychids (100 kg), paled in comparison with the tyrannosaurs (2,000 kg) that were ten times larger than the largest of the early mammal carnivores and possessed offensive weapons that would have terrified and destroyed them. Even the later mammalian carnivores—the wild cats and dogs (10–200 kg)—would have been no match for their dinosaur counterparts. Similarly the mammalian herbivores were also small and relatively docile. Compare the early condylarths (1–100 kg), the coryphodontids (400 kg), the perissodactyls (10–1,000 kg), or even the uintatheres and titanotheres (up to 5,000 kg) with the brachiosaurs (40,000 kg). It is significant that the larger early mammal herbivores, such as the titanotheres, did not survive the end of the Eocene (35 myrs BP), because they lacked the necessary finesse to adapt to changing environmental conditions. Even the more modern herbivores (such as elephants, hippos, and rhinoceroses) cannot compare with the size, ferocity, and armor of their dinosaur counterparts. True, there was a brief mammalian experiment with a megafauna consisting of giant herbivores (mammoths, rhinos, marsupials) and giant carnivores (saber-toothed tiger), but none of these relied on dinosaur-like weapons and all of them proved to be more vulnerable to large-brained hominids than their smaller counterparts. They were all driven to extinction by that master of finesse, *Homo sapiens*, between fifty and one thousand years ago. Once we get to the modern era, some 90 percent of all mammal species weigh less than 5 kg. In the mammalian world force is no match for finesse.

We come now to those experts in finesse among the mammals, the primates. Early primates were small, tree-dwelling animals that hunted insects on narrow branches high in the canopies of trees. Even today most primates fit this description. Interestingly, primates are most closely related to tree shrews and bats, which are also tree-dwellers.

It would appear that primates continued where premodern mammals left off—living obscurely by their wits in the forests far from the main action on the savanna, where the larger herbivores and, hence, carnivores roamed. Who could have imagined that these insignificant creatures would ever move to, let alone dominate, center stage? The sight of an unarmed primate was enough to make even a hyena laugh. The emerging characteristics of the primates, from which the anthropoids—the higher primates including man—arose, are: a larger brain size than other mammals owing to the more rapid growth of brain relative to body at the fetus stage; the development of considerable dexterity in the use of hands and feet as they leapt from branch to branch; the use of hands (the "opposable thumb") to grasp rather than to lunge at their prey and risk injury from falling; the development of stereoscopic color vision from the hunting of small insects; and the production of small numbers of offspring that develop slowly and live for a relatively long time, thereby enabling the creation and propagation of knowledge from generation to generation. All these characteristics were important in the employment of dynamic strategies that enabled anthropoids to cope effectively with their changing environment and, eventually, in discovering the technology option.

The story of the origin and development of primates down to modern man changes significantly from decade to decade as new fossils are discovered. At the time of writing it was generally thought that primates emerged in Africa during the Paleocene (65–56 myrs BP) and thereafter migrated to Asia and North America (where they died out). It is also thought that North Africa, this time during the Eocene (56–35 myrs BP), was the location of the origin and initial radiation of the anthropoids (monkeys and apes). Surviving anthropoids include Old World monkeys (cercopithecoids), New World monkeys (ceboids), and hominoids, which include gibbons, apes, and humans. The likely ancestor of Old World monkeys—and of *Proconsul* which led to the hominids—is *Aegyptopithecus*, a small anthropoid from the late Eocene discovered near Cairo. The primates that migrated to South America in Oligocene times (35–23 myrs BP) eventually gave rise to the New World monkeys but not to any apelike species.

Living hominoids include the hylobatids (gibbons), pongids (only the orangutan, or Asian ape, survives), panids (African apes, gorillas, and chimpanzees), and hominids (only *Homo sapiens* survives). According to DNA evidence, hominoids split off from the Old World monkeys during the Oligocene at about 33 myrs BP, the gibbons departed at about 22 myrs, the orangutans at about 16 myrs, and the hominids somewhere between 10 myrs and 6 myrs BP. The sparse fossil evidence, particularly between 30 myrs and 18 myrs BP, suggests later dates for all these speciation events. Some (Easteal, Collet, and Betty 1995) argue that hominids and panids (chimps and gorillas) may have parted company as recently as 3.5–4 myrs ago. Certainly we share 98 percent of our DNA with the panids.

The two earliest apemen yet discovered are *Ardipithecus ramidus* and *Australopithecus anamensis*, who emerged in Ethiopia (4.3 to 4.4 myrs BP) and Kenya (4.2 myrs BP) respectively. They had brain capacities about the size of a modern chimpanzee. *Ardipithecus ramidus* was probably the more primitive of the two, living in the forest, while *A. anamensis* was almost certainly bipedal. Both were displaced by *A. afarensis* ("Lucy") around 3.7 myrs BP. In turn *A. afarensis* succumbed to the forces of extinction about 2.5 myrs ago and was succeeded by *A. africanus* in South Africa from about 2.8 myrs BP and by *A. garhi* in East Africa, who is known to have been a meat-eating scavenger employing crude stone tools at about 2.5 myrs BP (Cullotta 1999: 572–3). Both species of apemen appear to have existed for about 1 myrs.

While there are particular difficulties with the fossil evidence at 3–2 myrs BP, it appears that at this time either *A. africanus* or *A. garhi* gave rise to the first true human, the large-brained *Homo habilis*, sometime before 2.4 myrs BP, followed by *H. ergaster* ("working" man) and *H. erectus* at about 2 myrs (or so) BP. They were the first humans to migrate out of Africa into Eurasia, possibly as early as 1.9 myrs BP (Balter and Gibbons 2000: 948–50). In the Middle East and Europe these species gave rise to *H. sapiens* (archaic) about 300,000 years BP and *H. neanderthalensis* about 120,000 years BP; in China to Peking man about 500,000–300,000 years BP; and in Southeast Asia to Java man.

Finally in this genealogy, *H. sapiens sapiens*, or modern man, emerged in South Africa about 150,000 years ago from *H. ergaster/erectus*, and possibly separately in Asia as well. We do know that modern humans moved out of Africa and encountered their distant relative, Neanderthal man, in the Middle East at about 100,000 years BP. The genetic difference between the two is about four times greater than we would expect between individuals in our own species. Then around 40,000 years BP, modern humans moved north into Europe and within 15,000 years Neanderthal man was extinct. Indeed, each stage in the emergence of modern humans appears to have been accompanied by the elimination of earlier stages. Intelligence was employed as an offensive weapon in the line of primates called hominid.

The important issue here is not the precise timing of the emergence of new primate species—the story changes slightly with each new fossil find—but the tendency for each new genetic strategy to be characterized by larger brain capacity. By 18 myrs BP, for example, a small species of *Proconsul*, which is thought to have been the ancestor of the higher apes, had a body weight of only 9 kg but a baboon-sized brain of about 167 cc. It was a tree-climbing, fruit-eating animal that may have spent some time on the ground, and have been able to stand upright when required.[5] Once again this larger brain must have provided the flexibility to cope with the exchange of animals both northwards and southwards when Africarabia docked with Eurasia during the Miocene (23 myrs to 5 myrs BP). From the north came giraffes, advanced deer, and other ruminants (cattle, etc.), and from the south departed elephants and hominoids.

It is possible to provide a quantitative outline of the growth of intelligence over the past 18 myrs. Reflected in table 9.1 and figure 9.7 are a number of fascinating features regarding the change in hominoid brain size. The first of these is the transition from constant *rates* of brain growth between *Proconsul* and *Ardipithecus ramidus* (18 myrs to 4 myrs ago) to exponential *rates* between apeman and early man (4 myrs to 2 myrs ago), and back to constant rates between early man and modern man (2 myrs to 150,000 years). Certainly nothing had prepared the world for this sudden acceleration of growth *rates*, nor for the increase in brain complexity in these add-on parts, from the appearance of the first apemen to the first appearance of modern humans. The past 150,000 years is also interesting because, after the emergence of modern humans, the rate of brain growth, for the first time since the extinction of the dinosaur dynasty, actually fell to zero.

Table 9.1 Growth rates of hominid brain size, past 18 myrs

Period	Brain size	Absolute increase in brain size per generation	Compound growth rate
(myrs BP)	(cc)	(cc)	(% per myrs)
18	167	—	—
4	400	.00042	6.4
3	450	.00125	12.5
2	650	.00500	44.4
1	925	.00688	42.3
0	1,350	.01063	46.0
Entire period			12.3

Source: Figure 9.7 and text.

The second interesting characteristic is that, while sounding impressive, these growth rates are based on time units of one million years each. If we think about what this means in terms of increases in absolute brain power per generation the achievement appears much less impressive. By arbitrarily imposing our concept of a generation—of about twenty-five years—on the data, we can see from table 9.1 that brain size increased by only 0.0004 cc per generation during the 14 myrs before the emergence of the first apeman. While this absolute increase for each generation is very small, the important point is that it continued to accumulate generation after generation for some 560,000 cycles. Yet even marginal increases of this amount gave each generation of hominoids an edge over other mammals in their determination to survive and prosper. And of course the generational increase in brain size grew significantly during the four million years from apeman to modern man: a 25.3-fold increase from 0.0004 cc to 0.01 cc per generation. It is also suggested in figure 9.7 that the increase in hominid brain size took place through a series of escalating steps as each new species replaced its progenitor.

Figure 9.7 Ascending the intellectual staircase—the past 5 myrs

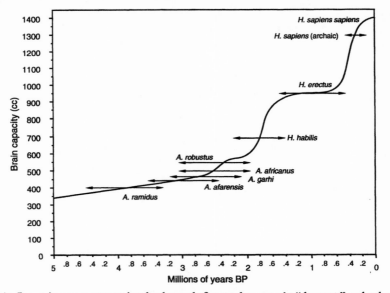

This figure is not meant to imply the path for modern man's "descent," only the stages achieved by the various hominid representatives in the fossil record.
Source: Drawn from data in Eccles 1989, White et al. 1994, Groves 1989, and Asfaw et al. 1999.

What accounted for this sudden increase in the size and complexity of hominid brains from about 3–2 myrs ago? And why did this growth rate level off at about 2 myrs ago and decline to zero from about 150,000 years BP? The acceleration in intelligence was due to the transformation of apemen's family-multiplication strategy. Until this time their existence both as individuals and as a species was precarious owing to their small and scattered populations. At about 3 myrs BP *Australopithecus* probably numbered no more than a few thousand individuals located in the rift valley of East Africa and in coastal South Africa. If they were to increase the probability of their survival as individuals and families, it was essential to be able to extend their territorial range, increase their populations, and improve their control over resources.

To achieve this would require a revolution in their family-multiplication strategy. While hominoids had always attempted to improve their prospects through improved intelligence—the reason for the slow but steady increase in brain size—from about 3 myrs BP their intelligence level appears to have passed the threshold that enabled them to make major advances in implementing their dynamic strategy. In addition, greater strategic flexibility was required owing to the changing climate (cooler and drier) and vegetation (advancing savanna) caused by a new ice age at about 2.5 myrs BP.

The breakthrough in the family-multiplication strategy was achieved by transforming it from a highly specific to a very general instrument of survival and prosperity. The higher primates had become very specialized in their diet and habitat, living on fruit, leaves, and nuts in tropical forests. With the emergence of apemen there is evidence that this diet was supplemented with high protein foods such as crabs and crocodile eggs to be found around lake shores in those fairly wooded, well-watered regions. But even this extension of diet and habitat would have limited the range over which apemen could have safely migrated. Certainly there was no possibility of their migrating to the cooler, drier, grassland regions, which were later occupied by *Homo* (Reed 1997).

This existential fact changed forever when the decision was made by some families of apemen to take up meat eating. Meateaters were not limited by a specific diet and associated habitat, because they were able to consume those animals that were highly specialized consumers, wherever they might live. In other words, the family-multiplication strategy based on meateating is a far more general strategy than one based on the eating of specific nuts, fruit, and leaves. This revolution in diet liberated hominids from their birthplace and enabled them to spread rapidly throughout the world.

But it did even more than this. By revolutionizing the basis of their family-multiplication strategy, some apemen were transformed into true men. What I am arguing is that the genetic change underlying the leap in intelligence was a response to the strategic demand generated by the family-multiplication strategy. The raison d'être for genetic change is always the attempt to gain greater access to natural resources, and it is pursued either as a strategy in its own right when resources are abundant following a mass extinction, or as an input—a **strategic instrument**—into another dynamic strategy, as in this case. This genetic change was required by apemen so that they could break out of their original primate environment and pursue their family-multiplication strategy in the rest of the world. In doing so they took their first great step in asserting their independence from the environmental constraints of life.

The essential point here is that the sudden increase in brain size was a response to systematic changes in strategic demand and not to ad hoc changes in supply as suggested by some scholars. It has been recently argued that hominids were able to increase their brain size, without increasing overall energy intake, by releasing energy constraints through meat eating, either by reducing the size of the gut to allow a transfer of resources for brain building or by enabling mothers to supply high-protein nourishment through the placenta before birth and through breastfeeding after birth (Gibbons 1998: 1345–47). In other words, meat eating, by reducing the constraints on brain size, was the driving force in the sudden increase in intelligence. Others prefer the idea that cooked tubers rather than meat "prompted the evolution of large brains" (Pennisi 1999: 2004). Similar supply-side arguments have been advanced in earlier publications (Crawford and Marsh 1989).

As usual these ad hoc supply-side arguments are incapable of explaining the forces and timing behind any increase in the use of resources. The problem is that supply-side forces are not treated as part of a general dynamic theory. These arguments are the biological equivalent of the catastrophe hypotheses of the physical scientists and paleontologists—involving ad hoc forces external to individual life forms. Ironically these supply-side arguments are also non-Darwinian and sound very Lamarckian to my ear—an increase in protein in mothers' milk allows an increase in brain size that can be passed on to future generations. Apart from the random reference to "evolution" or "natural selection" these scholars have in reality abandoned Darwin. Certainly they do not, indeed cannot, explain how mother's milk is related to natural selection.

In contrast my argument about the adoption of a new diet by the apemen is part of the general dynamic-strategy theory that has been used to explain the fluctuating fortunes of life over the past 3,500 myrs. Strategic demand called into being larger brains and all other anatomical changes—a smaller gut and richer mothers' milk—required to facilitate it, together with the associated technological ideas (fire, weapons, tools) and socioeconomic organization. Where there is a strategic demand there will always be a supply response, owing to the higher returns offered. Yet, while strategic demand creates its own supply, supply cannot create strategic demand (Snooks 1998b; 1999; 2000).

The importance of the decision of some apemen to generalize the family-multiplication strategy by meat eating/high intelligence/technological change can be seen more clearly against a backdrop of the majority of apemen—such as A. robustus (1.8–1.5 myrs BP)—that stuck to the old specialized family-multiplication strategy based on nut eating/lower intelligence/technological stagnation. The chewing power involved in grinding down hard nuts and tubers required small front teeth but massive molars, which in turn required a robust-shaped face with large jaws and a ridge on the top of the skull to anchor strong jaw muscles (McCollum 1999: 301–5). Although A. robustus had a brain slightly larger than A. africanus, a contemporary, they did not require one as large as that developed by the meat-eating generalists. Hence, the different sub-strategies of family multiplication led to very different physical (including brain) changes. This is a very important conclusion because it shows how genetic change is an outcome of the strategic pursuit—of choices made by strategists. Unfortunately for *A. robustus* they made the wrong choice and disappeared from the fossil record around the time that the large-brained *H. habilis* appeared. The "big-brains" outsurvived the "big-jaws." Finesse triumphed once again over brute strength.

But exactly what path did this strategic revolution take? Essentially the decision to adopt meat eating in order to transform the family-multiplication strategy required greater organization and technological abilities than formerly possessed by apemen. This transition took place in a number of stages. The first stage was based on the use of scavenging techniques. Scavenging required teamwork and primitive weapons, such as sharpened sticks and heavy clubs, to "persuade" carnivores to surrender a large carcass, and to keep other scavengers at bay. Also as apemen were not well equipped with the necessary teeth and claws to butcher a carcass, they needed to fashion sharp rocks for this purpose. They also required tools to extract marrow from bones, and they had to master fire to make meat eating easier and more palatable. There is clear evidence that scavenging was employed by *A. garhi* in East Africa at least 2.5 myrs ago (Cullotta 1999: 572–73).

This substrategy finally bore fruit at least by 2.4 myrs BP when a large-brained hominid suddenly emerged. This was *Homo habilis*, the first in our own line, who possessed a brain 50 percent larger than *A. africanus*, but still only about half the size of that of modern mankind. The structure of the economy and society of early man bore a close resemblance to that of modern man before the emergence of civilization (about 10,600 years ago) and of modern hunter–gatherer societies down to the present. They formed small kinship groups in which the males supplied meat through competition with other scavengers, and the females gathered fruit, nuts, and roots, and they bore and suckled the children (Leaky and Lewin 1992; Calder 1984: 140–1). These family groups foraged over increasingly longer distances searching for large carcasses that they brought back for butchering to central locations equipped with crude shelters protected by barriers made from thorn bushes.

Scavenging, however, is not a secure basis for globe-trotting. It is not always possible to find a freshly killed carcass just when required. To employ a viable generalized family-multiplication strategy it is necessary to be able to successfully hunt large animals on the open savanna. This placed significantly higher demands—strategic of course—on the intellectual capability of the African apemen. Hunting large and dangerous animals requires even more effective teamwork, particularly the abilities to plan and communicate clearly and precisely. No doubt this was the beginning of primitive but effective forms of language. Improvements were also needed in social structure—the emergence of leaders and greater specialization and division of labor—and in tools and weaponry. All of these matters established a strategic demand for higher intelligence.

A major breakthrough occurred about 2 myrs ago with the appearance of *Homo ergaster/erectus*. It involved a significant increase in brain size of about 37 percent over *H. habilis* and, just as important, this occurred particularly in the regions of problemsolving and language (Eccles 1989: 95). This growing intellectual capability led to a shift in mankind's economic role from scavenger to hunter—a shift as revolutionary in our history as that from hunter to farmer some 10,600 years ago, and from farmer to industrial worker about 200 years ago. For the first time in the entire history of life on Earth, a door had been forced open that might just lead to an escape from the eternal recurrence through the transcendence of the genetic option. The hunting (or paleolithic) revolution was the first technological paradigm shift in hominid—indeed life—history.

The new substrategy of hunting is reflected in the new stone tools (the Acheulean type) fashioned and employed by *H. erectus*. These included more sophisticated butchering, cutting, and scraping tools, together with deadly looking heavy axes and cleavers. It is thought that they also employed other hunting weapons fashioned from less durable materials, such as wood and bone, that have not survived (Cowen 1990: 414). And they used fire, both for cooking and, no doubt, for managing the savanna grasslands (Rowlett 1999: 741).

Because of this strategic breakthrough—which called into being greater intelligence, technology, and a more effective economic and social organization—*H. erectus* not only became the world's leading predator (displacing the saber-tooth tiger and eliminating other hominids) but also the first hominid to pursue the family-multiplication strategy outside Africa. This was the first **great dispersion** of mankind and it took *H. erectus* around much of the Old World, reaching the Middle East by 1.9 myrs BP, eastern Asia by at least 700,000 (possibly 1,800,000) years BP, and Europe sometime later.

To take the great dispersion to its logical conclusion—crossing great rivers and lakes, small seas, large mountain chains, and surviving extreme climates—the dynamic strategy of family multiplication generated strategic demand for even higher levels of intelligence. Consequently, by about 500,000 to 300,000 years BP Peking man in China had a brain size of 1,100 cc, about 80 percent of that of our own species, and by about 150,000 to 120,000 years BP both Neanderthal man and modern man had brain sizes of up to 1,400 cc. In early *H. sapiens sapiens* this greater intelligence can be seen reflected in the manufacture of more efficient stone tools (involving the Levallois technique by which long flints were extracted from within suitable rocks), a greater control over the animals they hunted, more effective management of their physical environment, improved methods of communication, and more sophisticated social, economic, and political organizations (groups) and institutions (rules).

There is considerable controversy over whether modern man emerged only in South Africa and spread throughout the world, replacing the descendents of *H. erectus* in a second great dispersion, or whether our species emerged in various parts of the world—including Africa, the Middle East, China, and Southeast Asia—at about the same time. I prefer the regional theory to the replacement theory because it is a natural outcome of the strategic story I am outlining in this chapter. My argument is that as human societies in different parts of the Old World—*H. erectus* did not get to the New World—struggled to extract the most from natural resources through family multiplication, they employed both genetic change and technological change as strategic instruments. There is no reason that this would have occurred only in South and East Africa. Neanderthal man, who had a slightly larger brain than modern man, demonstrates this. Also my theory suggests that Neanderthal man was eliminated by our forebears due not to their superior intellect but to a better substrategy. While Neanderthal man in Europe hunted wild cattle and lived in fixed settlements, our forefathers lived a more nomadic life by hunting reindeer and salmon which required the development of a new and more complex technology. The reasons that one succeeded and the other failed after a few thousand years of competition (prior to 25,000 BP) are no different to those that account for success and failure in competing civilized societies (such as ancient Rome and Carthage, Macedonia and Persia, or medieval Venice and Genoa). The issue here is strategic rather than genetic.

The strategic dynamic that led independently to both Neanderthal man and "African" man could also have led in other parts of the Old World to the independent emergence from *H. erectus* of modern man who subsequently migrated to the New World. The reason that the out-of-Africa, or replacement, theory is preferred by most natural scientists is that it seems to fit modern Darwinist theories about peripheral isolates—that genetic change occurs in a geographically isolated population, which on later contact eliminates the larger parent population (Mayr 1963; Eldredge and Gould 1972). But, as I have shown in parts I and II, the Darwinist theories cannot be sustained.

Before passing onto the next question, about the abrupt cessation of brain development about 150,000 years ago, we need to clarify the transitional period over the previous 2 myrs when *both* the genetic *and* the technology options were involved in human development. This transition began with the emergence of *H. erectus* and it ended with the appearance of *H. sapiens sapiens*. During these 2 myrs mankind employed both genetic change and technological change as dual instruments in the application of the family-multiplication strategy. Although the hunting (or paleological) revolution had begun about 2 myrs ago with the emergence of *H. erectus*, it could not be taken to its logical conclusion—the maximum population that could be supported by a hunter–gatherer lifestyle—without further joint increases in both genetic and technological change. After the emergence of *H. sapiens sapiens* about 150,000 years ago, the size and complexity of man's brain were sufficient not only to exhaust the paleolithic revolution but also to launch and follow through with the neolithic (or agricultural) revolution (10,600 BP) and the modern (or industrial) revolution (1780–1830).

It is likely that we are about to embark, after the lapse of 150,000 years, on a new interaction between genetic and technological change. This time it will be in support not of the family-multiplication strategy but of the technological strategy. In the process mankind will learn how to recombine both the genetic and technological paradigms.

The other major question about hominoid genetic change is: why in the past million years did the *rate* of growth of brain size slow and actually come to a halt about 150,000 years ago? This is usually glossed over by Darwinists, because natural selection is not the answer. Those who do attempt an explanation do so in non-Darwinian terms, such as the physiological constraints on fetus head-size imposed by female pelvic dimensions. This is not very persuasive as female anatomy changed considerably between *A. afarensis*—when females were only half the size of males—and *H. sapiens*, and could have changed still further in 150,000 years if strategic demand had so required.

The real answer is strategic in nature. One line of modern mammals had by 150,000 BP finally achieved what the dinosaurs had been unable to do: to replace their exhausting genetic paradigm with a technological paradigm. By 150,000 years ago our hominoid line had reached the level of intelligence required to pursue the hunting paradigm to the ends of the Earth. In other words, the strategic demand for genetic change generated by the family-multiplication strategy dried up after the emergence of modern man. The reason is that, when the paleolithic technological paradigm was finally exhausted—when all accessible hunting lands in the Old and New Worlds were finally used to capacity about 10,600 and 7,000 years ago respectively—it was found that human intelligence was sufficient to generate a new economic revolution or technological paradigm shift without any further genetic change. This neolithic (or agricultural) revolution involved the domestication of wild grass seeds and certain wild animals. And again, when the neolithic technological paradigm had been exhausted in the late eighteenth century in Western Europe—when all accessible agricultural lands throughout the world had been used to capacity—it was found that human intelligence was sufficient to generate the industrial technological paradigm shift known as the Industrial Revolution. Accordingly there was no further strategic demand in either of these revolutions for increases in intelligence, only for increases in technological ideas. As strategic demand creates its own supply in the long run, technological ideas were forthcoming and genetic "ideas" remained static.

In each of these technological paradigm shifts (or economic revolutions) the human race was able to transcend the old exhausted paradigm so as to release the hidden potential of the Earth's natural resources. And it was able to do so without making any further strategic demand for increases in brain size or complexity. The reason is that, once the threshold brain size has been reached, the technological strategy is a more precise, predictable, and precipitous instrument than the old genetic strategy. It is a strategic instrument that can be used to totally transform the world within the span of one lifetime rather than the hundreds of thousands of generations necessary under the genetic strategy.

In all other respects both strategies are very similar. For example, the new technological strategy, like the old genetic strategy, was employed in its own right only once scarcity had been removed. Once the paradigmatic breakthrough had been made, the technological strategy (like the genetic strategy in earlier times) was employed to develop all the technological styles (comparable to genetic styles) required to exploit natural resources in different circumstances. In the neolithic revolution, for example, these included wheat/barley and cattle in the Middle East; corn, pigs, and dogs in Mesoamerica; and rice, poultry, and oxen in Southeast Asia. Once fully developed these technological styles (like the old genetic styles) were exploited through the familiar family-multiplication strategy until they were exhausted. Further progress for individual societies in the premodern world depended on the successful pursuit of either the conquest or commerce strategies. This led to the **great wheel of civilization** that rose and fell without gaining traction—without generating any long-term increase in prosperity (that is, more intensive access to natural resources, measured in terms of real GDP per capita), only the short-term increase in monopoly profits that occurs in this zero-sum game as great conquest (Egypt, Assyria, Rome, Macedonia, Tenochtitlan) and commerce (Phoenicia, Carthage, Greece, Venice) societies rise and fall as part of the human eternal recurrence (Snooks 1996: ch. 12).

Just as mighty animal dynasties were doomed to the eternal recurrence until one dynasty was able to develop that rare commodity, intelligence, so the mighty human societies were doomed to the circular motion of the great wheel of civilization until one society was able to develop and adopt *industrial* technology. Without the **Intelligence Revolution** the mammal dynasty would have collapsed by now and have become extinct just like the dinosaurs; and, without the Industrial Revolution, Western civilization would have collapsed and become extinct just like the ancient Egyptian or Roman societies.

The patterns and dynamic explanations of both life and human society are very similar. Which is why any dynamic model that claims to be able to explain one part of life *must* be able to explain *all* parts of this great pageant. To be able to explain the collapse of the dinosaurs or the sudden increase in hominid brain size 3–2 myrs ago, it is necessary to employ a theory that can also explain the rise and fall of Rome, of Greek democracy, of any of the great religions, of the U.S.S.R., together with what the future may bring for our species and our society. These among many other matters both great and small. On the other hand, how do Death Stars, sundry catastrophes, cooked tubers, gut reduction, mothers' milk or, indeed, natural selection, stand up to this challenge? Obviously not very well at all. It is essential to realize that human society is neither artificial nor special as the Darwinists claim in order to rationalize their obvious failure. It is an integral part of the entire pattern of life on Earth.

One final point about the interaction between the genetic and the technological strategies is worth raising here. Having achieved the threshold level of intelligence necessary to substitute the technology option for the genetic option and, thereby, to break free from the eternal recurrence, our species was unable to achieve any higher level of intelligence. We are only as intelligent as was necessary to invoke the technology option. And the basic reason is this, that life does not attempt to maximize its IQ only to maximize the probability of survival and prosperity of its individual organisms. Only in the future might it be possible to maximize both through genetic engineering. But even this possibility may never be fulfilled because of the huge risks involved in changing the structure of our brains. It will be less risky and more economical to continue to focus on technological (AI) rather than genetic intelligence.

Conclusions

A number of quite radical issues have been explored in this brief survey of the drama of life on Earth. First, it is clear that the pattern of both life and human society is more complex than even the paleontologists recognize. Yet at the same time there is an endogenous regularity and predictability that can be persuasively modeled. But only if we abandon Darwinism in all its forms. Darwinism can be replaced using the inductive method to develop a single dynamic theory to explain the history not only of life but also of human civilization. It is a theory that can even be employed to make sensible predictions about the future of both.

What is radical about this? Simply that no other theory has been able to do so. While Charles Darwin claimed, mistakenly, that natural selection could explain life, it was clear even to him that it could not explain the dynamics of civilization. And while his contemporary Karl Marx claimed, mistakenly, that dialectical materialism could explain the past and future of human society, he made no such claims about the dynamics of life.

Second, it is clear from the dynamic patterns—the timescapes—of both life and civilization that Darwin's theory of natural selection (together with all its neo-Darwinian permutations and perversions) is, as a dynamic theory, a total failure. It is unable to explain not only the pattern and dynamics of life (which it confuses with "evolution") but also the transmutation of species. It is a theory in ruins.

Darwin predicted not only that the transmutation of species is a slow and continuous process, but that it occurs because of continuous population pressure (the "doctrine of Malthus") and the ever-present scarcity of resources. This is supposed to give rise to the survival of the fittest and the genetic advantages that they carry. Quite the contrary. As we have seen, speciation occurs only when resources are abundant and there is an absence of population pressure. Once speciation has taken place and the various genetic styles have been established, the dynamics of life is driven by nongenetic forces.

Darwin was also wrong about the reason that species, even entire dynasties, disappear. He incorrectly claimed that old species are eliminated by new species that possess even a slight genetic advantage over their "parents." In reality old species/dynasties go extinct because they have exhausted their genetic style/paradigm and either collapse by themselves or are swept away in this vulnerable state—they would eventually have collapsed on their own—by skillful opportunists.

Despite the obvious deficiencies of natural selection, even the more pragmatic paleontologists still defer to it. While they expertly *describe* the changes that have taken place in the history of life, they feel obliged to "explain" them by randomly repeating the Darwinian mantra—just as economists in the U.S.S.R. did with Karl Marx. But these paleontologists completely fail to show how natural selection can possibly explain a historical pattern stretching over 3,500 myrs that is demonstrably non-Darwinian. Instead they offer either mystical ("species speciate") or exogenous physical hypotheses without attempting to explain the inconsistencies with natural selection. Some merely suggest hopefully that Darwin is relevant to the microbiological level (but not how this could be), while exogenous physical forces are relevant to the macrobiological level. They are unable to say how these two levels interact. Unfortunately, paleontology is a descriptive rather than an analytical form of history.

Finally, it should also be clear by now that all those ad hoc exogenous, even extraterrestrial, "explanations" proposed by natural scientists, privately frustrated by the inadequacies of Darwinism, are far from the mark. There is just no way that these highly unsophisticated single-cause hypotheses will ever be developed into a general dynamic theory that could be used to explain and predict the dynamics of life or of civilization. The basic problem is that none of the authors of these shopping-list "explanations" of history has granted life forms the dignity of determining their own fortunes. What is required is an endogenous dynamic strategy of life.

In the remainder of this book I develop a formal endogenous dynamic theory that can explain, at both the micro- and macrobiological levels, the pattern of life outlined in this chapter. What we need is a single internally consistent theory that can explain, as parsimoniously and as completely (with *no* flying buttresses) as possible, the following matters:

- the micro- as well as the macrobiological patterns of life—the internal driving force and dynamic process as well as the biological outcomes;
- the need to focus on total biological activity in order to understand directional genetic change;
- why speciation occurs only in non-Darwinian circumstances;
- how the beginnings and endings of species and dynasties are related;
- why mammals succeeded where dinosaurs failed;
- the dynamics of human civilization as well as the dynamics of nature;
- why we are not smarter than we are; and
- the future of life as well as its past.

And many other empirical issues canvassed in this chapter.

No natural or social scientist has ever been able to build such a model, largely because no one has attempted to systematically observe and explain anew the entire pattern of life and civilization over the past 3,500 myrs. Instead they have adopted and patched up the existing, fatally flawed, and discipline-specific deductive theories inherited from their own narrowly based intellectual professions. It is time to break out of these limiting intellectual traditions.

Chapter 10

The Dynamic-Strategy Theory of Life

It is time to set out formally the vision of life developed in this book. In the foregoing chapter a new story of life was outlined to show how a simple version of my dynamic-strategy theory could be used to interpret the fossil evidence that has been painstakingly uncovered and presented by paleontologists over the past few centuries. It is a vision that arises from a systematic examination of the historical evidence concerning both nature over the past 3,500 million years (myrs) and human society over the past 2 myrs. It is, therefore, an inductive—what I call an existential—rather than a deductive dynamic theory. Indeed, it is argued that a realist dynamic model can be arrived at only inductively (Snooks 1996; 1998b; 1999).

To construct a realist dynamic model of life it has been necessary to go back to the metaphorical drawing board and to start again at the very beginning. We have seen how the neo-Darwinists began with Darwin's flawed farmyard theory and turned it inside out by arbitrarily and secretively substituting the sociological concept of "reproductive success" for Charles Darwin's materialist concept of "struggle for existence."[1] By contemptuously ignoring the historical evidence and by taking this viewpoint to its logical conclusion they have inadvertently revealed the absurdity of the deductivist approach. The historicists on the other hand have taken a more realistic approach by closely examining the fossil record. But by refusing to go back to the very beginning—by insisting on taking Darwinian baggage with them—they have failed to see the real message in their hard-won evidence. They have either forced the historical record into a distorting Darwinian straitjacket or they have attempted to explain it as the outcome of ad hoc physical forces. Or both. All the while chanting the Darwinian mantra. The outcome is incoherent.

It is time to cut through this incoherence by developing a realistic alternative to the flawed Darwinian and physical-science approaches to the dynamics of life. In this chapter I formally discuss, but in nontechnical terms, a general dynamic theory of life and show how it can be combined with the timescapes outlined in chapter 9 to isolate and explain the dynamic mechanisms operating over the past 3,500 myrs, which have translated the driving force of individual organisms into the **great waves of life**. And in the chapters that follow I discuss the main elements of this model in greater detail, extract the underlying laws of life, and consider what the future might bring.

The General Dynamic Theory of Life

What are the characteristics of a truly dynamic theory? Unlike the Darwinian model it will focus on *processes*, or mechanisms, as well as *outcomes*. These processes will involve the interactions between individual organisms and their physical and social environments, the **dynamic strategies** they adopt to survive and prosper, the manner in which these strategies operate at the micro- and macrobiological levels, and the impact that these strategies have upon the physical and social worlds.

An endogenous dynamic theory will be self-starting and self-sustaining, subject only to *passive* inputs (sunlight and various chemical substances) from the physical world. And the dynamic life system will respond positively and creatively, rather than passively and reactively, to outside shocks. Exogenous shocks are likely to be of two types: those emanating from the physical world, such as shifting landmasses, changing climates, and rising and falling seas; together with those generated by other life forms such as disease and invasion. Such a theory will be applicable to both the natural and human worlds. A brief summary of the general theory is followed by a more detailed analysis.

The dynamic-strategy theory consists of four interrelated elements and one random force:

1. the competitive driving force of individual organisms to survive and prosper—the concept of the materialist organism—provides the theory with its self-starting and self-maintaining nature;
2. the dynamic strategies—including genetic/technological change, family multiplication, commerce (symbiosis), and conquest—are employed by individual organisms through the process of strategic selection to achieve their objectives;
3. the strategic struggle is the main "political" instrument by which established individuals/species (old strategists) attempt to maintain their control over the sources of their prosperity, and by which emerging individuals/species (new strategists) attempt to usurp such control;
4. the constraining force operating on the dynamics of a species/dynasty is the eventual exhaustion of the dominant dynamic strategy/paradigm pursued by its members, which leads to the emergence of internal/external conflict and to collapse;
5. exogenous shocks, both physical (continental drift, volcanic action, asteroid impact, climate) and biological (disease and invasion), impact randomly and distortingly on this endogenously driven dynamic system.

In the dynamic-strategy model, therefore, the very long-run driving force arises from the universal motivation of all living organisms—to survive and prosper at any cost—and the wavelike process of biological activity by which this is achieved is due to the creative exploitation and exhaustion of these strategies. The constraints on life, in other words, arise from the very sources of its expansion and growth, which are internal to the model. Exogenous forces, in the form of physical and organic shocks, do *not* drive or systematically shape the dynamic process of life, but they do occasionally disrupt the systematic pattern in a random way. The internal dynamic always rapidly reasserts itself following any such shock, which, at most, temporarily distorts the endogenous dynamic pathway. These issues need to be discussed in greater detail.

The Driving Force

The endogenous driving force in life is the **strategic desire** of all individual organisms to survive and prosper. A systematic examination of the history of life undertaken in this book and elsewhere (Snooks 1996; 1997; 1998a) suggests that organisms attempt at all cost to survive *and*, having survived, to prosper—to maximize their consumption subject to the prevailing physical, social, and genetic/technological constraints. This is the dynamic concept of the **materialist organism**. It is the essence (and definition) of life, which has shaped our genetic and technological structure from the very beginning. Organisms that fail to feel this "desire" strongly, overwhelmingly, do not last long in this world.

To achieve their fundamental objective, organisms adopt the particular dynamic strategy that is expected to maximize their probability of survival and prosperity. This trial-and-error procedure is strongly influenced by the prevailing degree of external competition. We have seen that, contrary to Darwinian convention, intense competition leads to the adoption of the conquest strategy and to death and extinction, whereas minimal competition leads to the genetic strategy and to prolific speciation, and more "normal" competition generates the family-multiplication and commerce strategies that in turn lead to rapid population growth. This is not a haphazard affair. In each case a dynamic strategy is chosen because it is the most effective one available in the prevailing physical and social environments.

But how do individuals from the lowest to highest life forms make these choices? There are two existing extremist answers to this question provided by the neo-Darwinists and the neoliberal (economic rationalist) economists. The neo-Darwinists, as we have seen, insist that it is our genes that decide behavior, while the neoliberal economists insist that it is our rational faculties or ideas (Snooks 1997: chs. 4 and 5). Some readers might be tempted to say that the neo-Darwinists are right about animals and the neoliberals are right about humans. This, however, would only maintain the unsupportable dualism that exists in our treatment of life—a dualism that Darwin (1871) and, particularly, Wallace (1871) tried to eliminate more than a century ago.

Neither the neo-Darwinists nor the neoliberal economists are correct in this matter for any life form, high or low. Essentially the same process of decision-making is employed by all life forms and that is determined neither by genes nor by ideas. Rather it is determined by desire—the desire to survive and prosper. Both genes and ideas, combined in varying proportions by different life forms, merely facilitate desires. They encode and translate the methods by which desires are successfully achieved.

If genes and ideas do not drive animal society but merely facilitate the desires of its members, we need to replace both the neo-Darwinian genetic model and the neoliberal rationality model with a realist model of decisionmaking. Through the inductive method employed in this and earlier books it is possible to derive a model relevant to both low and high life forms and to finally eliminate the present dualism. I have called it the strategic-imitation model of decisionmaking (Snooks 1996; 1997; 1998b; 1999).

In reality, decisionmaking is based on the need to economize on nature's scarcest resources—intelligence. This is clearly the case with lower life forms, and it is one reason that neo-Darwinists have opted for the "genetic-leash" argument (the other is that they are professional geneticists of one sort or another and have a vested interest in claiming that genes rule life). But what of our own species? The neoliberal rationalists have adopted a totally unrealistic view of human decisionmaking: that it involves the construction of mental models about the way the world works, the collection of vast quantities of economic benefit–cost information, and rapid intellectual processing abilities.

The strategic-imitation model rejects both the anti-intellectual view of the neo-Darwinian geneticists and the super-intellectual view of the neoliberal rationalists. In the real world, rather than the fantasy worlds of these game-playing deductivists, individuals do make choices. But rather than collecting vast quantities of economic information on a large range of alternative options for processing through an intellectual model of the way the world works, the great majority of animals (including our own species) merely imitate those individuals and those activities that are conspicuously successful. This means that the only information the vast majority of decisionmakers seek is that needed to answer the key, but simple, questions: Who and what are materially successful and why? In this way the many follow the successful few.

The few are the strategic pioneers and the many are the strategic followers. Hence, the only information decisionmakers require is relatively costless **imitative information**, not the prohibitively expensive benefit–cost information; and the only intellectual faculty needed is that required to determine that someone else is more successful than the rest. Even the strategic pioneers do not employ rationalist techniques when seeking new ways of exploiting strategic opportunities. Rather than sophisticatedly calculating the best strategy from the large numbers available, the pioneers believe that their intuitively chosen strategies are best. Only a few of these will meet their expectations, but it is the fewest of the few that are slavishly imitated by the many. While the tendency for animals to imitate each other has received widespread recognition, the strategic-imitation model—to be discussed further in chapter 11—is the first to employ it as part of a general dynamic model.

But what of life forms without any intellectual capacity at all? Organisms in pursuit of survival and prosperity control their dynamic strategies through what I call **strategic instruments**. These strategic instruments include brains in organisms that possess them and special genes in those that do not. I have called these instruments the **strategic gene** and the **strategic cerebrum**. As shown in the foregoing chapter, early life forms were able to pursue the full range of dynamic strategies without the use of central nervous systems or brains. As discussed in detail in chapter 12, simple life forms possess a gene or genes that switch dynamic strategies on and off accordingly to the availability of nutrients and the degree of competition. Considerable scientific research has recently been conducted on this topic. The only matter in dispute is who is in control of this process, the selfish gene or the selfish organism. There are good empirical and theoretical reasons supporting my view that it is the selfish organism which controls its dynamic strategies through the strategic gene. Genetic structure, as we shall see in the next few chapters, has been shaped by the organisms themselves through their strategic pursuit. This is why organisms that developed central nervous systems quickly substituted the strategic cerebrum for the strategic gene in order to supervise their dynamic strategies. The reason is that the brain is a far more flexible and imaginative instrument of strategic control than the strategic gene.

The Dynamic Mechanism

The theory's endogenous driving force is provided by the dynamic concept of the "materialist organism." It is a self-starting and self-sustaining force that drives a dynamic mechanism which has at its center the **strategic pursuit**. The strategic pursuit involves the adoption and exploitation of the most effective available dynamic strategy or substrategy by the materialist organism—in the case of human society **materialist man**—to achieve its objective of survival and prosperity. A dynamic strategy begins as an individual or family activity which, if successful, is adopted by successively wider social groupings at first local, then regional, and finally global. This aggregation process—the progression from the micro- to the macrobiological levels—takes place through the **strategic imitation** mechanism by which conspicuous success is copied. In this way a successful dynamic strategy becomes widespread throughout a population, a species, even a dynasty.

The Sequence of Dynamic Strategies
The choice of strategy—from four possibilities including genetic/technological change, family multiplication, commerce (symbiosis), and conquest—depends on the underlying material conditions, such as the relative abundance of natural resources and the degree of external competition. The important point to realize is that organisms "invest" energy in each of these dynamic strategies at different

times and under different conditions to achieve the same objectives—survival and prosperity.

Typically, organisms in a species will pursue a sequence of strategies from the time they begin to diverge from the parent species until they finally go extinct. The sequence prior to the emergence of human society 2 myrs ago was typically genetic change, family multiplication or commerce, and conquest. Each dynamic strategy is exploited until it is exhausted, which leads to a temporary crisis until a new strategy can be developed and employed in the organisms' strategic pursuit. If, in a normally competitive environment, a new strategy is not adopted following the exhaustion of an old strategy, that species will collapse and go extinct prematurely. Accordingly, this **strategic sequence** leads not to a linear development path but to a series of waves consisting of phases of expansion, stagnation, crisis, decline, and renewed expansion.

As chapter 12 is devoted to examining the dynamic strategies of life in detail, only a brief outline is provided here. Imagine a non-Darwinian world characterized by minimal competition and abundant resources that emerges following the dramatic extinction of an earlier animal dynasty. Contrary to the predictions of Darwin's natural-selection theory, this is precisely the time when surviving organisms invest energy in the **genetic strategy**, because they have both the time and opportunity to exploit abundant resources. By creating a new **genetic style** they are able to gain more intensive access to already employed resources and/or new access to unused resources. This is equivalent to economic growth generated by technological change in human society. Essentially it involves a greater biological output from a given input of energy and other factors.

How—we need to ask—are organisms able to manipulate genetic change and turn it to the advantage of themselves and their families? While it is discussed in detail in a later chapter, it basically involves the process of strategic imitation. A beneficial mutation that improves an organism's access to resources will attract the attention of those with similar abilities. They will cooperate and/or mate with each other, thereby improving the prospects of the existing generation and the genetic characteristics of the next. Selection by the organism at the phenotypic level, therefore, shapes the genotype. It is what I call **strategic selection**—a form of self-selection at the "societal" level—which replaces the "divine selection" of the creationists and the "natural selection" of the Darwinists. Over very long periods of time it is responsible for the upsurge in speciation that always follows a major extinction.

Strategic selection is also the answer to the question that has always been a great embarrassment to the neo-Darwinists: Why sex? Neo-Darwinists are unable to satisfactorily explain the "popularity" of sex, because their central dogma about "reproductive success"—that organisms attempt to maximize copies of their genes in the gene pool—predicts that asexual reproduction, which enables individuals to pass on copies of all their genes, not just half of them, should be the norm. Clearly it is not. The dynamic-strategy argument is that **selective sexual reproduction** provides individuals with greater control over their dynamic strategies. In the first place it enables individuals to choose partners that display characteristics, both physical and instinctual, most required in the particular dynamic strategy they are pursuing. For genetic change one requires physical and instinctual characteristics that provide better access to natural resources; for family multiplication, those that provide greater fertility and mobility; for commerce, those that enable the monopolization and exchange of strategic resources; and for conquest, those that enable military success. Organisms are able to pursue and change dynamic strategies more effectively by being able to choose between the different physical and instinctual characteristics embodied by potential mates. Second, sexual reproduction increases the rate of mutation and genetic variety in the family, both of which are the primary material for genetic change. And third, sexual reproduction provides the basis for specialization and division of labor along gender lines in family groups, thereby increasing the probability of strategic success. Essentially all these matters provide organisms with greater control over life's strategic pursuit.

Once a new genetic style (or species) has fully emerged, the dynamic strategy of genetic change will have exhausted itself. It will, therefore, be more cost-effective in terms of metabolic energy use for an organism to switch investment from the genetic to the **family-multiplication strategy**. This will lead to an exclusive focus by organisms on the procreation and migration needed to fully exploit the new genetic style. In this way the members of our new species can outflank their parent and sister species by rapidly increasing their populations and spreading throughout the accessible world. This strategy, which leads to expansion (more resources accessed at the same degree of intensity) rather than growth (more intensive resource use), occurs during periods of "normal" competition—competition that is neither minimal nor intense.

This expansion phase is characterized in some species by the individual's pursuit of the more specialized dynamic strategy of commerce, or symbiosis. It is a strategy that involves the interaction between individuals in two different species, each specializing in access to different resources and "trading" with the other for mutual gain. These relationships exist between organisms in different species of plants, different animals and plants, and different animals. They characterize not only nature but also human society. The great commerce societies have included the Phoenicians, the Greeks, the Carthaginians, the Venetians, and the British.

In both the family-multiplication and the **commerce strategies**, the choice of and cooperation between associates and sexual partners is based not on the ability to use existing resources more intensively but on the characteristics required to procreate and migrate on the one hand and to monopolize and trade scarce resources on the other. In other words, through the operation of "strategic selection," benign mutations in the expansionary phase will be ignored unless they assist in promoting the family-multiplication or commerce strategies. It is for this reason alone that the genetic profile of a species will approximate what has been misleadingly called "punctuated equilibria."

By the time a genetic style has been exhausted, resources available to individuals in this species will be scarce, resulting in intense competition for them. This produces a crisis, because the earlier family-multiplication and commerce strategies will have generated levels of population and consumption that can only be maintained by a continuous inflow of natural resources. To prevent going under during such a crisis, individuals search for a new dynamic strategy to replace the old exhausted one. The only possibility, however, is the **conquest strategy**.

With the exhaustion of any given genetic strategy, individuals in that species will battle fiercely with each other for the diminishing supply of resources as overpopulation damages their ecosystem. In the process they will turn upon each other. This type of conquest strategy, which can be likened to civil war, renders a species vulnerable to a takeover of their ecosystem by a closely competing species or to any adverse change in their physical environment. Extinction of the species is highly probable. This may also happen in groups of species, as can be seen in the fossil record, when the exhaustion of their genetic styles coincides. And when the wider **genetic paradigm** of an era is exhausted, the entire dynasty resorts to the conquest strategy and becomes involved in a world war, which ultimately leads to the extinction of the entire dynasty. But as this world war is waged over a period of hundreds of thousands of years, many species employ genetic change to support the conquest strategy by developing offensive and defensive biological weapons. This add-on technology is a response to changing strategic demand as the conquest strategy unfolds. Organisms do not have the time or resources to effect a complete genetic transformation.

Hence, the Darwinian world of scarce natural resources and intense competition leads not to "evolution" through natural selection but to a short-term (a relative concept!) zero-sum game, in which the victor gains temporarily at the expense of the vanquished, and, ultimately, to the extinction of the entire dynasty. Only once the old dynasty has collapsed and the non-Darwinian world of minimal competition and resource abundance has been ushered in does directional genetic change and speciation occur once more. Darwinism has collapsed just as surely as the dinosaurs.

A Dynamic Form for the New Theory

As individual organisms seek to exploit their physical and social environments, which sets in train a mass movement orchestrated through strategic imitation, the dominant dynamic strategy unfolds. By this I mean that the inherent opportunities in the strategy are progressively exploited and, finally, exhausted. There is nothing automatic or inevitable about this unfolding process, which is merely an outcome of dynamic strategists exploring and investing in existing strategic opportunities.

But why does the expansion of a species lead ultimately to strategic exhaustion and stagnation? Strategic exhaustion is, as shown in chapter 15 (Law #7), the outcome of the "law of diminishing *strategic* returns." There will come a point in the expansionary phase of a species when each *additional* unit of metabolic energy invested by individuals in the dominant dynamic strategy leads to a decline in the *additional* units of resources accessed. Eventually the extra or marginal unit of energy cost will be driven down to equality with marginal resource returns. At this point the dynamic strategy is exhausted and it will be abandoned, because any further expenditure of energy will not pay for itself. No life form can afford to expend more energy than it is able to access through hunting and gathering food. Stagnation of the species sets in, and this in turn leads to a general decline as some individuals are unable to obtain resources to satisfy basic requirements.

Figure 10.1   Great waves of economic change in England—the past millennium

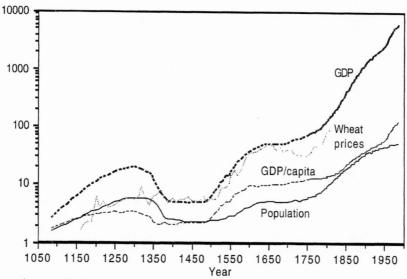

Log scale on vertical axis.
Source: Snooks 1993: 257.

The rise and fall of a dynamic strategy pursued by individuals in a species or a dynasty traces out a distinctive strategic pathway measured in terms of biomass, which would be seen in the fossil record if the evidence was sufficiently detailed and continuous. Because our evidence for human society is more complete, the strategic sequence of conquest → commerce → technological change can be seen quite clearly in figure 10.1, showing the great waves of economic change for Britain over the past millennium. It is this rise and fall of dynamic strategies that accounts for the systematic fluctuations in the numbers of animal families shown in figure 9.1.

What this means is that the dynamic-strategy theory expresses itself through a series of biological waves within waves. The **great waves of life** (of geometrically declining duration) generated by the rise and fall of dynasties encompass the long waves (of about 30 myrs) generated by the rise and fall of groups of species, which in turn contain the shorter waves (of up to 6 myrs or so) generated by the rise and fall of individual species. At all levels these waves are generated by the expansion, exhaustion, and replacement of dynamic strategies, genetic/technological styles, or genetic/technological paradigms. This pattern of waves within waves, therefore, constitutes the dynamic form of our model.

Strategic Demand and Strategic Confidence

The unfolding dynamic strategy, driven by the competitive energy of the materialist organism, plays a central role in our theory. Not only does it provide a realistic dynamic form, but it also gives rise to two important concepts that I call **strategic demand** and **strategic confidence**. These concepts are important in explaining the investment of energy made by organisms in the various dynamic strategies.

Strategic confidence, which rises and falls with the dominant dynamic strategy, is the element that binds any society together. It is responsible for the dynamic order that underlies the trust and cooperation between individuals in any group. Rather than having trust in each other—which both sociobiologists and economic rationalists are unable to convincingly explain—individuals have confidence in their successful dynamic strategy owing to the flow of resources that it provides. It is this confidence that gives rise to the decision made by organisms to invest energy in a particular dynamic strategy. When a dynamic strategy is exhausted, the evaporating strategic confidence contributes to the emerging crisis in the species by fracturing the relationships between individuals. Only when a new strategy arises does strategic confidence and, hence, order, stability, and trust return.

Strategic demand also rises and falls with an unfolding dynamic strategy. It comprises the demand generated by the **strategists**, or individual decisionmakers, for a range of inputs required in the strategic pursuit. To exploit strategic opportunities individual organisms need to invest in infrastructure (the construction of shelters, burrows, nests, beaver dams, farms, and cities) and "ideas" (both genetic and technological); to pass on acquired knowledge to the younger generation; and to develop and enforce social conventions and organizational arrangements.

Strategic demand, therefore, is the central active principle in life. Naturally the supply response of population change, infrastructure construction, and genetic/technological change—which in turn is influenced by relative scarcity of these factors (that is, by relative factor "prices")—will contribute to the way in which strategic opportunities are exploited, but they do so passively. It is strategic demand that creates its own supply, *not* the other way around as most scientists, both natural and social, claim. Take, for example, the recent claim that the apparently fortuitous act of meateating (or the eating of tubers—the experts cannot decide which) triggered the dramatic increase in the brain size of hominids. In contrast, the dynamic-strategy model suggests that meateating was a response to the strategic demand for greater strategic flexibility in the pursuit of family multiplication. Strategic selection is a response to strategic demand.

Strategic demand also undermines the neo-Darwinian concept of the "extended phenotype." The Dawkins argument that infrastructure (such as beavers' dams) and even the bodies of parasitic hosts are a "phenotypic expression" of the genes contained within another individual is both unrealistic and absurd. Changes in all supply inputs, including infrastructure, are a response to strategic demand that changes as the dominant dynamic strategy unfolds.

The Strategic Struggle

A mechanism central to the "politics" of the strategic pursuit is the **strategic struggle**. It is engaged in by individuals attempting to maintain or gain control of their "society," and by groups, populations, and ultimately species attempting to replace old genetic styles/paradigms with new ones. In the process these individuals and groups employ the **dynamic tactics of order and chaos**. The tactics of order aim at maintaining and exploiting the status quo by insiders and include the threat of punishment or ostracism and the enforcement of customary "rules," while the tactics of chaos aim at disrupting the existing order by outsiders and include attempts to undermine the authority of the leader or even to challenge him to combat. In each case the aim is to maintain or to hijack control of the dominant dynamic strategy, because it is the source of survival and prosperity.

The common occurrence in many species of males intimidating and fighting with each other is not, as the neo-Darwinists claim, primarily about sex. Males battle for supremacy, we are told, to maximize the presence of their genes in the gene pool by mating with as many females as possible. But in reality males battle with each other to gain control over the sources of their dynamic strategy, which are territories that provide access to food and shelter. They battle, in other words, to become the leading strategist in their group. The conflict between them is part of the struggle for "political" control of their "society." Having maintained or achieved this strategic control, which ensures his survival, the **strategic leader** is in a good position to maximize his prosperity, which involves the consumption of food, sex and, for the time being, leisure. While procreation assumes greater significance when it is part of the dynamic strategy of family multiplication, even then it is a means to a more important end.

The strategic struggle also plays a central role in the rise and fall of genetic styles and paradigms. With the exhaustion of a genetic style, individuals within a species battle with each other for control over the depleting sources of their dynamic strategy. Like warlords in a world of conquest, they attempt to become the strategic leader in order to improve their prospects of survival and prosperity at the expense of others in their species.

More significant is the strategic struggle when an entire genetic paradigm approaches exhaustion. In these circumstances the struggle takes place between the old strategists—the leaders of the old dominant dynasty—and the new strategists—the leaders of the newly emergent or resurgent dynasty. Because the old strategists have been weakened by the exhaustion of their genetic paradigm, the strategic struggle eventually favors the new strategists, such as the triumph of the archosaurs (protodinosaurs) over the therapsids (protomammals). But, as we have seen, this may involve many regional battles throughout the world over tens, even hundreds, of thousands of years. In these battles the old strategists employ the dynamic tactics of order, while the new strategists employ those of chaos.

Strategic Exhaustion

The force constraining life is not a limited supply of natural resources but rather the limitations of its dynamic strategies. Many people have trouble with this concept. To their minds natural resources are finite, so that, once a species or, on a wider scale, a dynasty has "used them up," its ecosystem will crash. This is the thinking behind the currently popular concept of "sustainable development." Fortunately this type of thinking is not correct, otherwise life would have either ended or stagnated about 3,500 myrs ago when heterotrophic life forms exhausted the supply of free-floating organic molecules to create the world's first energy crisis.

The whole history of life is the outcome of organisms using dynamic strategies to gain access to natural resources. This proceeds in two ways, one involving a more intensive use of natural resources and the other a more extensive use thereof. To gain more intensive access to resources, organisms employ either the genetic or the technological dynamic strategy. This involves the creation of new genetic/technological styles and, at the wider level, of new genetic/technological paradigms. With these new genetic/technological skills, organisms can employ other nongenetic/technological dynamic strategies—family multiplication and commerce—to migrate around the world to access underused resources. Once these skills can no longer access any further natural resources throughout the known world, strategic exhaustion occurs. Further progress requires the development of new styles and paradigms through further genetic/technological change. Hence, the supply of services that can be extracted from natural resources is limited only by the effectiveness of our dynamic strategies. In other words, the supply of resources can only be defined relatively (rather than absolutely) in terms of the dynamic strategies available to exploit them.

Strategic exhaustion operates at a number of different levels. The first of these is at the level of the species or individual genetic style. A new genetic style, which provides improved access to natural resources, is exploited by individuals in that species until it is exhausted. With exhaustion a crisis occurs as population presses heavily on natural resources. The response by individuals in the species is to opt for the internalized conquest strategy (civil war), which is a short-run zero-sum game that ultimately leads to the collapse of both the genetic style and the species. This accounts for the "background extinctions" for any dynasty that have been noted by paleontologists.

A more widespread type of strategic exhaustion occurs at the level of the dynasty or genetic paradigm. A new genetic paradigm, or collection of closely related genetic styles, is exploited by a large number of closely related species—such as the Permian reptiles or the protomammals, or dinosaurs, or modern mammals—until that paradigm is exhausted. Approaching paradigmatic exhaustion is reflected by a growing background extinction rate as groups of species exhaust their genetic styles, and by an increasingly depressed origination rate as unaccessed global resources decline. The outcome is a growing dominance of the prevailing dynasty by one or two species. In turn this is a sure sign that the dynasty is approaching exhaustion which will inevitably lead to crisis, world war (as a result of adopting the conquest strategy), collapse and, finally, extinction. While this is a terribly destructive process, it clears the way for the emergence of a new dynasty that will create a range of new genetic styles within this recent genetic paradigm, in order to gain even greater access to the Earth's natural resources. *The meaning of life, therefore, is all about unlocking a growing stream of services from the natural resources that were inherited when the Earth was formed some 4,500 myrs ago.*

External Shocks and the Dynamic-Strategy Theory

The broad outlines of the dynamic-strategy mechanism are summarized in figure 10.2, which is a simple diagrammatic model developed to illustrate the dynamics of human society. It can be employed with only a few substitutions of terms—such as "materialist organism" for "materialist man" and "biological

Figure 10.2 Diagrammatic model of the Dynamic Society

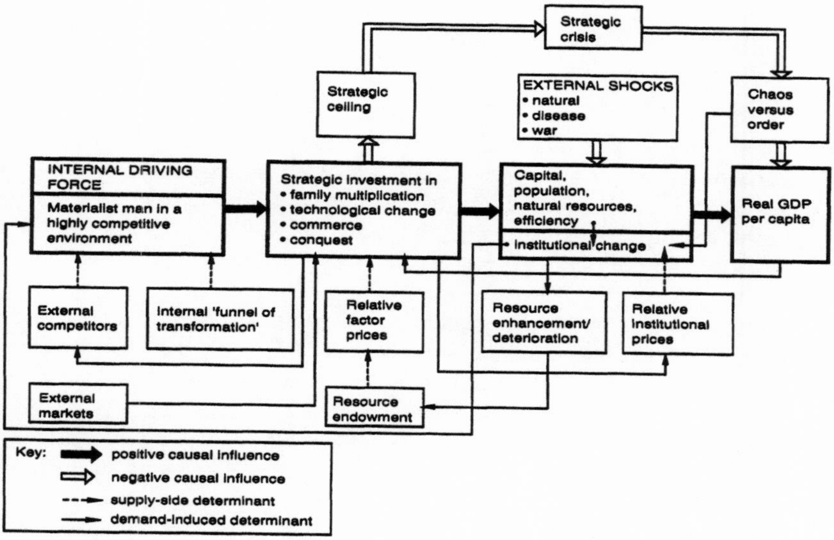

Source: Snooks 1996: 400.

output" for "GDP per capita"—to illustrate the dynamics of life. While there are a number of induced feedback effects, the overwhelming causal influence arises from the objectives of the materialist organism and the way in which they are pursued through its dynamic strategies. This great dynamic pursuit leads to the rise and fall of species (genetic styles) and of dynasties (genetic paradigms). It is all an outcome of the endogenous dynamic-strategy mechanism, and it would occur in the absence of any exogenous shocks.

Exogenous shocks, however, will impact on and distort this dynamic process in a random way. Major changes in sea levels, climate, volcanic action, and asteroid impact can be expected to have, at most, an *unsystematic* influence by accelerating, distorting, or delaying a dynamic process that is driven by forces internal to life itself. They do this by decreasing the amount of natural resources available to life forms at a given level of genetic/technological change. Timing is of the essence in this matter. If a dynasty is enjoying the fruits of its vigorous expansion phase, even major external shocks are unlikely to do more than cause a blip in the strategic trajectory. But if these shocks occur during a major internal crisis, such as the late exhaustion phase of a dynasty's genetic paradigm, they may hasten the dynasty's extinction. Yet at the same time other minor species that are experiencing a different strategic phase and are not caught up in world war will remain largely untouched. This is why there are so many "inexplicable" (to the natural scientist) survivors of the great extinctions in life. The essential point to realize is that exogenous physical shocks have no *systematic* influence on the dynamics of life. It is high time that natural scientists took a more sophisticated approach to this centrally important issue.

The Historical Dynamic Mechanisms

It is now possible to employ both the timescapes presented in chapter 9 and the general dynamic-strategy model outlined above to identify the complex interlocking mechanisms that have been driving life over the past 3,500 myrs. They are the great genetic (or technological) paradigm shifts and the great wheel of life. These dynamic mechanisms underlie the great waves of life illustrated in figure 9.3. Only a brief outline is given here as they are discussed in more detail in chapter 13.

The great genetic and technological paradigm shifts (figure 13.1) give the appearance of a flight of stairs—the **great steps of life**—that has changed exponentially in two dimensions: increasing in height but decreasing in depth. This reflects the accelerating impact of genetic change on life between 3,500 myrs and 2 myrs BP, and of technological change thereafter. There have been six paradigm shifts, the first three were genetic and the last three were technological.

The first genetic paradigm shift was an outcome of the Prokaryotic Revolution driven by blue-green algae from about 3,500 myrs ago; the second arose from the Eukaryotic Revolution driven by primitive plants and animals (including reptiles) from 900 myrs BP; and the third had its origin in the Endothermic Revolution begun by the protomammals from 245 myrs BP. While the next revolution—the **Intelligence Revolution**—was genetic in nature it led not to a genetic paradigm shift as one might have expected but to a series of technological paradigm shifts: the paleolithic paradigm shift (the hunting revolution) beginning about 2 myrs ago; the neolithic paradigm shift (the agricultural revolution) beginning about 10,600 years BP; and the modern paradigm shift (the Industrial Revolution) beginning in the late eighteenth century. Each revolution followed the exhaustion of the previous paradigm, and each made possible a more intensive access to natural resources that generated a higher level of biological or economic activity than ever before. In other words it was the sequence of genetic and technological paradigm shifts that enabled the increasingly energetic surging of the great waves of both life and human society.

The Intelligence Revolution differed from the three earlier genetic revolutions because it enabled the substitution of the **technology** option for the long-exhausted **genetic option**. It was because of this **strategic substitution** that the fourth and final genetic revolution generated a technological rather than a genetic paradigm shift. While the enabling condition for this strategic substitution was the achievement of a threshold level of brain size (in the range 700 cc to 1,000 cc), the driving force was provided by the strategic desire of one small branch of the mammal dynasty—the hominids—to acquire more precise and precipitate control over the means of intensifying their access to natural resources. Initially this was achieved through genetic change (before 3 myrs BP), then by a combination of genetic and technological change (3–0.15 myrs BP), and finally by technological change alone (since 150,000 years BP).

The technology option liberated life from sole dependence on the very slow-acting dynamic strategy of genetic change. The transition period between 3 and 0.15 myrs BP, when both the genetic and the technological strategies were employed in an interacting fashion, is particularly interesting. To increase their mobility through the pursuit of a more generalized family-multiplication strategy, the apemen needed, as we have seen, to change their diet from nuts to meat. But to become meateaters, these defenseless primates had to develop tools and weapons. Although the time-honored way was to develop biological appendages through genetic change, the apemen had, owing to their relatively large brains, a comparative advantage in producing detached stone/wood/bone tools and weapons, if only they could increase their intellectual capability. For the next 2 myrs or so an increase in brain size enabled the hominids to improve their tools and weapons, which in turn helped to improve the effectiveness of the family-multiplication strategy, which increased the strategic demand for greater intelligence to improve their tools and weapons. And so on. Genetic and technological change interacted in a joint response to strategic demand. In the process a branch of apemen was transformed progressively into modern man as its members changed from scavengers to highly skillful hunters that migrated to all parts of the globe, wiping out the megafauna as they went.

Over the past 150,000 years, since the emergence of modern man, our species has pursued the technology option exclusively, because from that time our brains were finally large enough to enable us to negotiate the neolithic and industrial revolutions with ease. Owing to the final liberation of the technology option, the development of brain size was terminated but the pace of life accelerated. And the driving force in life became materialist man rather than the materialist organism.

The genetic option exhausted itself once the full potential of the endothermic paradigm had been fully exploited. This was at about the time the dinosaurs were at their peak during the late Cretaceous—say 80 myrs ago. This meant that from then on life forms would only be able to employ genetic change to create *different* genetic styles from their warm-blooded predecessors, not more *innovative* genetic styles. Accordingly there would be no further dramatic increases in global biological output or productivity as new dynasties such as the mammals emerged from the ruins of the dinosaur dynasty. Any improvements in this respect would merely be the result of fine-tuning. Actually, the writing was on the wall of life—for those with eyes to see—when the endothermic paradigm was pioneered by the protomammals during the early Triassic (245 myrs ago). Even the succeeding dynasties of warm-blooded archosaurs and dinosaurs merely employed genetic change to realize the full potential of the endothermic paradigm that had been introduced by the protomammals. That was to be the last genetic paradigm shift in life on Earth.

The exhaustion of the genetic option meant that from the extinction of the dinosaurs the only things to change would be the participants. In the future new dynasties would replace old dynasties eternally without being able to substantially increase the amount and productivity of global biological activity. Each subsequent great wave of life would peak at about the same level as that achieved by the dinosaurs. This is what I have called the **eternal recurrence**. In terms of our historical dynamic mechanisms we can show that the **great wheel of life** replaced the great steps of life—the genetic paradigm shift—by the late Cretaceous (about 80 myrs ago). Thereafter the great wheel turned slowly without gaining upward traction.

I have called this the eternal recurrence because one can easily imagine life forms on other planets reaching this stage and not being able ever to escape from the great wheel of life as it rises and falls eternally. Each new rotation would differ from the last only in terms of the type of life forms it carried into inevitable extinction. In the case of our own planet the eternal recurrence was broken—life managed to leap from the great wheel—only because the dynasty of mammals followed the strategic pursuit with finesse rather than force (relatively speaking).

By using finesse rather than force, the dynasty of mammals eventually gave rise to a large-brained species that was able to generate the Intelligence Revolution. It was this revolution that introduced the technology option to replace the exhausted—some 80 myrs before—genetic option. But it was not inevitable. Indeed it should be regarded as a remarkable and highly improbable accident that may never be repeated on any other life-bearing planet. While intelligence is the scarcest resource on Earth, it may well be nonexistent in the rest of the universe.

With the Intelligence Revolution of 3 to 2 myrs ago, life began moving forward again rather than just repeating itself. A process of more intensive resource-use by life was reintroduced after a lapse of nearly 80 myrs, this time by a series of technological paradigm shifts (see figure 13.3), driven by the Paleolithic (hunting) Revolution about 2 myrs ago, the Neolithic (agricultural) Revolution at 10,600 years BP, and the Modern (industrial) Revolution, 1780 to 1830. These technological paradigm shifts, which generated the progressively surging great waves of economic change (measured in terms of real GDP per capita), were driven by materialist man in pursuit of the fourfold series of dynamic strategies. Details can be found in my global history (Snooks 1996; 1997; 1998a) and social dynamics (Snooks 1993; 1998b; 1999) trilogies. The point to emphasize here is that the dynamics of human civilization is just part of the dynamics of life, and it can be explained using the same dynamic-strategy theory. Human civilization is neither unique nor unnatural as the neo-Darwinists claim.

Social Organization and Change

Sociobiologists, as shown in part II, attempt to explain the social interactions between related and unrelated individuals by employing the genetically grounded kin-selection model. Even if we assume, only for the sake of argument, that their model is correct, sociobiologists are still unable to "explain" social grouping beyond the extended family. In particular they have failed to explain the structure of human society or the way it changes over time. And as the kin-selection model is merely a flight of deductive fantasy, they are in fact unable to explain social relationships even in the extended family.

Social organization is a response not to the alleged attempt by genes to maximize their presence in the gene pool, but rather to changes in strategic demand. Strategic demand varies in any species according to the genetic style within which individuals operate, the type of dynamic strategy they are pursuing, and the stage reached in the unfolding of that strategy. The basic social unit is the family, where the relationship between males and females is based on an exchange of food, shelter, protection, and status for sex, support, and comfort. Both partners contribute to a cooperative relationship by specializing according to their comparative advantage in the pursuit of their chosen dynamic strategy. The basis of mate choice will also vary according to the dynamic strategy being pursued: the physical characteristics on which choice is based will be of a resource-accessing nature under the genetic strategy, of a reproductive nature under the family-multiplication strategy, of a "trading" nature under the commerce strategy, and of a "military" nature under the conquest strategy. This is the basis of **strategic selection**, the key concept of the dynamic-strategy theory that, in response to strategic demand, determines not only genetic change but also macrobiological growth.

The structure of wider social groupings will also depend on the type and stage of the dynamic strategy, which are communicated through strategic demand. While in lower life forms this social structure will be relatively simple, in the higher forms (particularly in mankind) it will achieve considerable sophistication. In lower life forms, group activities can be seen in colonies of insects (ants and termites) and in the swarming of birds and various flying insects; and in the higher forms it can be seen in the complex structure of groups of non-related individuals as well as of families. In human society for which we have considerably more detailed information, the entire organizational and institutional structure is geared to the dynamic strategies that humans pursue: such as the war and imperial organization of a conquest society; the trade and colonial groups of a commerce society; the family and migration organizations of frontier society; and the industrial production and distribution organizations of a technological society (Snooks 1997). While none of this can be explained by neo-Darwinist kin-selection theory, it is rendered intelligible by the dynamic-strategy theory.

But how do we account for the coherence of animal—including human—society? The dynamic-strategy theory contains a model of animal inter-action that I have called the **concentric-spheres model** (Snooks 1994; 1997; 1998b; 1999). It is based on the notion of genetically determined desires, but allows for varying degrees of independent action. It shows that the way the self relates to other individuals and groups depends on their potential contribution to maximizing the probability of the self's survival and prosperity. This potential contribution is measured in the concentric-spheres model (see figure 14.1) by the **economic distance** between the self and all other individuals and groups who occupy positions on a set of concentric spheres that radiate outwards.[2] Underlying this model are two balancing sets of forces, one centrifugal and the other centripetal. The centrifugal force is the incessant desire of the self to survive and prosper, which leads typical individuals to consistently pursue their own self-interests. This is the life-energizing force, which continually threatens to disrupt social relations. The centripetal force, which can be thought of as the economic gravity holding animal society (from the lowest to the highest forms) together, is generated by the self's need to cooperate with other individuals and groups in order to achieve its own objectives through the joint pursuit of a common dynamic strategy. While this model was developed primarily to explain animal behavior it can also be applied to plant and earlier life forms (bacteria and viruses).

It is through the interaction of competition and cooperation in the concentric-spheres model that individuals maximize the probability of their survival and prosperity. And it is in this way that their "societies" also prosper. But it is important to realize that the underlying condition for the "trust" required for cooperative activity is not "altruism," genetic or otherwise; rather it is (as we have seen) strategic confidence. This is the confidence that arises from the conspicuous success of the dynamic strategy being pursued jointly by these cooperating but self-interested individuals. Because Darwin's theory of natural selection does not allow for cooperation—individuals are constantly locked in a struggle for existence—neo-Darwinists were forced to invent an absurd hypothesis about "altruism," which is based on alleged genetic links, to "explain" social relationships.

But what of the dynamics of social organization? Systematic changes in strategic demand provide the incentives, opportunities, and imperatives for the changing structure of animal and human society. The strategic cycle of adoption, expansion, exhaustion, stagnation, and decline—the complete unfolding process of the dominant dynamic strategy—has a characteristic impact on observed changes in the organizational and institutional structure of society. Institutional change, despite the claim of evolutionary institutionalists, has no life of its own. Nor is it the outcome of the independent actions of genes plotting from within their "lumbering robots," "survival machines," or—as you and I recognize them—individual organisms. Institutional change is reactive, not proactive. It has no "evolutionary" logic of its own. Animal society, whether simple or sophisticated, is merely a vehicle for achieving the basic "desires" of life forms. While the dynamic process is eternal, the social institutions and organizations of life are ephemeral.

Conclusions

Life has an observable pattern and an existential meaning. The rise and fall of species and of dynasties, the great genetic and technological revolutions, the great dispersions, civil wars, world wars, and extinctions are all part of a whole. They are the outcome of individual organisms attempting, through the pursuit of a range of dynamic strategies, to gain access to resources so as to survive and prosper. And this whole can be understood from within life itself through the dynamic-strategy theory.

The dynamic-strategy theory is briefly outlined in this chapter. To begin with, a general model is derived from a systematic study of the fluctuating fortunes of life in all its forms, including human society, over the past 3,500 myrs. In essence it involves the employment by individual organisms of dynamic strategies to maximize the probability of survival and prosperity. At the model's center is the concept of strategic selection, which places the responsibility for the dynamics of life as well as **biotransition** (the transmutation of species) firmly in the hands of individual organisms responding to strategic demand. Accordingly there is no need for the metaphysical concept of "divine selection" or the fanciful farmyard concept of "natural selection."

By applying the general model to the timescapes discussed in chapter 9, it is possible to isolate and analyze the dynamic mechanisms that have operated in the various historical eras. These dynamic mechanisms—the great paradigm shifts and the great wheel of life—show how the actions of individuals, which are coordinated through the process of strategic imitation, generate the great waves of life and how they led to the replacement of the genetic option that operated until 80 myrs ago with the technology option. This enabled life to break free from the eternal recurrence that probably dominates the rest of the universe.

In the next few chapters the main features of the dynamic-strategy model are examined in detail. Once this has been completed, our model is employed to derive the laws of life and to consider the future of life on Earth and in the universe.

Chapter 11

The Driving Force

At the center of the dynamic-strategy theory lies an endogenous driving force. Its vitality is generated by the **materialist organism** rather than the selfish gene because only the individual life form can engage in the strategic pursuit that is life. The organism is the dynamic **strategist**, who employs nongenetic as well as genetic strategies to achieve its objectives of survival and prosperity. Indeed, over the past one hundred and fifty thousand years the genetic strategy has been sidelined by life, because mankind has found it a less effective means of gaining access to resources than the technological strategy. Only now that we have been able to map the human genome will the genetic strategy be brought back to contribute to the strategic pursuit. Yet the only difference between the future and the deep past (prior to 150,000 BP) is that the control exercised by **materialistic man** over genetic change will be more direct, precise, and quick acting than that exercised by materialist organism. Genes are not, and have never been, our masters as the neo-Darwinists claim; rather they are merely one of the instruments employed by organisms in their strategic pursuit.

The Materialist Organism

Some 4,000 million years (myrs) ago the attribute that first separated living from nonliving cells was not the ability to reproduce systematically but the **strategic desire** to survive and prosper. This desire translated itself into an urgent need to gain access to chemical and organic materials—or "natural resources"—in order to convert them into the metabolic energy required to fuel the life process. This desire—called here **dynamic materialism**[1]—is the driving force in life. Without it, life would not exist. Hence, this driving force is to be found at the very center of the dynamic process we are exploring in this book.

While there is no escaping the fact that life is driven endogenously, most natural scientists attempt to explain major changes in its historical patterns—such as the extinction of the dinosaurs—in terms of exogenous physical forces. But these ad hoc arguments are totally unconvincing because they cannot be shaped into a general dynamic theory and because they take no account of endogenous forces. In this chapter I argue that the dynamics of life can be successfully analyzed only if we develop a theory that is self-starting and self-maintaining.

But what is the source of this internal driving force? Prior to the emergence of neo-Darwinism it would not have been necessary to ask this question. Systematic historical observation made it quite clear that the driving force has its source in the organism. But historicism is a method despised by neo-Darwinists. In chapter 8 we discovered that sociobiologists like Edward Wilson are contemptuous of the attempt to understand human nature through observation. It is an issue that cannot be understood, we are told, without an expert knowledge of the genetic building blocks of the brain.

The neo-Darwinists tell us that, despite our observations to the contrary, the driving force in life is the "selfish" gene. Genes are driven by a determination to maximize their presence in the gene pool, and organisms are merely the instruments by which this is achieved. The selfish gene, we are told, is "the prime mover of all life" (Dawkins 1989: 264). A variety of metaphors are employed to express this master–slave relationship. We are told by some (Dawkins 1989) that organisms are no more than the "survival machines" and the "lumbering robots" inhabited and manipulated by the "immortal gene" to achieve its own ends; and by others (Wilson 1975; 1998) that genes hold organisms on a tight "leash." In effect the neo-Darwinists "see the gene as sitting at the center of a radiating web of extended phenotypic power" (Dawkins 1989: 265).

A major problem for this interpretation—which is metaphysical because neo-Darwinists have been unable to show how it can be tested empirically—is that there can be no doubt that in the twenty-first century mankind will be in total control of the human genome. Even Edward Wilson (1998: 273–74), author of the "genetic-leash" concept, admits that "hereditary change will soon depend less on natural selection than on social choice" and that "humanity will be positioned godlike to take control of its own ultimate fate." He also concedes that, since the emergence of *Homo sapiens* some one hundred and fifty thousand years ago, and particularly since the rise of neolithic societies ten thousand six hundred years ago, "cultural evolution sprinted ahead at a pace that left genetic evolution standing still by comparison" (Wilson 1998: 157–58). Even the high priest of sociobiology indirectly concedes that human beings have been able to turn genes on and off in the strategic pursuit over the past one hundred and fifty thousand years. During this time (at least), we humans, rather than our genes, have had a major influence over the dynamics of life.

Neo-Darwinists attempt to shrug off this damaging fact by arguing that, owing to our unique achievement—our large brain—humans are different from the rest of nature in that we have the power to "rebel against the tyranny of the selfish replicators" (Dawkins 1989: 201). But by claiming that mankind is unique, the neo-Darwinists are also admitting that their selfish-gene theory is unable to explain the dynamics of human society. And it will certainly not be able, on their own admission, to explain the future of our species or the rest of nature under our biotechnological control. Clearly even the neo-Darwinists do not believe that they have been able to construct a *universal* theory of life.

But if neo-Darwinism is not a universal theory of life, why should its irrelevance be restricted to the time following the emergence of man? While the burden of proof lies heavily on the neo-Darwinists, they have been able to provide no proof at all. Indeed, the neo-Darwinists have only been able to create the evolutionary myth because even educated people are willing to believe quite strange things about other species—things that they are unlikely to believe about their own. This is why the critical test for any theory of life is whether it can explain not only the dynamics of nature but also that of human society.

It is my intention to banish the selfish gene from the entire history of life on Earth by developing a far more persuasive dynamic theory on both logical and empirical grounds. The role of the great majority of genes is to provide a blueprint or, as Dawkins suggests, a recipe for making copies of body parts. Genes exist merely to store, recover, and transmit the biological information necessary to maintain the effective operation of existing organisms and to create new organisms. But this role as "replicator" does not automatically grant genes the more fundamental role as "the prime mover of all life" as Dawkins (1989: 264) glibly asserts. That role is reserved for those entities engaged in the strategic pursuit. And those entities driving the dominant dynamic strategy of any society or species are individual organisms. Genes are merely strategic instruments which respond not to their own "desires" but to strategic demand.

Recently it has been suggested by some natural scientists that while the great majority of genes play a replicating role there are some more specialized genes that turn these replicating processes on and off. For example, Adam Arkin, a physical chemist at Lawrence Berkeley National Laboratory (LBNL) in California, has been working on the way viruses "decide" whether to replicate inside their bacterial host or to lie dormant, waiting for a better opportunity (Service 1999: 83). By employing experiments and computer modeling techniques Arkin and his colleagues have concluded that there are five genes to inform the virus whether to replicate or to lie dormant, and that these are switched on and off by six other genes: four "promoters" that turn on gene transcription and two "terminators" that turn it off. The presence of "smart" genes that switch other genes on and off is also recognized in bacteria (Maynard Smith 1993: 21).

These "smart" genes play an important role in my dynamic-strategy theory, where they are called **strategic genes**. They are used by organisms to respond to the changing physical and social environments, which determine the availability of nutrients and presence of competitors, and to activate the most appropriate dynamic strategy. When nutrients are abundant this will be the family-multiplication strategy; when competition is intense it will be the conquest strategy. In lower life forms, therefore, these strategic genes play the same role as central nervous systems in higher life forms—they are the strategic instruments that select the most appropriate dynamic strategy to fulfil the desires of the materialist organism. As such they do not drive life; they merely facilitate it. It is essential to realize that these strategic genes emerged in response to, and were shaped by, the strategic pursuit of their organisms. They do not have an independent existence or any driving ambitions of their own as claimed by the neo-Darwinists.

Genes of all types only attract strategic attention from the materialist organism when they fail to do their jobs properly—when they make a mistake either in faithfully copying existing body parts or in selecting the right time to turn a dynamic strategy on or off. The first of these—copying errors or replication mistakes—lead to changes, usually called mutations, in the body parts of existing organisms. In most cases mutations create problems for organisms and lead to their premature death (which, by the way, is not a very effective way for genes to maximize their presence in the gene pool). But even when a mutation is not life endangering it will have little effect on future generations unless it has the potential to facilitate the dynamic strategy being pursued by an organism and its family and friends. If, for example, an organism is successfully pursuing the family-multiplication strategy and the benign mutation cannot contribute to either fertility or mobility, it will be ignored. In other words, it will not be adopted through the process of **strategic selection** (to be discussed in detail in the next chapter). But if it has the potential to contribute to the family multiplication it will be adopted.

Take another example. If an individual organism is pursuing the dynamic strategy of genetic change and a benign mutation emerges with the potential to assist the process of speciation then, through strategic selection, it will be adopted and passed on to future generations. Hence, the genetic replicator plays no part in the strategic pursuit, as that role is reserved for the materialist organism—life's dynamic strategist. Only organisms possess the desire to survive and prosper.

In the second place, any errors made by strategic genes in turning dynamic strategies on and off in lower life forms have a number of obvious and not so obvious outcomes. Quite clearly a major mistake in strategic timing can lead to the death of a simple organism. Less clearly, a stream of more minor errors in the detection of environmental change and of strategic timing would lead to the emergence of a strategic demand by the materialist organism for a more efficient and flexible strategic instrument. This would be most urgent in the case of organisms attempting to gain access to radically different sources of nutrients. The prospects of the organism would be greatly enhanced if the selection and control of dynamic strategies were more sophisticated. In particular it would be preferable if the selection and control device—the strategic instrument—could not only just respond to environmental change but if it could actually anticipate change, if it could design substrategies that were more effective than those of its competitors, and if it could make fundamental changes more quickly. All these improvements and more would occur if it were possible to substitute the **strategic cerebrum** for the strategic gene.

The strategic desire of early vertebrates (with elementary nervous systems) to exploit the resources of the oceans and to invade the land appears to have been an important reason for the development of brains. Clearly it was beyond the capability of the strategic gene. This first occurred in primitive fish during the Ordovician period from 510 myrs to 440 myrs ago. From this time, therefore, brains began, through the process of strategic selection, to replace genes as the materialist organism's instrument of strategic control. And whenever strategic demand increased for even more sophisticated employment of dynamic strategies—of an increase in finesse rather than force—brain size also slowly, very slowly increased. Eventually at some indeterminate time, brain size passed a threshold level at which it experienced consciousness and began to regard itself as the materialist organism. Since then the conscious mind has been involved in a struggle between its idea of the rational self and the desires of the materialist organism that it was designed to serve.

Herein lies the origin of the conflict between ideas and desires. While the conscious mind believes that it is in control of the individual, in reality that role is played by the desires of the materialist organism or, more recently, of materialist man. Desires drive and ideas facilitate. The materialist organism, therefore, is still in control of the strategic pursuit, even in large-brained hominids. The conscious mind is just the latest method of strategic supervision.

In the strategic gene and the conscious mind we have the sources of the two extreme models of animal behavior. The neo-Darwinists have placed their faith in the strategic gene (not that they recognize it as such), while the neoliberal rationalists have embraced the conscious, rational mind. What neither group realizes is that they have focused on the shadow rather than the substance—on the strategic instrument rather than the dynamic strategist (or materialist organism). Our model of animal behavior in this chapter is based on the substance rather than the shadow.

The distinction made here between the materialist organism and its strategic instruments makes it clear that neither the strategic gene nor the strategic cerebrum (even less the conscious mind) has universal significance. The strategic gene was the materialist organism's instrument of choice for at least 3,000 myrs, and it continues to be exclusively employed by some lower life forms, such as bacteria, viruses, and plants. Plants in particular demonstrate that there was nothing inevitable about the emergence of intelligence. Some 1,500 myrs ago their eukaryote forefathers made the fateful "decision" to give up generating energy by engulfing other organisms and fermenting and oxidizing them—the animal way of life—in favor of photosynthesis—the life of plants. Because of this decision plants never had the opportunity to develop central nervous systems or brains. They continued to rely, therefore, on the strategic gene to supervise their strategic pursuit. The descendents of eukaryotic life forms that remained true to their animal origins, however, eventually developed a more effective instrument of strategic supervision—the brain. And with this instrument one insignificant representative of the animal kingdom—mankind—took over the world.

The Dynamic Strategist

Our interest in organisms is in their role as dynamic strategists: in their attempt to survive and prosper through the exploitation of strategic opportunities. These opportunities are exploited through the employment of a limited range of dynamic strategies. While these dynamic strategies are discussed in detail in the next chapter, we focus here on strategic choice.

How do organisms make strategic choices? As readers will have anticipated from the last section, there are two ways in which the materialist organism can adopt and supervise its strategic choice: the strategic gene employed by lower life forms, which switches dynamic strategies on and off as circumstances change; and the strategic cerebrum adopted by higher life forms, which is able to do so in a more effective and flexible manner. Both methods are briefly outlined.

Essentially the method of strategic choice by animals economizes on the world's scarcest resource—intelligence. This is as true of humans as of all other animals. Despite mankind's relatively large brain, we employ the same basic method of decisionmaking as that used by all other animals. As our species began to emerge about 2.4 myrs ago, it carried with it the decisionmaking methods inherited from earlier mammals. This is because rationalist decisionmaking—the collection of large quantities of benefit–cost information, the use of intellectual models of reality, and rapid processing techniques—is very costly in terms of intellectual and energy resources, and the probability of its success is not very high (Snooks 1997: ch. 4). Instead, we, like all other higher animals, employ a more pragmatic and effective method of decisionmaking.

In this book a realist approach to decisionmaking, based on a systematic examination of the history of both nature and human society, is adopted. Observation suggests that the decisionmaking process is imitative rather than rationalist or genetic and that it is dualistic rather than holistic. I call it the **strategic-imitation** model of decisionmaking. In this model the decisionmaking process can be divided into the pioneering and routine phases, with each part of the process being dominated by a different type of decisionmaker—the strategic pioneer and the strategic follower.

The Pioneering Phase

The investment of energy and resources in new genetic ideas or new activities is risky because little information of any kind about likely outcomes is available to strategic pioneers and because the probability of failure for new ventures is high. It is a matter of trial and error, with the pioneers operating largely on intuition about outcomes and on faith in their own abilities. It is a faith often misplaced.

The essential point to realize is that the strategic pioneers do not make their choices in a vacuum. They work within the boundaries of an unfolding dynamic strategy, and they respond to the changing incentives it generates. In other words the pioneers operate at the leading edge of the unfolding strategy and they actively explore its material potential. There is, therefore, an interaction between the pioneers and the strategy that gives direction and meaning to their activities. Yet, while these activities are informed by the changing dynamic strategy, many pioneers misread the signs and fail. While only the few succeed, these provide the example for the many to follow. And it is this mass following that provides the driving energy for the strategic unfolding.

The pioneers are few in number and even less significant in proportion to the total population. Paleontologists have referred to them as "aberrant" individuals (Cowen 1990: 155). They are more ambitious, possess greater skills, and have more faith in themselves than do their "normal" peers. The few are more driven than the many. When conspicuously successful the pioneers are followed by large numbers of imitators. Yet the pioneers, as already suggested, are not always successful. Owing to the risks that attend untried ways, only a small proportion of the pioneers achieve their objectives. But as the rewards, in terms of favorable access to natural resources, are great, there is always a steady stream of "aberrant" individuals willing to be tested.

What is the mechanism by which successful pioneering decisions are made? The pioneers invest most of their time not in the collection of "benefit–cost" information about alternative ventures but in pursuing the particular ventures that appeal to them and which intuition tells them will bring windfall gains. In essence the pioneers provide the drive and enthusiasm to propel their ventures into the marketplace. And it is the marketplace of competing ideas and activities, shaped by the dominant dynamic strategy, that determines the outcomes.

That small proportion of pioneers who get it right are provided with the material encouragement to continue. Successful feedback in terms of access to better and/or more resources leads to further exploration along the same lines. For as long as individual pioneers are achieving the outcomes they desire—survival and prosperity—there is no need to explore the alternatives. That will only happen when their current strategy has been exhausted and a crisis has emerged.

The individual pioneers between them embody the alternative possibilities that the rationalist approach assumes will be evaluated by each and every one of them. Each pioneer directs his energy to the exploration of a single strategic possibility. Naturally, many of these strategic possibilities will be duplicated or will overlap with others. What the pioneers require to enter into this competitive process is not rationalist abilities and techniques (which even modern humans do not employ) but great curiosity, energy, and enthusiasm. Curiosity rather than applied intelligence is the hallmark of the strategic pioneer. And once they stumble on a new and better way of accessing resources, they do not compare this with all other possible ways as the rationalists insist. Rather than shopping around for an even better way, the pioneers proceed as if their way is best. And in some cases it will prove to be so.

The Routine Phase

The successful pioneering few initially earn "monopoly profits" on their strategic activities. They flourish and provide the first positive information for the many about "rates of return" on the investment of energy in alternative ventures. The pioneers provide, in other words, **imitative information**, which satisfies the key questions: Who and what are successful and why? Armed with this simple, almost costless information, the vast number of risk-averse organisms, ranging from lower to higher (including man) life forms, are also able to invest scarce energy in the successful strategies and to share the rewards of better access to natural resources. This is the central mechanism of **strategic imitation**, which transforms the individual rashness of the "aberrant" few into the dynamic strategies of the "normal" many—of populations, species, and dynasties.

Dynamic-strategy theory, therefore, is the very antithesis of game theory developed by neoliberal economists and later adopted by neo-Darwinists such as John Maynard Smith. Individuals are not continually second-guessing each other and adopting *different* and highly competitive "game strategies" (actually they are just "tactics"). Rather they adopt *similar* dynamic strategies and substrategies—the dynamic strategies of the successful few. It is not surprising, therefore, that game theorists are unable to develop either realist models of the interaction between individuals or macroeconomic or macrobiological models. In contrast the dynamic-strategy theory can do both (Snooks 1998a; 1999).

In the strategic-imitation model, therefore, choice is exercised not by calculating the benefits and costs of energy/resource use, but by imitating those who appear to be materially successful in life. In animal society, young males will imitate older successful males who battle with each other to maintain or gain control over territories (economic resources), and who monopolize the attention of females by trading food, protection, and status for sexual and other personal services. Females on the other hand will select their mates by observing their strength and other skills (food gathering, shelter building, and sexual) in male to male and male to female interactions. Their choice will depend on the attractiveness, vitality, and popularity of the males. They are attracted to those males that most other females find attractive.

In human society, imitative information is acquired easily not only from the direct observation of those around us as in animal society, but also from books, magazines, "how-to" manuals, the Internet, the media, and professional advisers. The main form of imitation in the past was through the direct observation of local celebrities. Even now we are well aware of the nouveaux riches in our midst. Modern humans still possess a primitive and unquenchable need to display the objects of material success. Indeed the intense desire to imitate the successful is matched only by the desire to display the fruits of success.

The only difference between primitive and sophisticated societies, apart from the scale of material success, is the way in which this interactive mechanism of display and imitation is effected. In fact, with the recent development of the Internet, the ancient role of word of mouth has made a spectacular comeback, this time on a global basis. Today we also gain a large proportion of our imitative information from the media, which specialize in watching and reporting the rich and successful, often in the most crudely voyeuristic form. We are treated to a continuous parade of the successful to be imitated, and of the failures to be avoided. This is a fascinating subject that has been discussed in greater detail elsewhere (Snooks 1997: 40–46).

The method of imitative decisionmaking, therefore, involves no more than a minimal amount of intellectual capacity. All animals, including humans, have a remarkable ability to mimic those around them. And mimicry is all that is required. Animal studies are replete with examples of the way mimicry is employed in discovering new food sources and new mates (Trivers 1985). In animals with small brains this decisionmaking procedure is simplified by the more narrowly defined boundaries of their existence. The narrower the genetically programmed parameters of existence, the less brainpower is required. Nonetheless, animals do make nonprogrammed choices. They are not genetically determined automations as the sociobiologists would have us believe. To argue that animals and humans are programmed by genes is to miss the central fact of the history of life that the strategic cerebrum replaced the strategic gene some 500 myrs ago.

What then is the role of genetic influences in the animal and human worlds? The genetic structures of animal organisms have been shaped by the strategic pursuit, rather than by the "selfish gene." This is particularly so in the case of the dynamic strategy of genetic change pursued when conditions are right by the materialist organism. This has been achieved by individuals through the mechanism of strategic selection. Genetic structure has been shaped by choices made by the organisms themselves. Genetically speaking we have created ourselves.

The question here is: How does this *strategically determined* genetic structure of animals influence the actions of individuals in any species? The dynamic-strategy theory suggests that:

- it *defines* the *range* of possible physical and "intellectual" responses to external stimuli;
- it *influences*, along with the physical and cultural environments, both individual tastes and personality;
- it *influences*, along with the physical and cultural environments, the ways animals learn.

Yet, even where genetic inheritance determines an organism's range of capacities, there is scope for nongenetic manipulation. Genetic structure has been shaped by the strategic pursuit and exists merely to serve the materialist organism. Ideas, both genetic and technological, are slaves to desire. Not the other way around. Hence, even fundamental genetic controls—those strategic instruments of the materialist organism—can be transcended by determined individuals. And genetically constructed capabilities, both physical and intellectual, can be extended through the acquisition of learned skills and, in the future, by genetic engineering.

What genetic inheritance does *not* do, therefore, is to program the behavior of individuals in the way that sociobiologists claim. As the brains of animals have been constructed by the materialist organism in the strategic pursuit, albeit by employing genes, the behavior of animals is determined by the organisms themselves and not by the instruments they employ. While it is quite clear to the careful observer that this is true in human society, the evidence marshaled by naturalists suggests that it is also true for other animals. Recent experiments with animals such as chimpanzees, monkeys, dolphins, sea lions, dogs, and parrots suggest that animals possess not only cognitive skills but also a conceptual grasp of symbols and words imparted to them by human trainers, together with the ability to learn a large number of tasks (Hauser 2000). Not surprisingly the capacity to do so varies between species.

Chimpanzees, for example, display far greater perception and understanding of "ideas" than rhesus macaques, similar to the way that four-year-old human children do in comparison with three-year-olds. Indeed, it is now clear that chimpanzees can formulate plans and make tools to achieve their objectives. They also employ elaborate tactics to manipulate and even deceive those around them to attain social dominance and favorable access to food or sexual partners (Hauser 2000).

Why should any of this surprise us? Just like humans, other animals exercise choice subject to their genetic, physical, and economic constraints. Obviously the genetic parameters within which individuals operate vary across species at any given time. Humans have a wider range of physical and intellectual options than any other species. While there is a hierarchy in this, the available evidence on behavior suggests that all animal life forms exercise some real choice. And as these choices affect their/our genetic structure, the "leash" operates not from genes to organisms but from organisms to genes. Sociobiology is utterly confused about who is taking who for a walk around the block called life.

That may be all very well about animals, you might say, but what about plants, bacteria, viruses, and other life forms that do not possess central nervous systems and brains? How do organisms in these species make choices? How do they adopt appropriate dynamic strategies? As outlined above, recent research shows that organisms without brains are able to respond to their environment by "employing"—I use the word advisedly because they can also be made redundant—specialized genes to do what the brain does in higher life forms. These "smart" genes—or strategic genes as I have called them—work on behalf of the materialist organism by monitoring the outside world and switching on and off appropriate dynamic strategies, such as genetic change (for example, the creation of siderophores in bacteria to attract and trap iron atoms whenever iron levels fall below a critical level), family multiplication (to replicate and colonize as in bacteria and viruses), commerce (to monopolize and exchange resources with other species as in lichen), and conquest (to eliminate competitors to gain access to resources as bacteria do with their waste products and siderophores).

Strategic genes are also employed by organisms to monitor their competitors, which allows them to participate in a successful dynamic strategy through the mechanism of strategic imitation. This would account for the periodical explosion of bacterial and viral infections. Such a high degree of coordination could not be achieved by the "selfish genes" all making their own calculations about when to replicate and to invade new tissue. It would also explain how some two hundred bacterial forms developed different iron-seeking siderophores some 2,500 myrs ago.

Strategic imitation also enables simple organisms to economize on scarce DNA, just as higher animals use it to economize on the even scarcer commodity of intelligence. This resolves the problem for paleontologists who are aware that the small amount of DNA in bacteria and viruses is not sufficient to carry out the detailed "if . . . then . . ." genetic instructions so beloved by sociobiologists. The reality is that organisms provide the driving force, while genes facilitate their desires in a passive way. There is no such thing as a prime-moving "selfish gene." That is merely a product of neo-Darwinian fantasy. It is the organism that fights for survival and prosperity, not the gene, even the strategic gene—which is merely a strategic instrument. And when that simple genetic instrument was found to be inadequate in gaining better access to natural resources during the Eukaryotic Revolution it was replaced with the strategic cerebrum. In life, desires drive and ideas facilitate.

Conclusions

The driving force in life is the universal desire of organisms to survive and prosper. It is this desire that drives the materialist organism in a strategic pursuit to gain ever greater access to natural resources. This pursuit is undertaken through the mechanism of strategic imitation, which is managed by organisms through either the strategic gene (in the case of viruses, bacteria, and plants) or the strategic cerebrum (in the case of animals and humans). Both devices, which were shaped by the organism through the strategic pursuit, enable the relevant materialist organism to monitor the changing physical and cultural environment and to adjust its dynamic strategy accordingly. Hence both the strategic gene and the strategic cerebrum are merely instruments employed by the materialist organism to achieve its objective of survival and prosperity. When the strategic gene proved inadequate for this purpose by more aspiring life forms it was replaced by the strategic cerebrum. The "selfish" gene has no independent role to play in the strategic pursuit.

Chapter 12

Strategic Selection

To achieve its objective of survival and prosperity, the **materialist organism** must find a way to gain consistent, long-term access to the resources required for generating metabolic energy and for providing shelter. It must pursue a dynamic strategy that will deliver a reliable return on the energy it expends in its participation in life. The most appropriate dynamic strategy will depend on the physical environment—on the availability of natural resources—and on the degree of competition with other organisms in their own and neighboring species. At the center of this strategic pursuit is the mechanism of **strategic selection,** which is examined in this chapter.

There are only four dynamic strategies, which have been employed by life forms over the past 3,500 million years (myrs) and by human society over the past 2.4 myrs, although there are many more ways (substrategies) by which they can be implemented. Owing to their universality, the pursuit of these dynamic strategies depends not on the intellect but on the strategic-imitation mechanism. Life, or the **strategic pursuit**, is driven not by ideas, whether genetic or technological, but by desires.

The strategic pursuit, however, is not just about finding the best dynamic strategy. We all live in a dynamic world, which is affected by the choices we make and the strategies we pursue. A successful dynamic strategy eventually will be exhausted and general environmental conditions will change. This creates a crisis for any species or society, which can only be overcome by the development of a new dynamic strategy. The strategic pursuit, therefore, involves a sequence of dynamic strategies that generate a succession of waves of biological or economic development.

But this dynamic sequence of strategies cannot continue indefinitely. Eventually a species, society, or dynasty will be unable to find a replacement strategy and, when this occurs, it will stagnate, decline, and eventually collapse. In the natural world this will usually occur once the **genetic style** of a species, or the **genetic paradigm** of an entire dynasty, has been exhausted. The history of life, therefore, is an account of the process by which organisms gain access to the Earth's natural resources through a sequence of dynamic strategies, by which those strategies are progressively exhausted, and by which species and eventually dynasties collapse and are extinguished to be replaced by eager new aspirants. Life, in other words, is an eternal strategic pursuit, and its organisms are ephemeral dynamic strategists.

What Are Life's Dynamic Strategies?

The dynamic strategies of life are, as we have seen, fourfold: genetic/technological change, family multiplication, commerce (symbiosis), and conquest.[1] These strategies are "dynamic" because they are employed by organisms (including man) to maximize the probability of survival and prosperity *over the lifetime of the individual*, and because, when successful, they lead to the transmutation of species, to population increase, to geographic expansion of the species or dynasty, and to biological or economic growth (an increase in "output" per unit of input). Dynamic strategies therefore are central to the "progress" of life in general and human society in particular.

The externalities generated by individual strategies, however, differ significantly. Genetic/technological change and family multiplication increase not only the survival and prosperity of the species/society employing them but also, fortuitously, that of life/humanity as a whole. There are, for example, spin-offs or externalities from genetic or technological change that are of benefit to other species or societies through strategic imitation. This is merely fortuitous because the dynamic **strategist** is attempting only to maximize its own survival and prosperity and not the survival and prosperity of other organisms. As these strategists are unable to confine the benefits of their actions to themselves and their families they are responsible for generating positive externalities for life as a whole.

Conquest, in contrast, is a zero-sum dynamic strategy, because the gains made by the victor are at the expense of the vanquished. There are no positive externalities. And even these exclusive gains are "short-term" (possibly hundreds of thousands of years!), because in nature they lead ultimately and inevitably to extinction. The commerce (symbiotic) strategy is intermediate in terms of externalities. While the exchange of monopolized resources benefits both parties involved, although not equally, the externalities are limited because they arise not from a more intensive use of resources but from restricted access to resources.

The four dynamic strategies are pursued by different species/societies in different ways, reflecting the different physical and social environments and the different stages reached in the unfolding of the encompassing genetic/technological paradigm.[2] This gives rise to different substrategies or genetic/technological styles. Take, for example, the case of the industrial technological strategy introduced into human society by the Industrial Revolution. Since the 1780s five different substrategies (or technological styles) have emerged in the industrialized world: the first, employed by Britain from 1780 to 1830, involved small-scale, labor-intensive production of textiles, basic iron and steel, and coal; the second, by France and Germany from 1830 to 1870, focused on larger-scale, more capital-intensive production of more sophisticated engineering and chemical products; the third substrategy was pioneered by the United States from 1870 to 1970 and involved mass production and mass distribution of consumer durables for its own and, later, the world's megamarkets; the fourth, by Germany and Japan from the late 1960s, was based on a new microelectronic technology; and the current substrategy involves the development of mega-states—using large-scale electronic/biological technology—including the European Union, China, and Russia to compete with the superpower, the United States of America (Snooks 1997: 422–25).

Or take the era of the dinosaurs, made possible by the Endothermic Revolution, in which a series of genetic styles rose and fell. Examples include the herbivores that were either high feeders (the stegosaurs and brontosaurs) or low feeders (nodosaurs and iguanodonts) or the highly specialized carnivores (such as the tyrannosaurs). These were the substrategies of the broader dynamic strategy of genetic change. Or take the apemen who appeared as a result of the early stages of the **Intelligence Revolution**: some (*A. robustus*) based their family-multiplication strategy on a diet of nuts and tubers, whereas others (*H. habilis*) based theirs on a meat-and-marrow diet. These different substrategies had very different outcomes. Or take the species of modern man during the final stages of the same revolution: *H. neanderthalensis* specialized in the more settled substrategy of hunting wild cattle, while *H. sapiens sapiens* specialized in the more nomadic pursuit of hunting reindeer and salmon. And as we have seen some substrategies were more successful than others: both *A. robustus* and *H. neanderthalensis* were unable to compete with our ancestors and went extinct.

A distinction must also be made between the dominant dynamic strategy that is the primary focus of a species/society, and a subsidiary strategy that is called forth by the **strategic demand** generated by the unfolding dominant dynamic strategy. Take the conquest strategy, for example, in which the main focus is on maximizing the probability of survival and prosperity through forcibly acquiring the natural (and other) resources held by other species/societies. To achieve these gains it is usually necessary to develop special weapons of offense and defense. In the dinosaur dynasty some species developed huge crushing jaws, tearing teeth, and slashing claws, while others developed armor plating, deadly horns like lances, and war-club tails; and in Roman society the conquest army employed body armor, handheld weapons, heavy cavalry with lances, siege machines, military roads, warships, and so on. In both dynasties, genetic or technological change was a subsidiary strategy to the dominant conquest strategy; and in both cases these military weapons were a response to a changing strategic demand generated by an unfolding conquest strategy.

Employing these dynamic strategies requires considerable investment. Individuals in nonhuman species invest metabolic energy in the strategic pursuit, while those in human society invest not only energy but also financial resources (stored-up energy). My point is that dynamic strategies do not just happen in some sort of costless way. They are the outcome of a creative exploration by individuals of their environment in order to survive and prosper. This is a positive act that requires abstinence from present consumption (including leisure) in order to make this investment of personal energy and other resources, and it is undertaken in the expectation of a positive return. The argument is that non-human species of life are able to shift stores of energy between different strategic activities just as do human societies. The decisionmaking process, as discussed in chapter 11, is one of **strategic imitation**.

The dynamic strategies employed by organisms and species/societies are responsible not only for biological/economic expansion and growth, but also for its ultimate and inevitable downturn. In the initial stages of an unfolding dynamic strategy a species/society will experience increasing returns to scale. A new dynamic strategy will bring greater specialization and better ways of accessing resources, which in turn will increase the strategic return on a species'/society's investment and lead to increasing rates of biological/economic growth.

But there will come a point in the expansion of any species/society when its dominant dynamic strategy is beset by diminishing returns, owing to an increasing investment within a *given* genetic/technological style. This will cause growth to slow. While it may be an extended process, once the returns on an extra unit of investment are no greater than the energy/resources expended—when the marginal return is equal to the marginal cost—growth will cease and may even become negative. It is at this point that the dominant dynamic strategy is exhausted. There will be no further investment of energy/resources because expenditures will be greater than returns. Hence, in the dynamic-strategy theory, exhaustion is a function of diminishing returns not to scarce natural resources but to finite dynamic strategies. If this were not so, a species/society would never be able to recover from a crisis of this nature. And we know that many do. Further biological/economic growth will depend on a species'/society's ability to adopt new dynamic strategies. Only if this cannot be achieved will the forces of chaos overwhelm those of order and will the species/society collapse.

The remainder of this chapter focuses on each of the four dynamic strategies in turn. It also discusses the central role of strategic selection in this process. As these strategies have been dealt with in relation to human society elsewhere (Snooks 1996; 1997; 1998a; 1998b; 1999; 2000), the discussion here relates mainly to nonhuman species. It should be recognized that there is nothing automatic or inevitable about the unfolding of a dynamic strategy; rather it is the outcome of individuals exploring and investing in strategic opportunities. If these individuals make major errors of judgment, or if they are not as competent as a close competitor, then their dynamic strategy will fail prematurely.

The Genetic Strategy

The most controversial feature of my dynamic-strategy theory is its treatment of genetic change. Rather than viewing the emergence of species as the outcome of either "divine selection" or "natural selection," I see it as the result of **strategic selection**, or self-selection writ large. Genetic change in other words is a dynamic strategy that is deliberately pursued by organisms—similar to the role of technological change in human society—when the circumstances are favorable. Organisms select those mutations that assist their dominant dynamic strategy and they ignore the rest. They do this by cooperating and mating with those individuals who have an edge in accessing natural resources as demonstrated by their material success. This occurs not just when pursuing genetic change as a dominant dynamic strategy, but also sometimes under the nongenetic strategies in response to the strategic demand that they generate for military weapons together with highly aggressive instincts (conquest) and for fertility and mobility aids together with sexual and adventurous drives (family multiplication). Accordingly, the concept of strategic selection is far more general in its application than Charles Darwin's concept of natural selection, to which he was forced to add "sexual selection" as a separate and puzzling process. This is why the dynamic-strategy theory, in contrast to Darwinian evolution, is a *general* dynamic theory.

As a Dominant Strategy

When natural resources are abundant and competition is minimal, organisms—and hence populations, species, and even dynasties—invest much time and energy in the pursuit of genetic change. The evidence is discussed in chapter 9. Organisms respond to these conditions by creating new genetic styles—or species—in order to reap "monopoly profits" by gaining sole access to new resources or more intensive access to old resources. Speciation, therefore, occurs merely because organisms attempt to gain better access to natural resources. It is significant that speciation occurs *only* in a non-Darwinian world, because monopoly profits (guaranteed returns) and vast periods of time are required to reap the rewards of investment in the drawn-out process of genetic change. A Darwinian world of scarce resources and intense competition cannot provide either of these essential conditions.

Genetic change is the outcome of deliberate and sustained decisionmaking. It involves a change in the way metabolic energy is employed by organisms. Investment in genetic change requires a deliberate transfer of internal energy from an existing biological activity to an entirely new activity. This, for example, can be seen in the genetic shift from apemen (*A. africanus*) to early humans (*H. habilis*) about 2.4 myrs ago. What was involved was a change in diet from tough nuts and tubers to meat and marrow. To achieve this change in family-multiplication substrategy, it was necessary to divert metabolic energy from maintaining a massive jaw, large teeth, and extended gut to developing and maintaining a large brain. This was achieved by reducing the size of the jaw, teeth, and gut. As argued earlier, this redirection of energy was in response to a changing strategic demand as the unfolding family-multiplication strategy was transformed from a specialized to a general basis. The resulting larger brain of early man made it possible for them to greatly increase their geographic range by using new weapons and tools to gain reliable access to supplies of meat and marrow.

This brings us to the centerpiece of my dynamic-strategy theory: **strategic selection**. Strategic selection is a dynamic process in which organisms are themselves responsible for selecting or rejecting benign (nonlethal) mutations. But what does this radically novel mechanism involve? Essentially it operates through the strategic imitation process. If an individual experiences a beneficial mutation that enables better access to natural resources—within the context of the dynamic strategy it is pursuing—that individual will increase its prospects of survival and prosperity. This success will attract the attention of others. Those with similar abilities will cooperate with each other to improve their joint prospects. If they are of the opposite sex they will also mate together and produce offspring that will, on average, carry the new successful genes. I have called this **selective sexual reproduction**. Success attracts and, literally, breeds further success.

Mate choice is an important part of the strategic pursuit. Individuals select mates possessing the phenotypic characteristics that they hope will improve their family's access to natural resources. When pursuing the genetic strategy they will reject individuals possessing other characteristics—such as greater fertility or better military weapons—that are not relevant to their success in developing a new genetic style. In fact, they will find these aberrant individuals unattractive, even freakish. Similarly, strategists following nongenetic strategies will find other individuals who are marginally better able to access new resources unappealing and unhelpful, because they possess "deviant" physical and instinctual characteristics. They will be attracted to those who display characteristics associated with greater fertility, mobility, or military abilities on the one hand and greater sexual and aggressive drives on the other. Strategic selection not only involves selective sexual reproduction, it also leads to the culling of offspring that do not possess the strategically required characteristics. This is widespread in animal and pre-neolithic human societies. In this way, selection by the organism at the phenotypic level shapes the genotype.

We are now in a position to revisit the question briefly touched on in chapter 10, a question that has always puzzled the Darwinist: Why sex? One of the obvious failures of neo-Darwinism is that it has no convincing explanation of why reproduction occurs other than asexually, because the mixing of genes from two parents reduces the probability of maximizing the presence of the genes of any one of them in the gene pool. Under sexual reproduction any individual must have twice the number of offspring to achieve the desired outcome from asexual reproduction. Clearly this is highly inefficient, and in species such as mammals where the number of lifetime offspring is small (say two to six) some genes may not be passed on to the next generation (often called "genetic drift"). To maintain their selfish-gene hypothesis, neo-Darwinists such as John Maynard Smith have resorted to ad hoc arguments of the following kind: the short-term genetic costs of sexual selection *must* be outweighed by some sort of (unspecified) long-term benefit; genetic mixing provides protection against parasites;

and it provides protection against harmful mutations in one parent. Essentially, these ad hoc arguments (flying buttresses) are an admission that the Darwinists cannot explain sexual reproduction using the theory of natural selection (Wilmut, Campbell, and Tudge 2000: 63).

The dynamic-strategy theory, however, can explain sexual reproduction without resorting to ad hoc arguments. Genetic mixing through sexual reproduction is undertaken to gain greater control over genetic change—and hence greater access to natural resources—than is possessed by life forms limited to asexual reproduction. It allows individuals to influence the genetic structure of their offspring by choosing sexual partners who possess the physical and instinctual characteristics required in their strategic pursuit. It is for this reason that asexual reproduction is rare in vertebrates and nonexistent (except for *identical* twins) in mammals.

Strategic selection, therefore, is a response to strategic demand. This means that an unfolding genetic/technological dynamic strategy is responsible for driving the adoption of "ideas," whether genetic, technological or, as in the case of early man, some combination of both. Remove strategic demand for new ideas and the dynamic process will break down, even if environmental change causes a steady stream of mutations. This is why there is a vast chasm between inventions—the discovery of new ideas—and innovations—the application of new ideas. On the other hand, if the supply of new ideas were to be cut off, the materialist organism would merely adopt another dynamic strategy—such as family multiplication, commerce, or conquest—and the dynamic process would continue. The bottom line is that the dynamics of life is demand driven. This is the essence of my strategic theory.

The success of the strategic pioneers in the pursuit of genetic change will soon attract the attention of other families, which will attempt to imitate the feeding and sheltering activities of their successful neighbors. Those with the abilities to adopt these new methods will join in and, through appropriate mate choice, will select for those physical and instinctual characteristics that support the successful genetic strategy. This marks the beginning of a new genetic style. After the passage of many generations, possibly taking tens or hundreds of thousands of years, this imitative process of strategic selection will lead to the emergence of a new species.

The great advantage of my strategic-selection model is that it can actually explain why and how speciation occurs in a non-Darwinian world. Darwin claimed that speciation is the outcome of the "struggle for existence"—that intense competition for resources resulting from an incessant increase in population. But, as shown in chapter 9, speciation only occurs when resources are abundant and competition is minimal. Only strategic selection within the context of the wider dynamic-strategy theory can explain speciation in these circumstances. The Darwinian world, as we have seen, gives rise not to directional genetic change and speciation but to conquest and extinction. Reality is the very opposite of the outcome predicted by Darwin's theory of natural selection. It is just staggering to realize that the so-called Darwinian revolution is based on a theory that bears no relation to reality. We are indeed fortunate that biological and economic success depend not on ideas but on desires. Had our theories about life been critical to our future success as a species and a society, either Western civilization would have collapsed or the theory of natural selection would have been aborted when Charles Darwin first conceived the idea.

As a Subsidiary Strategy

Genetic change has an important, if secondary, role to play as a supportive or subsidiary strategy to the dominant nongenetic dynamic strategies. In these circumstances it is a response to changing strategic demand as the dominant dynamic strategy unfolds, and it is used by organisms to facilitate, support, or protect the dominant strategy. Once again it is an outcome of strategic selection. A number of examples will illustrate this concept and will distinguish it from genetic change as the major strategy.

The conquest strategy provides some interesting examples. Probably the most dramatic, but certainly not atypical, era was the age of the dinosaurs. In any new dynasty the usual strategic sequence is genetic change → family multiplication → conquest. While military weapons and aggressive instincts will be useful under any strategy in the struggle between carnivores and herbivores, they are absolutely essential during the conquest phase because of the intense competition for scarce resources. In the conquest phase not only are carnivores pitted against herbivores, but herbivores against herbivores and carnivores against carnivores. Genetic change is employed in these circumstances as a supporting strategy rather than as a strategy in its own right, because add-on technology can be achieved more quickly and economically than can speciation. During a war a species/nation has the time and resources to produce armaments but not to generate a biological/economic revolution.

As shown in chapter 9, the carnivores in the dinosaur dynasty developed in size from the relatively small coelophysids (200 kg) to the frighteningly large but agile allosaurs (3,000 kg), ceratosaurs (3,000 kg), and tyrannosaurs (2,000 kg) with their powerful crushing jaws, tearing teeth, and slashing hind claws. In order to defend themselves against these killing machines, and to struggle against each other for overcrowded territories, the herbivores developed not only defensive armor but also offensive weapons of their own. The stegosaurs (3,000 kg) had the body size to stand toe to toe with these carnivores and they employed offensive weapons consisting of large (about a meter long) triangular plates along the spine and a powerful war-club tail armed with four to eight spikes, each up to a meter in length. Then came the heavily armored ankylosaurs (2,000 kg) with their war-club tails, the dome-headed stegoceras (55 kg) that cannoned into their opponents like flying projectiles, and the ceratopsians (2,000 kg) with their giant horned heads. Only the giant diplocids (17,000 kg) and brachiosaurs (40,000 kg) had no use for armor as they contemptuously swept aside their smaller opponents with massive tails.

In order to develop these military weapons it was also necessary to build skeletons and muscle systems that could deploy them effectively. The stego-

saurs, for example, had to develop powerful, flexible spines to effectively operate heavily spiked war-club tails. To do this they had to eliminate the bony ten-
dons initially employed to stiffen dinosaur tails, to develop enlarged shelves of
bone to anchor powerful tail muscles, and to develop massive deltoids, or shoulder muscles, to provide the thrust that they needed to pivot on their hind legs so
that their menacing tails always faced the agile allosaurs and ceratosaurs. It is
interesting how similar this is to the strategic demand and response—with wood,
leather, and metal replacing bone and muscle—under the conquest strategy employed by human society.

Another fascinating example is the strategic demand–response mechanism
for genetic change generated by the family-multiplication strategy of the hominids. As already outlined in detail in chapter 9, the unfolding family-
multiplication strategy pursued by apemen (*Australopitheci*) led in some families and clans to a strategic demand for genetic change that would enable these
individuals to expand their geographical range. This required a shift from their
specialized and regionalism diet of nuts and tubers to the generalized and global
diet of meat and marrow. While suitable nuts and tubers could be found only in
restricted localities, meat could be found in all habitable areas. But for the apemen this change in diet required a rapid increase in brain size to develop the
weapons and tools required by this defenseless species to scavenge and, later, to
hunt large animals, to butcher their carcasses, and to crush their bones to extract
marrow; and it required the management of fire to extend grazing lands and to
cook the meat, thereby making it easier to chew and digest.

A final example is the strategic demand for genetic change required to overcome damage done to the environment by the waste products of organisms pursuing the family-multiplication strategy. The species I have in mind is the blue-
green algae, which at various stages suffered the adverse effects of increasing
levels of oxygen in the seas and atmosphere. Increasing oxygen levels, as shown
in chapter 9, not only had a toxic effect on this life form but also caused the essential mineral iron, which was held in suspension in seawater, to precipitate. To
combat these problems and to continue to pursue their family-multiplication
strategy throughout the world, blue-green algae developed an antitoxin known
as superoxide dismutase, a respiration process to utilize oxygen in generating
energy, and a molecule (a siderophore) that could hunt and capture iron atoms.

Without these genetic changes, blue-green algae would not have been able
to continue their strategic pursuit and their dynasty would have stagnated and
collapsed some 1,500 myrs earlier than it did. In turn this would have halted the
growing oxygenation of the atmosphere, together with the developing ozone
layer that provides protection to life forms against UV radiation. This would
have at least delayed, possibly even terminated, the development of more complex life forms. Once again this is an example of ideas being slaves to desire.

We now need to examine the mechanism by which strategic demand generated
by the dominant dynamic strategy is able to call forth appropriate genetic
change. Once again strategic selection is at the center of this process. Benign
mutations are selected by organisms and passed on to future generations if they

meet the requirements of strategic demand—if they assist or support the dominant dynamic strategy—and they are ignored if they do not. This, as we now know, is achieved as follows. Those organisms experiencing a favorable mutation will display greater success and will, thereby, attract the attention of those around them. Through cooperation with others possessing similar abilities this group will improve their prospects of survival and prosperity. These beneficial physical and instinctual attributes will be enhanced and passed on to the next generation through selective sexual reproduction. Other groups noticing this success will be on the look out for any mutations that provide similar benefits. This is the mechanism of strategic imitation.

The strategic imitation concept can be extended to include the creation as well as the adoption of favorable mutations. There is evidence in bacteria that, under conditions of stress, organisms can actually increase their rate of mutation. Some researchers (Cairns, Overbaugh, and Miller 1988; Hall 1988) are convinced that the increase in mutation rate will occur mainly in genes that can provide the assistance required by the organism. Their research suggests that mutation as well as selection responds to changes in strategic demand. Even neo-Darwinists such as John Maynard Smith found these initial results interesting if controversial:

> By itself, this finding is interesting, but does not challenge the idea that mutation is non-adaptive. It would be explained if a cell which is in difficulties, and cannot grow, increases the mutation rate of all its genes. This would be a sensible thing to do: if in trouble try anything. However, Cairns and Hall go further and claim that the mutation rate increases only, or at least mainly, in those genes which, if they mutate, will help the cell to resume growth. . . . This claim is still highly controversial. (Maynard Smith 1993: 2–3)

Nevertheless, Smith is willing to suggest a mechanism by which this could occur, involving genes that switch other genes on and off. Here is our **strategic gene**.

Over the decade since then, the pioneering research of Cairns and Hall has generated a large and supportive empirical literature (see the bibliography in Foster 1999). It now appears to be widely accepted that beneficial mutations can be generated by simple organisms when subjected to "nonlethal selective pressure." This has come to be called "adaptive mutation," and has even been defined as "a strategy to overcome adversity"; also, these simple cells are considered to "have diverse ways of achieving this goal" (Foster 1998: 1453).

This research is still in its early stages. As it develops I expect that it will demonstrate that Maynard Smith's belief that mutation is random is wrong. Even now these mutation researchers are referring to bacterial response as a "strategy." Also it will be found that the strategic response of these simple cells is made not just to artificially induced stress but to a wider range of real situations outside the laboratory. If I was going to test my strategic demand concept at this level, subjecting bacteria to artificially induced stress would be a rather crude and limited procedure. But the results are promising.

Evidence that organisms can influence the beneficial mutation rate of their genes in response to changed circumstances finds a comfortable home in the dynamic-strategy theory. Conversely, it provides the dynamic model that mutation researchers are looking for to provide realistic direction to their empirical work. This research also provides micro-evidence to stand alongside the abundant fossil macro-evidence that seriously challenges the Darwinist concept of natural selection. Certainly in human society some major inventions—which are the source of technological ideas just as mutations are the source of genetic ideas—as well as innovations—the application of ideas to production—respond to changes in strategic demand. It should not surprise us, therefore, that in nature mutations, as well as their selection by organisms, also respond to the same demand-side forces.

The Family-Multiplication Strategy

By the time a new genetic style has fully emerged, the dynamic strategy of genetic change will have been exhausted by that species. Individuals within the new species will discover that the struggle for survival and prosperity is more cost-effective if they switch strategic investment from genetic change to family multiplication. Also this new strategy generates quicker results than genetic change: a species can migrate around the known world to access new resources in a fraction of the time it takes to construct a new genetic style. But, of course, it can only migrate around the world once the new genetic style has emerged.

A slowly expanding population will, therefore, suddenly increase rapidly, greatly extending its geographical range. It is an outcome of an exclusive focus—earlier population increase was subsidiary to genetic change—on procreation and migration to exploit the new genetic style. In this way members of the new species can "outdistance" their parent and sibling species. It is a strategy that occurs, as suggested earlier, during periods of "normal" competition—neither minimal as under the genetic strategy nor intense as under conquest.

Family multiplication is a strategy pursued by plants as well as animals. Plants are expert at enlisting natural elements—such as wind, rain, and tides—together with animal species—such as insects, mammals, and birds—to assist in pollination and the distribution of seeds. Once they had developed a range of genetic styles, gymnosperms and angiosperms spread rapidly around the globe from about 300 myrs and 135 myrs ago.

It would be wrong, however, to assume that the family-multiplication strategy is pursued continuously and without fluctuation. The most rapid phase of population expansion is associated with global migration, which proceeds in a fluctuating manner as physical barriers and reversals of various kinds (disease and conflict) are encountered. In a mature and stable environment—one in which resources are fully exploited and not yet subjected to intense competition—the strategist responds to changes in the environment by turning the family-multiplication strategy on and off. One interesting case study is the Murray

River ecosystem in the southeastern corner of Australia. This ecosystem consists of massive River Red Gums (*Eucalyptus camaldulensis*), fish, birds, and marsupials. The signal that leads various species to turn their family-multiplication strategies on and off in unison is the rise and fall of the river. When the Murray breaks out across its floodplains, bringing with it an increased supply of nutrients, the entire ecosystem bursts into life. The giant trees begin to grow rapidly and to set seed, and the river birds, animals, and fish begin to breed. When the river falls again the family-multiplication strategy is turned off. In a stable ecosystem the populations of plants and animals fluctuate according to the changing availability of natural resources.

As well as entire ecosystems, individual species have developed ways of regulating their populations to prevent mass starvation. These include the estrous cycle in mammals, which is typically short so that the opportunity for conception is limited (Raven and Johnson 1992: 1138); the diapause—or manipulation of reproductive activity—by which the life cycle is synchronized with appropriate environmental conditions (such as temperature, humidity, light), food supplies, and population density (Hickman, Roberts, and Larson 1993; Hodek 1983; Pener and Orsham 1983; Lee and Cockburn 1985).

Investment in the family-multiplication strategy involves a shift of energy by organisms from increasing the intensity of access to natural resources to improving fertility rates and migration skills. Population increase is not, as Darwinists believe, beyond the control of individuals—not something that all individuals are forced to maximize at all times and in all places as required by the "doctrine of Malthus." There is, as we have seen, a considerable literature of the self-regulation of populations within a wide range of plant and animal species. Rather it is the outcome of a deliberate dynamic strategy that is only pursued under well-defined circumstances. In the course of pursuing the family-multiplication strategy, males in some species invest a greater proportion of their energies in eye-catching displays, attractive colors, nest building, and male-to-male combat. They do so in order to attract/snare mates that they associate with higher fertility. Females also view these characteristics to the same ends. This is the famous but misnamed and misconceived "sexual selection" that so worried Darwin, that Wallace warned Darwin about, and which the neo-Darwinists, fulfilling Wallace's prediction, elevated to the very center of Darwinian "evolution." There is, however, nothing evolutionary—to do with natural selection that is—about this matter. It is a response to strategic demand generated by the unfolding family-multiplication strategy.

The basis for mate choice, therefore, switches from more intensive resource-accessing abilities to family-forming abilities. This is not about sex for its own sake—that is always an important aspect of consumption once survival has been ensured—but rather about gaining greater family control over natural resources through spatial expansion into unused or underused regions of the world. By strategic selection the embodied physical characteristics that promote greater fertility and mobility together with the instinctual sexual and adventurous drives are adopted and passed on to future generations through selective sexual reproduction; and other physical and instinctual characteristics, such as those highly prized under the former genetic strategy, are ignored. In fact, what appeared attractive in others under the genetic strategy may now seem repulsive, even freakish. "Mutants" will be shunned, ridiculed and, in human society, even placed in sideshows to provide ghoulish entertainment as well as examples of what must be rejected.

Strategic selection, therefore, is a mechanism that transcends the genetic change strategy. It involves the selection of physical and instinctual characteristics that support and enhance the prevailing dynamic strategy, whether it is genetic or nongenetic. And by strategic imitation this new strategy will spread rapidly throughout the species, even the entire dynasty. Unlike Darwin's natural selection hypothesis which, at best, was only applicable to genetic change, my strategic-selection concept is applicable to all aspects of biological activity. It is, in other words, a general theory.

The macrobiological outcome of this strategy-switching once a genetic style approaches completion will be clear. The focus of mate choice, family formation, and other cooperative activities will shift from genetic change to propagation and migration. Mutations will continue to appear, but they will not be adopted through strategic selection unless they assist the new nongenetic strategy. Hence, the earlier relatively rapid expansion in genetic change, expressing itself through speciation, will level off, giving the "punctuated equilibria"—or steplike—shape to the genetic profile of species and dynasties.

While the steplike shape of the genetic profile has "recently" been rediscovered and renamed by Niles Eldredge and Stephen Gould (1972), it is no closer to being understood by modern Darwinists than it was in Darwin's day. It cannot be explained by the comparative-static concept of convergence to equilibrium as suggested by the inventors of "punctuated equilibria." It can only be explained by the development of a truly dynamic and general theory of life. The dynamic-strategy theory—which possesses the required theoretical properties—suggests that the steplike genetic profile is the outcome of the exploitation, exhaustion, and replacement of dynamic strategies. It is the strategic sequence of genetic change → family multiplication/commerce → conquest that underlies the steplike genetic profile of species and dynasties.

This brings us to the forces motivating the family, which is the organization responsible for the family-multiplication strategy. Needless to say, the interpretation of the family based on the dynamic-strategy theory is very different to that based on neo-Darwinist theory.

The neo-Darwinist concept of family—as we saw in chapter 7—is based on genetic arguments about "altruism" and "kin-selection." The degree to which individuals are prepared to risk their own welfare to ensure the welfare of others, we are told, depends on the strength of the genetic relationship—measured by "generation distance"—between them. The shorter the generation distance, the more willing individuals are to assist one another, because they will be assisting copies of their own genes to increase their presence in the gene pool. In other words, "altruism" at the organism level is an outcome of "selfishness" at the gene level. The family, therefore, is a social organization dedicated to maximizing the survival and propagation of its shared genes. There are, however, far too many empirical difficulties with this theory, such as the close bonds between genetically unrelated individuals at the center of the family (between spouses), between adopted children and adopting parents, between homosexual couples, together with the abuse and violence in genetically close families.

Family relationships, as I have attempted to show elsewhere (Snooks 1994; 1996: ch. 8), depend not on "generation distance" between organisms sharing the same genes, but on **economic distance** (see chapter 14) between organisms—the most important of whom (spouses) are not related—sharing the same objective function. Individuals promote the interests of the group and cooperate with each other (this is not "altruism" but mutual self-interest) in order to maximize their own material advantage. And this applies to social groups other than families, a matter that kin-selection theory cannot even pretend to explain. The best test of the "altruism" hypothesis is to examine situations in which cooperation is not mutually advantageous: the association between individuals always breaks down.

More specifically the family is the social organization central to strategic selection by which cooperative groups and mating couples make arrangements to select those physical and instinctual characteristics that assist in the strategic pursuit. Clearly the family plays a central role in the dynamic strategies of genetic change and family multiplication. In the case of the genetic strategy the family focus is on attracting (through mate choice), exploiting (through cooperative action), and enhancing (through selective sexual reproduction) those physical and instinctual characteristics that are needed to develop and drive a new genetic style (to provide more intensive access to natural resources). The focus of the family here is on maximizing biological change while maintaining a modest family size.

In the case of family multiplication the focus is on attracting, exploiting, and enhancing those physical and instinctual characteristics that are required to maximize the propagation and resettlement of offspring. The focus here is on fertility rather than biological change, with family size approaching maximum levels. This can be illustrated from case studies in human society. Under the family-multiplication strategy in the United States and Australia in the early nineteenth century, average family size ranged from nine to eleven people, whereas under the technological strategy it has fallen progressively to about three people—less than one-third of the former (Snooks 1994: 63–65; 1997: 411). Similar differences can be expected in the family size and function of other species pursuing those different strategies.

The Commerce Strategy

The dynamic strategy of commerce, or symbiosis, is a more specialized means of achieving prosperity practiced by a variety of plants, bacteria, viruses, insects, and other animals, and, most spectacularly, by some human societies. It can be thought of in nature as an alternative dynamic strategy to family multiplication and in human society as an alternative to conquest. Under this strategy an organism/society that manages to gain a monopoly over an important resource or location may exchange it for a resource or locational access held by a different organism/society. This exchange is to their mutual advantage.

I have called it the commerce strategy rather than the symbiotic strategy largely because it reached its peak in human society. The great societies of Phoenicia, Greece, and Carthage in the ancient world, and Venice, Holland, and Britain in the medieval and early-modern worlds were highly successful commerce societies. In their hands this strategy led not only to the emergence of wealthy merchants and mercantile communities, but also to the development of great trading and maritime empires (Snooks 1996: ch. 11; 1997: chs. 8–9).

These societies show that the commerce strategy is not just a matter of trade but also of the ability to extract large surpluses from other societies that need to trade. They can do so because of their monopoly over certain highly desirable tradable goods, key trading routes, or transport facilities. But because these monopolies are difficult to acquire and maintain, the commerce strategy can be pursued only by a fortunate few. Fortunate because this strategy in human society leads to great wealth, to a wider distribution of that wealth among the population, to a richer culture, and to a less severe way of life than the conquest alternative. And because of these advantages, commerce societies are irresistible to the conqueror. For example, the Persians and Macedonians harried the Greeks, the Assyrians and Macedonians made life difficult for the Phoenicians, the Romans utterly destroyed the Carthaginians, and the Turks and, finally, Napoleon pressured the Venetians. For this reason, as well as to "persuade" reluctant societies to trade, commerce societies found it necessary to employ conquest as a supporting strategy in their commerce ventures.

In a no less spectacular way a variety of animal and plant life forms have successfully adopted the commerce strategy. Natural scientists usually call this process symbiosis. A symbiotic relationship is one from which two different organisms derive mutual benefit by specializing in the exploitation of natural resources and in trading surpluses. This is a very economical use of resources by simple organisms. It must be distinguished from a parasitic relationship, which is a conquest rather than a commerce strategy because it usually ends in the death—or at least the poverty—of the host. Probably the most remarkable role played by symbiosis occurred during the Eukaryotic Revolution when eukaryotes, or multicellular organisms, seem to have developed from the close relationship formed between several prokaryote cells. This is known as the endosymbiotic theory which, however, is not without its critics. Another example is the symbiotic relationship forged between plants emerging from the seas and pre-existing soil fungi that facilitated nitrogen uptake. By pursuing the commerce strategy, therefore, early plants were able to colonize the land (Redecker, Krodner, and Graham 2000: 1920–21).

A few other examples will suffice, because symbiosis is a well-known life plan. In coastal regions an ancient symbiotic relationship exists between algae and coral polyps. Algae employ the waste products of polyps, which include carbon dioxide, in the production of metabolic energy using photosynthesis, and in turn they provide oxygen and other waste products that are required by the polyps (Black 1988: 116). This highly successful commerce strategy was essential to the initial emergence and maintenance of coral reefs, together with their reemergence after various widespread extinctions.

Algae on the land have also had a long-standing symbiotic relationship with fungi, forming what are commonly known as lichens. There are at least thirteen thousand species of fungi, belonging to five hundred genera, that live in lichen relationships with algae (Hawksworth 1988). In this type of relationship a fungus completely covers the algae, thereby protecting the algae from the sun and drying winds and providing it with necessary waste products, while the algae in turn provide the fungus with organic compounds. This is a very successful example of the commerce strategy in the plant world, which enables algae and fungi to live in regions—such as the Himalayas, northern Scandinavia, and under rocks in the Antarctic—too inhospitable for either species on its own. The success of this commerce strategy is reflected in the great age, up to nine thousand years old, of living lichens found in northern Sweden (Johns 1983). This evidence also shows that the commerce strategy has been a successful vehicle for the migration of these (and other) species around the globe.

There are many other examples of the commerce strategy in nature, with specialist readers being able to nominate them by the score. Some of these include plants (such as acacias) that house and feed ant colonies in return for protection from grazing animals and other insects; ant species that culture fungi for food; bacteria that live in the digestive systems of many animal species, thereby providing enzymes to break down cellulose that cannot be digested by the host's own chemicals; and bacteria that assist the life struggle of marine organisms, such as the luminous bacterium (*Vibrio fischeri*) that colonizes the mantle cavity of the Hawaiian squid (*Euprymna scolopes*) activating a light-emitting organ, and the sulphur bacteria that enabled early marine animals (such as trilobites) to counter the toxic effects of widespread sulphur-rich sediment beds (Redecker, Krodner, and Graham 2000: 1920–21; Fortey 2000: 6574–78).

Some neo-Darwinists, such as Richard Dawkins, have developed totally unrealistic arguments to explain these symbiotic (and parasitic) relationships. They assert that the genes of symbionts, which are supposed to be maximizing their presence in the gene pool, exercise influence not only over their host organism but also over any other organism in the symbiotic relationship. This is achieved, Dawkins claims quite absurdly, by genes exercising "mind control" over these organisms and having "phenotypic expression—a type of genetic colonialism I suppose—in their bodies." Needless to say, neo-Darwinists can only assert such science fiction scenarios because they are completely untestable.

In the real world symbiosis is a matter of two organisms that are in control of their own genes profiting mutually from an exchange of resources over which they independently exercise monopoly control. This mutual gain is an outcome of the economies flowing from specialization according to comparative advantage. It is just the same as human societies pursuing the more familiar commerce strategy. For some plants and animals the dynamic strategy of commerce is an effective alternative to either family multiplication or conquest, provided they can gain exclusive control over resources or locations that are valued by other organisms.

The Conquest Strategy

The conquest strategy has played a major role in the dynamics of both life and human society. In life, conquest has been responsible for the elimination—the extinction—of those species and dynasties that had exhausted their genetic styles and/or genetic paradigms. By doing so they provided other species and potential dynasties with the opportunity to show what they could do. If the dinosaurs, for example, had not eliminated themselves through their world wars, the mammals would never have had the chance to show what their passion for the intellect could do on life's stage. And in human society, conquest was responsible for taking the agricultural revolution around the world—at first the Old World and then the New World—thereby ultimately exhausting the neolithic technological paradigm (by the mid-eighteenth century in the Old World) and making way for the Industrial Revolution (1780 to 1830). Needless to say this transforming role could only be identified in retrospect.

In prospect, conquest has always been adopted because, once a species' family-multiplication strategy had been exhausted, it was the only way that organisms could hope to survive and prosper. While conquest is a zero-sum game—in that the victor's gains are never greater than the vanquished's losses—it does provide fabulous returns to successful individuals, societies, and species. In the plant and animal worlds the victors gain better access to food sources, favorable habitats and locations, and sexual partners; and in the human world the successful conquest societies acquire additional agricultural land, slaves, equipment, infrastructure, treasure, and tax revenue. On the basis of this inflow of resources, the conquest metropolis and empire grow to unimagined sizes (Rome at its peak was a city of one million people and an empire of fifty million) that could never have been built using agricultural surpluses (supporting a city of thirty thousand people at most). Conquest was an extremely lucrative dynamic strategy for those species and societies that were successful in war (Snooks 1996: ch. 10; 1997: chs. 6 and 7).

To achieve this prosperity, species and societies had to invest their energy and accumulated surpluses in the weapons, infrastructure, and organizations required to conduct systematic warfare and to occupy invaded territories. In animal species, as we have seen, individuals, through strategic selection, invested energy in the development of powerful weapons of defense and offense; in plant species individuals invested energy in the production of thorns, toxins, strangling vines, suffocating canopies, or sheer size; and in human society individuals and the state (the strategic leader) invested accumulated surpluses in military equipment, war machines, chariots, warships, transport and communication facilities, together with the infrastructure of empire. While the dinosaurs, for example, employed genetic change to develop organic weapons, mankind has used technological change to develop inorganic weapons. To a visitor from another world, the conquest warriors and the great battles of both dynasties would have looked much the same.

These conquest species/societies are always in a state of flux, always dynamic. As the conquest strategy unfolds, strategic demand for these weapons changes gradually. And in response to this changing demand, the military hardware, strategies, and tactics are transformed. Individuals under threat of war in the animal world select mates with physical characteristics that help them to survive. These characteristics include size, strength, massive crushing jaws, tearing teeth, slashing claws, armored skin, huge horns, and war-club tails. And individuals in human society select companions, overlords (this usually means giving up economic independence and freedom) and spouses that possess the military hardware, personnel, or connections to protect them from their enemies.

In the natural world, strategic selection is responsible for a shift of metabolic energy from procreation to military hardware (in animals) or chemical warfare (in plants) once the family-multiplication strategy is exhausted. This involves a shift from the largely genetic-neutral outcome of family multiplication to the demand-led genetic change required to develop add-on military technology and to enhance aggressive instincts. Military technology does not increase the intensity of resource-accessing capabilities as does the dynamic strategy of genetic change; rather it provides conquest warriors (both animals and plants) with a military advantage over their rivals. Clearly it does not lead to the transmutation of species, merely to death and extinction. There is just no further scope or potential following genetic style/paradigm exhaustion to employ genetic change as a dominant dynamic strategy. The genetic sponge for this species/dynasty has been wrung dry. Only with their extinction can the dynamic strategy of genetic change be revisited, and only then by a new species/dynasty. The reason is, as we have seen, that the Darwinian world of intense competition provides neither the time nor the returns to make the huge investment in speciation feasible. Only the add-on technology required by conquest can be achieved in these circumstances. And this is the outcome of strategic selection in response to strategic demand.

Clearly the Darwinian world of scarce natural resources and intense competition leads not to "evolution" through the mechanism of natural selection but to death and extinction through the conquest strategy. Only once the old dynasty has collapsed and become extinct, thereby creating a non-Darwinian world of minimal competition and abundant resources—a world of monopoly profits and vast time horizons—will speciation resume again owing to the dynamic strategy of genetic change. And this is the outcome not of natural selection but of strategic selection in response to strategic demand. While Darwinian natural selection is at a complete loss to explain either the entire dynamics of life and human society or even of transmutation, the dynamic-strategy theory can cover all these bases.

Conclusions

At the heart of the dynamics of life and of human society can be found the four dynamic strategies of genetic/technological change, family multiplication, commerce (symbiosis), and conquest. In this fundamental sense there is nothing unique about human civilization. Our species employs exactly the same type of dynamic strategies that were employed by the first protocells to emerge in the primeval seas some 4,000 myrs ago and which have been pursued by all life forms ever since.

What we share with all life is the universal desire to survive and prosper. And while we are better able to facilitate this desire owing to the fact that we possess a brain and that this brain is much larger than that possessed by any other species, we approach the problem of life in the same way as they do. Like other animal life forms we participate in life's strategic pursuit through our facility for mimicry rather than that for intelligence. We choose our dynamic strategies by imitating those who are successful in our community rather than by making millions of complex benefit–cost calculations. This is what life forms have always done and will always do.

The macrobiological and macroeconomic patterns in the history of life and of human society are outcomes of the dynamic strategies we have always pursued. Strategic selection is central to this process, for it is through this mechanism that organisms adopt and pursue the dynamic strategies of life. This is achieved by individuals selecting comrades and sexual partners on the basis of the physical and instinctual characteristics that are required by the dominant dynamic strategy. Strategic selection, therefore, is a response to strategic demand.

The dynamic strategies are adopted through the investment of biological or economic resources, and they are exhausted when the biological or economic returns no longer exceed the costs of that investment. At this stage crisis emerges. Only if and when a new dynamic strategy is successfully launched does a species or society recover its vitality. This strategic sequence imparts a characteristic wavelike pattern to the development path of life or society. In this way we observe the rise and fall of species and of dynasties. It is to the dynamic mechanisms underlying these historical patterns that we turn in the next chapter.

Chapter 13

Life's Dynamic Mechanisms

The great stumbling block for all existing theories of "evolution" is that they are unable to satisfactorily explain the historical patterns of life over the past 3,500 million years (myrs). Darwin's theory of natural selection assumes that the transmutation of species occurs very slowly but continuously, with new species gradually replacing their predecessors. Worse still, the necessary conditions for "evolution," in Darwin's theory—the "struggle for existence"—lead in reality not to speciation but to extinction.

Neo-Darwinist theory, which focuses on "reproductive success" rather than the "struggle for existence," suffers from the same problems. Even the historicists, or paleontologists, who recognize the rise and fall of dynasties, are unable to explain these patterns satisfactorily because they insist on clinging to the fatally flawed Darwinist theory. What we need in order to explain the dynamic mechanisms by which the great dynasties of life have risen and fallen is a theory that can account for both individual choice and macrobiological/macroeconomic dynamics. It is argued here that the dynamic-strategy theory can provide that explanation.

In this chapter the dynamic-strategy model is employed to analyze the dynamic mechanisms that underlie the historical patterns of life. This is done by bringing together the timescapes outlined in chapter 9 and the general dynamic model discussed in chapter 10. This approach identifies and explains two important dynamic mechanisms that in turn can be used not only to analyze the past but also to make sensible predictions about the future. These backward and forward predictions avoid the well-known problems of historicism by being based not on the historical patterns—the timescapes—but on the mechanisms *underlying* these patterns, mechanisms that are based on a general dynamic theory. These two mechanisms are the great steps of life—the great genetic and technological paradigm shifts—and the great wheel of life. One leads to biological and economic progress, the other to eternal recurrence. Both play a central role in life's dynamics.

The Great Steps of Life

The dynamic structure of life—reflected in figures 13.1 and 13.2—is defined by a series of genetic and technological paradigm shifts, which I have called the **great steps of life**. These great steps, which outline the genetic/technological potential for biological/economic development, are driven by the major genetic/technological revolutions examined in chapter 9. Owing to the nature of the dynamics of life, these great steps have been changing exponentially in two dimensions: increasing in height while reducing in depth. What this implies is that the dynamics of both life and human society have been accelerating over the past 3,500 myrs. In other words each great step is not only larger but takes place more quickly than the last. There have been three genetic paradigm shifts and three technological paradigm shifts over that vast period of time, and they are involved in a systematic geometric relationship. This suggests that a new technological paradigm shift will occur soon (in the twenty-first century) and rapidly (within a generation) (Snooks 1996: ch. 13).

Figure 13.1 The Great Steps of Life—the past 4,000 myrs

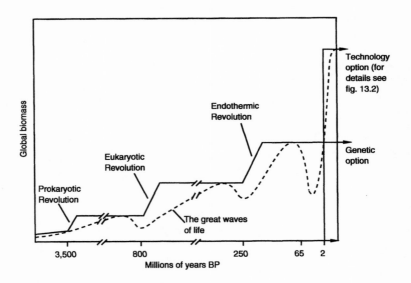

Source: From the dynamic-strategy theory—see text.

The genetic paradigm shifts are part of what I call the **genetic option** and the technological paradigm shift are part of the **technology option**. With each paradigm shift the capacity of the genetic option was progressively exploited until it was finally exhausted by the time the dinosaurs were in their prime (80 myrs ago). The technology option, however, only displaced the defunct genetic option when the intellectual capability of the hominids enabled them to employ the technology strategy (sometime before 2.4 myrs ago). During the almost 80 myrs between these events, nature was dominated by the **great wheel of life** discussed below.

Figure 13.2 The Great Steps of Life—the past 80 myrs

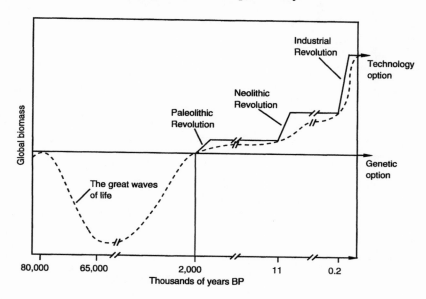

Source: From the dynamic-strategy theory—see text.

The great paradigm shifts were generated by a series of genetic and, later, technological revolutions. The genetic revolutions, as we have seen, included:

*   the Prokaryote Revolution by blue-green algae from 3,500 myrs BP;
*   the Eukaryote Revolution by multicelled plants and animals (including reptiles) from 900 myrs BP;
*   the Endothermic Revolution by warm-blooded animals and associated plants from 250 myrs BP; and
*   the Intelligence Revolution by the hominids from 2.4 myrs to 150,000 years BP.

The last of these genetic revolutions led not to a genetic paradigm shift but rather to a series of technological paradigm shifts. At this time genetic change by itself was unable to provide more intensive access to natural resources, only the ability to create and employ technology to do so instead. These technological revolutions (figures 13.2 and 13.3) include:

*   the Paleolithic (hunting) Revolution from 2 myrs BP;
*   the Neolithic (agricultural) Revolution from 10,600 years BP;
*   the Modern (industrial) Revolution from AD 1780; and
*   a future revolution—the Solar Revolution—which can be expected in the twenty-first century.

Figure 13.3 The Great Steps of Human Progress—the past 2 myrs

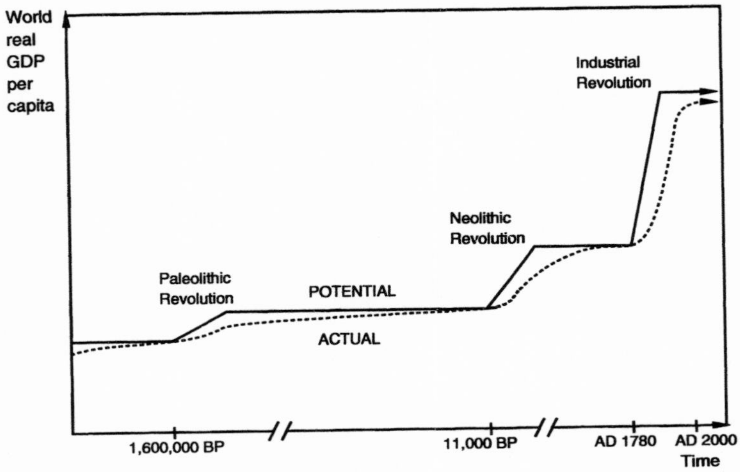

Source: Snooks 1996: 403.

What the great steps of life represent is the steplike progress that has taken place in biological and economic activity throughout the history of life on Earth. Each giant step forward, which was facilitated by a genetic or technological revolution, made possible more intensive access to natural resources. In turn this enabled a higher level of biological/economic activity—or life "output"—to be achieved. For the prehuman era this meant a growing number of more complex organisms that contributed to a higher level of global biomass. And in the human era it involved not only a rapid increase in human and domesticated animal and plant populations but also an exponential increase in human living standards, particularly regarding the consumption of services and nonperishable commodities.

The reason that the genetic/technology profile is steplike is that with each genetic/technological revolution there was a quantum leap in *potential* access to natural resources owing to the occurrence of a major innovation or cluster of innovations. The *actual* utilization of this potential, however, was a gradual and wavelike process driven by the materialist organism pursuing its nongenetic dynamic strategies. But in order to achieve this potential it was usually necessary to develop add-on genetic features, such as the organic devices used by blue-green algae to combat the adverse effects of their waste product, oxygen. This relationship between potential and actual outcomes is represented diagrammatically in figures 13.1 to 13.3.

In figures 13.1 and 13.2 the *actual* curves represent the biological development path that I have called the **great waves of life** (measured in terms of global biomass); and in figure 13.3 the *actual* curve represents the development path of global economic activity (measured by real GDP per capita). These were out-lined and discussed briefly in chapter 9. Once a particular genetic or technologi-cal paradigm has been exhausted, the *actual* curve presses persistently against the *potential* curve. Under the genetic option this overpopulation situation leads to the collapse and extinction of the prevailing dynasty of life due to world war, whereas under the technology option it leads to the adoption of a new techno-logical revolution by the same dynasty owing to the greater rapidity with which it can be introduced—a generation or less compared to millions of years.

Only once the intense competition for scarce resources has been eliminated is it possible to bring the slow-acting genetic strategy into play. Fortunately hu-man society is able to avoid the collapses that characterize earlier dynasties of life because the technology strategy can be applied more directly, precisely, and quickly, thereby keeping world war at bay. Hence, in human society the tech-nological strategy can be employed effectively during periods of intense compe-tition for scarce resources as a way of easing the pressure through economic revolution (providing the opportunity once more for monopoly profits that are necessary to encourage sustained technological innovation) without widespread resort to the conquest strategy and world war. Technology is the antidote to ter-minal global war.[1] If, however, we ever eliminate growth-inducing technological change in the mistaken hope that we are saving the environment, then conquest and collapse—possibly even extinction—will reemerge on our planet just as it did 65 myrs ago (Snooks 1996: ch. 13).

The interplay between genetic and technological change in that critical period between *H. habilis* and *H. sapiens*—between 2.4 myrs and 150,000 years BP—is fascinating. It constitutes what I call the **Intelligence Revolution**. As suggested earlier, the strategic demand generated by the hominid family-multiplication strategy led to a relatively sudden increase in brain size in order to create and employ the inorganic tools that were required to change their diet from nuts and tubers to meat and marrow.

The Intelligence Revolution, therefore, led to a more intense use of natural resources through better instruments of a technological rather than a genetic nature. Genetic change was employed only until technology could take over; then it was dropped because technological instruments are more direct, precise, and quick acting—in short, more economical. This, together with the much ear-lier exhaustion of the genetic option, is why the Intelligence Revolution led to a technological rather than a genetic paradigm shift.

As can be seen in figure 13.2, the genetic paradigm profile continues as a horizontal line from the time that the dinosaurs were at their peak, whereas the technological paradigm takes off vertically from about 2 myrs ago and thereafter traces out the familiar steplike pattern, except that it occurs in a greatly acceler-ated fashion. Even in the future when genetic engineering is undertaken on a large scale it will lead to paradigm shifts not of a genetic but of a technological nature. Genetic engineering merely serves the dominant technological strategy. The technology option has permanently replaced the genetic option.

The emergence of the dynamic strategy of technological change, therefore, set life free from the genetic option which, after 3,500 myrs, had finally been exhausted by the dinosaur dynasty some 80 myrs ago. Had the hominids not been able to adopt the technological strategy, the mammalian dynasty would have collapsed just as the dinosaurs had done.

The mechanism driving *actual* biological activity to meet the potential ceiling defined by the genetic paradigm is the **great dispersion**, which is an outcome of the dynamic strategy of family multiplication. Through procreation and migration a family is able to gain greater control over unused, or underused, natural resources and, thereby, increase its prospects of survival and prosperity. In the process, a species, group of species, even an entire dynasty disperses around the accessible world. This occurs with each species as it exploits its particular genetic style, which provides it with access to a particular set of resources. Ultimately it exhausts that style, turns to conquest, collapses and goes extinct. The same happens for an entire dynasty as their genetic paradigm is exploited to the point of exhaustion, after which the constituent species war with each other until the whole dynasty collapses and disappears forever from this world.

In the history of life on Earth, the great dispersion at the dynasty level has occurred on a large number of occasions. For example, blue-green algae dispersed around the world between 3,000 myrs and 1,000 myrs BP; reptiles did so only once between 350 myrs and 250 myrs BP, as they experienced only one wave of widespread expansion and extinction; the protomammals probably did so on at least three occasions between 250 myrs and 200 myrs BP owing to three rapid waves of expansion and extinction; the archosaurs did so between 225 myrs and 190 myrs BP before they finally went extinct; the dinosaurs, which generated a number of waves of expansion and extinction, also dispersed around the world on two or more occasions between 200 myrs and 65 myrs BP; and the mammals did so thereafter. Of course, the limits to these great dispersions were defined by the changing land-mass patterns generated by continental drift (Snooks 1996: 28–30).

The great dispersion also occurred many times at the species level as individual organisms sought to exploit their genetic styles. The most recent and momentous occurrence for the history of life on Earth was the great dispersion of mankind, by *H. erectus* and possibly *H. sapiens*, over the past 2 myrs. The great dispersion, driven by the family-multiplication strategy, is, therefore, the mechanism by which a genetic style/paradigm is exhausted.

The Great Wheel of Life

Many appear to regard the emergence of mankind—or the intelligent organism—as a natural, even inevitable, outcome of Darwinian "evolution." They believe that the emergence of intelligence was merely a function of "evolutionary" time. But they are wrong. There was nothing inevitable about the rise of intelligent life on Earth, or anywhere else. It was, I argue, purely a matter of chance. Had we the ability to rerun the dynamics of life on Earth over and over again, the chances that intelligent life would ever reemerge again are very small indeed. Or to put it another way, even assuming that life is commonplace throughout the universe, the probability that it would throw up an intelligent species is very low. Why? Because the emergence of intelligent life on Earth was the outcome of a long chain of improbable and unsought accidents.

Of course, this is not to say that the dynamics of life is unsystematic. As I have attempted to demonstrate in this book, species and dynasties rise and fall in a predictable way. This suggests that there is a dynamic mechanism in life that could account for the continual rise and fall of dynasties without the emergence of intelligence until the solar system itself expires. Indeed, what I am arguing is that such a dynamic mechanism should be regarded as the normal condition of any mature system of life anywhere in the universe. Intelligent life is an aberration, not the norm.

This mechanism of eternal recurrence, which I call the **great wheel of life**, comes into operation in a mature life system when the genetic option has been exhausted. As we have discovered, a genetic option is exhausted when it is no longer possible to gain further access to natural resources through the use of genetic instruments—when all conceivable habitats or niches have been as fully occupied as is possible through biological adaptation. At this point in time it is no longer possible through genetic change to increase the global level of the biomass of life. Of course this is not to say that minor increases or decreases in global "output" will not occur as the physical environment makes marginal adjustments (such as the rise and fall of sea levels), just that the available natural resources cannot be used any more intensively through genetic adjustment alone.

The exhaustion of the genetic option on Earth appears, as already suggested, to have occurred by the time the dinosaurs were at their peak about 80 myrs ago. Evidence for this comes from a comparison of the metabolic systems of the four great endothermic dynasties: the protomammals (therapsids), the archosaurs, the dinosaurs, and the modern mammals. Advanced dinosaurs possessed more sophisticated metabolic systems—higher body temperatures, higher calorie intakes, faster speed—than the protomammals, but they are similar to those of modern mammals (Bakker 1978: 137–40). What this implies is that the dinosaurs took the Endothermic Revolution, which had been initiated by the protomammals and carried forward by the archosaurs, a further and final step. Accordingly they were able to employ the Earth's resources more intensively than all preceding dynasties. Modern mammals, whose great radiation was not based on any genetic innovation, do not appear to have advanced beyond the dinosaurs in this central respect. Indeed, while the dinosaurs existed, the mammals were only marginal players in life, unable to move to center stage.

But, you may ask, what about the ability of mammals to generate more species than dinosaurs? Is that not a sign of greater effectiveness in accessing natural resources? The short answer is no. The explanation is that there is a three-way trade-off in life between the number of species, the size of individuals, and the number of individuals that can exist at a point in time with a given supply of resources. If members of a dynasty opt for large individual size and relatively larger numbers of individuals, then they will be limited to a few species. But if they opt for greater diversity—a larger number of species—together with a similar number of individuals, then they will be limited in individuals' size. My argument is that the dinosaurs opted for size and numbers rather than diversity, whereas the modern mammals did the reverse. The mammals were able to do so even though they were operating in the same genetic paradigm at the same level of **genetic competence** as the dinosaurs.

But, you may ask again, why and how were dinosaurs and mammals able to opt for different sizes and diversities? The answer arises from my "force versus finesse" argument. A dynasty, such as the dinosaurs, that favors force rather than finesse will opt for size and numbers rather than diversity. Large size and numbers are necessary for any dynasty that is accustomed to pursuing their dynamic strategies with brute strength. This was the case for all land dynasties before the era of modern mammals, and particularly for the dinosaurs. As we have seen, dinosaur herbivores reached sizes of 40,000 kg and carnivores, of 3,000 kg, while mammal herbivores—excluding the unsuccessful megafauna—reached 5,000 kg and their carnivores, little more than 200 kg. Mammals, which in the Mesozoic period had been forced to live by their wits on the margins of life for 150 myrs by the aggressive dinosaurs, favored finesse rather than force and, hence, opted for diversity rather than size. And greater diversity, which requires more regular genetic innovation, requires greater intelligence to control the mechanism of **strategic selection**. This is why mammal brains were relatively larger than dinosaur brains, and why they continued to grow.

From the time of the exhaustion of the genetic option—when the dinosaurs were at their peak—the great wheel of life displaced the great steps of life as the dominant dynamic mechanism underlying macrobiological activity. And while the great wheel rotated slowly in space–time no progress could be made in terms of intensity (or productivity) of resource use or of increase in global biomass. From this time, new dynasties only able to resort to genetic (and not technological) change would merely replace old dynasties without gaining upward traction, Thereafter, each great wave of life would peak at about the same level of biological activity as that achieved by the dinosaurs. This is the eternal recurrence of life.

The great wheel of life is represented diagrammatically in figure 13.4. When examining this diagram we should be aware of different but related types of motion: the rotation of a point on the circumference of the wheel around its axis I; and the forward movement of the wheel from I to II over a long period of time, say 100 myrs. Like any wheel travelling along a plane, the rotation leads to a directional movement. In other words the rotation of the wheel cannot lead to reversal of time as would be implied by a fixed axis. Also it should be noted that, as the genetic option has been exhausted, the plane along which the wheel travels through time is horizontal. This implies that there can be no progress, in terms of biological outputs or productivity (output per unit of input) between this noninnovating dynasty and its predecessor.

The above argument can be cast in terms of the great-wheel diagram as follows.

- A new, postgenetic-option dynasty begins its adventure in life at **a**, when competition is minimal and resources are abundant. Individuals are faced with the size/numbers/diversity options, which they resolve by pursuing the dynamic strategy of genetic change, with varying degrees of "preference" for force/finesse. This causes the great wheel to rotate upward, increasing biological output and moving forward in time.

Figure 13.4 The Great Wheel of Life—the rise and fall of a noninnovating dynasty

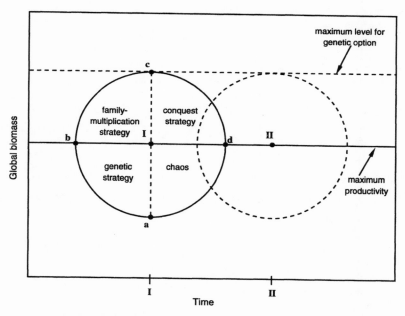

Source: From the dynamic-strategy theory—see text.

- Once they have arrived at **b** they will have exhausted the genetic strategy and have achieved the maximum level of "productivity" (output per unit of input) in the use of energy and other resources. Any further rotation of the wheel and forward movement through time requires the pursuit of the family-multiplication and commerce strategies to take their genetic styles around the globe. As the wheel turns from **b** to **c**, global biomass (biological output) increases but productivity (output per unit of total input) remains the same, because the wheel is travelling along a flat plane dictated by the exhausted genetic option.
- When the great wheel reaches **c** the family-multiplication strategy and the genetic paradigm for the dynasty as a whole will have been exhausted. The only option for individuals in this Darwinian world of intense competition

and scarce resources is to pursue the non-Darwinian strategy of conquest. While this leads to gains for the winners, it is at the expense of a growing loss of life, a reduction in diversity, a decline in global biomass, and a deterioration of the environment. Hence, the wheel rotates slowly but inevitably from **c** to **d**, at which point conquest gives way to chaos.

- Between **d** and **a** the wheel accelerates downward as the entire dynasty begins to collapse. By the time **a** is reached, the dynasty has gone extinct. During the time taken for one complete revolution of the wheel of life—say 100 myrs—in the case of this postgenetic-option dynasty, its forward motion has been along a horizontal plane **I–II**. In other words there has been no progress in terms of biological output or productivity beyond that of the former dynasty. That would require traction along an upward sloping plane as was the case before the genetic option was exhausted.

- With dynastic extinction the opportunity is presented to the surviving species, previously on the margins of life, to take over center stage where now competition is minimal and resources are abundant. As they do so the great wheel begins to rotate once more around the same horizontal axis. The cast of characters in the drama of life will have changed but the play and its ending are always the same for a dynasty trapped by an exhausted genetic option. The great wheel of life leads to the eternal recurrence, which should be regarded as the normal condition of life throughout the universe.

The only way to break out of the eternal recurrence once the genetic option has been exhausted is to adopt the technology option. Technology can be used to obtain more intensive access to resources once all possible habitats have been saturated through genetic change, and it is a process that appears to have no limits. Ideas are able to transcend biological limitations.

It is sobering to realize, however, that the technology option is the outcome of chance. The reason is that this option does not bestow any advantage on the *individual* decisionmaker—the dynamic strategists—only on the species as a whole. And species do not make strategic decisions. In reality it is only something that can be recognized in retrospect and, so far, only in this book. But why is the technology option not in the best interest of the dynamic strategists? Because all strategists die irrespective of whether one species (mankind) stumbles on the secret of perpetuating its existence. This is the organism/species paradox. Even had the apemen—or the mammal species before them—recognized the very long-run implications of adopting the technological strategy, would any one of its members have traded even a day's supply of food for the remote possibility that the species replacing their own would, some 2.4 myrs later, break out of the eternal recurrence? I suspect that most decisionmakers in this position would answer in the negative. The dynamic strategist is only concerned with his/her own immediate survival and prosperity. Hence, the technology option is merely the unintended consequence—and a highly unlikely one at that—of the individual struggle for a secure and better life.

What then is required to enable the inadvertent substitution of the technology option for the exhausted genetic option? Here are a number of important, but not necessarily exhaustive, conditions.

- Life forms need to develop central nervous systems. This eliminates all nonvertebrate life forms such as bacteria, viruses, soft-bodied marine organisms, and the extensive plant kingdom. As far as the development of the technology option is concerned, the development paths "chosen" by these life forms were all dead ends. Yet these highly successful life forms neither know nor care.

- For vertebrates to develop the large brains required to stumble on the technology option they needed to develop endothermic (warm-blooded) systems so as to control body temperature and keep it at a constant level. Just a slight change in body temperature in either direction renders our large and complex brains inoperative. This requirement rules out all ectothermic (cold-blooded) animals such as fish, amphibians, and reptiles. The only remaining candidates are the protomammals that emerged as recently as 240 myrs ago, the archosaurs from 225 myrs BP, the dinosaurs from 190 myrs BP, and modern mammals and birds from 60 myrs BP.

- It is also essential that warm-blooded vertebrates opt for finesse rather than force in their strategic pursuit. Without this vital ingredient there will be no pressure to increase brain size from the minimum required to effectively pursue the various dynamic strategies. This eliminates the protomammals, the archosaurs, and the dinosaurs, which were warm-blooded but pursued their dynamic strategies with force rather than finesse. Only a few species of small insignificant mammals together with birds, the only descendents of the dinosaurs, remain as possible candidates. Of these, birds have shown no sign of rapid brain development—hence "bird brain" is used as an insult by humans—and the survival of small Mesozoic mammals was purely fortuitous.

- Having barely survived the dinosaur world war, the mammals needed not only to engage in the strategic pursuit with finesse rather than force but also to do so at a pace that enabled the technology option to be stumbled on before the family-multiplication strategy had been exhausted and the terminal conquest strategy embarked upon.

- In retrospect it is clear that only one small branch of the mammalian family tree—the primates and among the primates only the hominids and among the hominids only man—had this potential. Had we all taken the path chosen by the apeman *A. robustus*—who pursued the more specialized substrategy of nut-and-tuber-eating rather than meat-and-marrow-eating—then this race of blindman's bluff would have been lost.

- And it was only just won because there are signs that the family-multiplication strategy for the entire dynasty was approaching global exhaustion. Indeed, the heroic attempt made by mankind to generalize their version of this dynamic strategy and to break out of their restricted and highly specialized environment reflects an exhausting strategy. Had they not succeeded against all odds or had a modest catastrophe wiped out their small (not more than a few thousand individuals some 3–2 myrs ago) and concentrated (in a few restricted areas of one continent) numbers, the great wheel of life would have continued to turn unwittingly in endless time without gaining any upward traction.

Owing to these extremely stringent requirements we can conclude that the probability of the technology option ever being discovered was very low. It was the outcome of a long chain of highly unlikely, unforeseen, and unsought events. The implication, therefore, is that intelligence is not only the scarcest resource on Earth but also the likelihood of it occurring anywhere else in the universe is not particularly high. Most likely we are alone in the universe and any life elsewhere is dominated by the "great wheel." If, in the highly unlikely event that intelligent life does emerge elsewhere, the probability is that it will lag considerably behind that on Earth owing to the remarkably fortuitous causal chain that we have experienced.

The Future Revolution

The dynamic-strategy model not only explains the dynamics of life and human society in the past, it also makes useful predictions about the future. It has been shown in this chapter that once a technological paradigm has been exhausted, intense pressures mount for a new technological paradigm shift. If a new revolution were to be blocked, either by a world government dominated by fanatical ecologists (Snooks 1996: ch. 13) or by the crippling influence of neoliberal (economic rationalist) policies (Snooks 2000), then population growth would press heavily on exploitable natural resources, widespread poverty would emerge, the environment would be irreparably damaged, and, ultimately, the conquest strategy would reemerge as the only option available to the dynamic strategists. In this way mankind would face the same destructive forces as the dynasty of the dinosaurs some 65 myrs ago.

But, assuming that common sense prevails—it rarely deserts us for long when it comes to material self-interest—a new economic revolution will definitely occur when the present technological paradigm has been exhausted. This will happen once most of the resources of the Third World have been drawn into the dynamics of the global strategic core (Snooks 1999). A sure sign that the future revolution is approaching is not only the growing economic development of the Third World but also the growing pressure being placed on natural resources. This is similar to the environmental degeneration that occurred in Europe in the mid-eighteenth century as the neolithic technological paradigm approached exhaustion (Snooks 1996). Just as the Industrial Revolution released the intense pressure on natural resources (marginal agricultural land and forests), so the future revolution will do the same. It is important not to confuse the signs of an approaching technological paradigm shift with those of a fictitious environmental collapse. But it is equally important to employ existing technology wherever possible to restore the environment.

When will the future revolution occur and what will it involve? Owing to the exponential nature of genetic/technological change over the past 3,850 myrs and to the growing signs of exhaustion of the industrial technological paradigm, the future revolution can be expected some time during the twenty-first century. And it will take place rapidly, probably within a generation or so. It is important to realize that there is nothing new about exponential change—the dynamic process has been accelerating ever since life first appeared on our planet. Indeed, it is highly likely that, beyond the next technological paradigm, economic revolution will become continuous.

What will the future revolution—called here the Solar Revolution (Snooks 1996: 429–30)—look like? This is difficult to say with any precision. Just imagine living in the mid-eighteenth century and attempting to predict what shape the world would take by the end of the twentieth century once the Industrial Revolution had largely run its course. It is unlikely we could have imagined motor vehicles, aircraft, spaceships, radio, TV, computers, or biotechnology. Yet we might have been able to say something sensible about the use of energy, which plays a central role in any technological revolution, and about the broad nature of the technology that would be employed.

Similarly, we can speculate sensibly about the use of energy in the future revolution. The first technological revolution saw the extension of human energy with the use of more efficient stone tools; the second revolution saw the partial substitution of animal, water, and wind energy for human energy; and the third revolution saw the substitution of thermal energy based on fossil fuels (later supplemented by nuclear energy) for animal and human energy. It is highly likely that the fourth or future revolution will involve the substitution of solar energy for fossil fuel and nuclear energy. From the fourth revolution, therefore, the physical constraints on economic (not to be confused with population) growth will be limited only by the flow of energy from the sun. It is for this reason I have called it the Solar Revolution.

While it is more difficult to predict the nature of the technology that the Solar Revolution will employ, there are a few signs of what is to come. Certainly the perceptive observer in the mid-eighteenth century could have guessed that the technology of the Industrial Revolution (which he could only have identified had he been in possession of my dynamic-strategy theory) would probably involve iron machinery and locomotives driven by steam power fuelled by coal, but he could not have foreseen the emergence of computerized factory robots powered by electricity or of the internal combustion, jet, or rocket engines. Similarly we can speculate about the use of nanotechnology—small, invisible machines—that will produce whatever we require through the manipulation of individual atoms, about artificial intelligence and personal robots, and about biotechnology. But how significant these technological developments will be or what lies beyond them we cannot say. Nevertheless, we can outline the dynamic mechanisms that will generate these technological and genetic details. This is the book's purpose.

Conclusions

The dynamic mechanisms that underlie the historical patterns—the time-scapes—of life are driven by the adoption, exploitation, exhaustion and, some-times, replacement of dynamic strategies, of genetic styles, of genetic paradigms and, ultimately, of the genetic option itself. The dynamic mechanism responsible for the macrobiological patterns of activity is the process of genetic paradigm shifts. It generates the steplike genetic profile and the great waves of life. But this mechanism ceased to operate when, finally, the genetic option was totally exhausted about 80 myrs ago. This process of exploitation and exhaustion of the potential available in the genetic option took about 3,500 myrs.

Once the genetic option was exhausted, the dynamic mechanism underlying the macrobiological pattern became the great wheel of life. Since it was no longer possible to gain more intensive access to the planet's resources through genetic change alone, life faced the prospect of dynasties rising and falling without any further biological progress. The great wheel of life began to turn slowly in time without gaining any upward traction. This is the eternal recur-rence that most life systems throughout the universe can be expected to experi-ence until their solar systems finally run out of energy. The remarkable, possibly unique, characteristic of life on Earth is that, through a long sequence of fortu-nate and unsought accidents, an Intelligence Revolution occurred that enabled the substitution of the technology option for the obsolete genetic option. This enabled mankind to break out of the eternal recurrence. But only just in time. What should be realized is that there is nothing inevitable or even very likely about such a breakout. We are probably alone in the universe.

What does the future hold? The dynamic-strategy theory predicts that, when the present industrial technological paradigm has been exhausted, a new revolu-tion driven by the insatiable appetite of materialist man will break out. Our the-ory also suggests that, owing to the exponential nature of the dynamics of life and human society over the past 3,850 myrs, the future revolution will occur soon (during the twenty-first century) and rapidly (within a generation or so). And it will release the increasingly intense pressure on our planet's natural re-sources and environment. We can also suggest, from observation rather than theory, that the future revolution will be a solar revolution employing nanotech-nology, artificial intelligence, and biotechnology. In the process of this revolu-tion we will change the nature not only of our world but also of ourselves. As we have always done.

Chapter 14

The Dynamics of Social Organization

The explanation of social organization is a critical test for any dynamic theory of life. As we have seen it is a test that Darwinism has failed. This failure in natural selection was starkly exposed when, in *The Descent of Man*, Charles Darwin attempted to employ it to explain the "evolution" of human civilization. Similarly sociobiology came unstuck when its supporters used neo-Darwinian genetic theory for the same purpose. Yet, undeterred by this failure some sociobiologists claim that they are about to rewrite the social sciences and humanities. They have yet—after some four decades—to show how this can be done.

In *Descent* less than two percent of its contents were devoted by Darwin to human society. As we discovered in chapter 4, this was a matter not of idiosyncratic preference but of necessity. Not being a realist dynamic theory, natural selection is unable to explain the changing social forms of life. Darwin merely states his *belief* that natural selection determines the formation and development of "social virtues"—largely "altruism"—that gave some nations an advantage in their struggle with each other for survival. But, as Darwin admitted, he could not explain how "altruism"—he means "cooperation"—triumphs over selfishness in successful societies or why this does not weaken them in their struggle with other societies. In any case, even a realistic argument about the adoption of "social virtues" such as "altruism" cannot explain either the structure or the dynamics of social organization, just its coherence.

This problem has led the neo-Darwinists to adopt a genetic explanation of "altruism." As shown in chapter 7 this is the much heralded but deeply flawed kin-selection hypothesis pioneered by W. D. Hamilton (1964). Supporters of this hypothesis claim that the relationships between individuals are determined by the degree to which they share genes in common—by, in other words, their "genetic relatedness." The reason is, they tell us, that this is the way genes manipulate organisms in order to maximize their presence in the gene pool. Hence, the "selfish" gene generates the "altruistic" organism.

Darwin's dilemma is neatly solved. Or is it? There are two problems with the neo-Darwinist position: first, this absurd hypothesis is denied by reality—by, for example, the strong nongenetic relationships with spouses and adopted children; and second, it cannot explain social organization beyond the family—even with Robert Trivers's "reciprocal altruism" concept—any more successfully than could Darwin's farmyard theory.

Kin-selection theory was widely adopted in biology during the last quarter of the twentieth century for two reasons. In the first place it appeared to fill a gaping hole in Darwinian theory. Not only did it lead to the development of the new discipline of sociobiology but it also generated the hope that it would become the cornerstone of all human knowledge. Second, kin-selection theory brought genetics to the center of modern Darwinism. It did not matter to these deductive thinkers that neo-Darwinism was genetically deterministic, totally unrealistic, or that it could be taken quite logically to absurd conclusions by popularizers like Richard Dawkins and Edward Wilson. They appeared quite content with the science fiction about the organism being a slave to the selfish gene. In this chapter a realistic alternative is explored.

A Realist Model of Behavior

To understand the structure of animal and human society we need to develop a realist model of individual behavior. By employing the historical (or inductive) method I have been able to construct such a model—the **concentric-spheres model of behavior**—which, allowing varying degrees of individual freedom, overcomes the fantasy of genetic determinism (Snooks 1994: 50–51; 1997: 28–32). The way an individual behaves in relation to other individuals in the concentric-spheres model depends not on "generation distance" (or "genetic relatedness") but on what I call **economic distance**, a measure of the importance to the self of other individuals in the maximization of the material interest—of survival and prosperity.

In the concentric-spheres model represented in figure 14.1—originally developed to explain human society—the self is at the center of a set of concentric spheres that define the varying strength of the cooperative relationships between it and all other individuals and groups in society. The strength of the social relationships between the self and others—which can be measured by the economic distance between them—will depend on how essential they are to maximizing the probability of the self's survival and prosperity. Those aspects of the self's objective function that require the greatest cooperation—such as the generation of love, companionship, and children—will be located on spheres with the shortest economic distance.

But even in the case of spouse and children, the economic distance will be greater than zero, implying that the average individual will discriminate between him/herself and even those closest to him/her. For the typical individual, spouse and children (the order will vary) will occupy the spheres closest to the central self, with other relatives, friends, workmates, neighbors, members of various religious and social organizations, other members of its socioeconomic group, city, nation, and group of nations occupying those concentric spheres that progressively radiate out from the center. As the economic distance—a measure of materialistic rather than genetic relatedness—between the center and each sphere increases, the degree of cooperation (*not* "altruism") between them diminishes. It is because economic forces transcend biological forces that the concentric-spheres model can explain what the kin-selection model cannot—the varying cooperative relationships between any given individual and all others in society.

Figure 14.1 The concentric-spheres model of human behavior

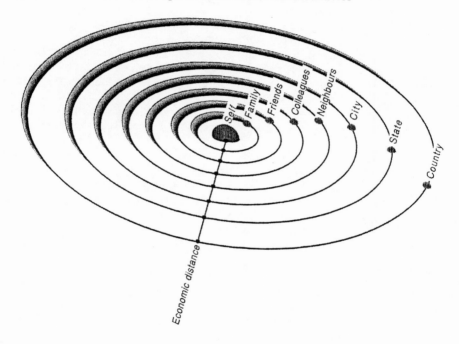

Source: Snooks 1996: 180–82.

There is always tension between the center and the periphery no matter how short the economic distance, because all personal relationships are developed by the self to maximize its utility. While one must cooperate with others to maximize individual utility, other cooperating individuals are still perceived as a constraint on what one can achieve. This accounts for the persistence of tension in all societal relationships. And the degree of tension appears to be inversely related to economic distance, with most individual (rather than organizational) conflict and violence occurring between those, usually genetically related, who are closely associated with each other. As shown in chapter 7, the greater conflict between genetically related individuals is another stumbling block for neo-Darwinian theory.

The concentric-spheres model is not a static theory of behavior, because individuals and groups on the various spheres, and the economic distance between these spheres, are constantly changing in response to **strategic demand**. Individuals form, dissolve, and re-form groupings at the various levels of the outwardly radiating concentric spheres in the struggle for life and prosperity. While not all competition occurs on an individual basis—groups compete against groups—all cooperation is part of the struggle by which individuals attempt to achieve their materialist objective of greater access to and consumption of natural resources. The **materialist organism** is always selfish, only the means adopted—involving varying degrees of competition and cooperation (*not* "altruism")—change with changing circumstances.

The concentric-spheres model, therefore, involves two sets of forces, one centrifugal and the other centripetal in nature. The centrifugal force is the incessant desire of the self to survive and prosper—a desire that leads, on average, to the individual placing itself above all others. It is the vital force in life. The centripetal force—the economic gravity holding society together—is the need to cooperate with other individuals and groups in order to pursue the dominant dynamic strategy more effectively. It is through both competition and cooperation that the materialist organism maximizes the probability of its survival and prosperity.

But exactly what is it that enables self-seeking individuals to build cooperative rules and structures? The neo-Darwinist claims that it is due to the genes they have in common, while the economic institutionalist argues that it is due to "trust." The genetic response has already been rejected, but the institutionalist response has yet to be considered. To the institutionalist, trust is the outcome of the "evolution" of informal (customs) and formal (laws) rules, which determine predictable and cooperative conduct by self-interested individuals. These rules are said to "evolve" through some sort of Darwinian mechanism involving survival of the fittest. The case for rejecting the institutionalist approach is both conceptual (it does not constitute an endogenous dynamic model) and empirical (it cannot explain frequent institutional/organizational *reversals*) and can be found detailed elsewhere (Snooks 1997: ch. 4). In contrast the dynamic-strategy theory suffers none of these deficiencies.

The dynamic-strategy theory provides a new analysis of societal change. It shows that the reason chaos does not overwhelm a society is not because of the infallible selfish gene or because of autonomous institutional rules but rather because the strategists in society are pursuing a *viable* dynamic strategy. It is this successful dynamic strategy that generates a network of competitive/cooperative societal relationships, together with the necessary supporting rules and organizations. The institutions and organizations, therefore, are a response to a changing strategic demand as the dominant dynamic strategy unfolds.

Individuals in society, therefore, relate directly to the successful dynamic strategy and only indirectly to each other. It is not a matter of mutual trust as such—of having confidence in the nature of other individuals through repeated interactions—but rather of having confidence in the wider dynamic strategy that they are jointly pursuing. This confidence depends on the materialist returns that are derived by individuals from the dominant strategy. Hence, it is not individual trust but **strategic confidence** that keeps society together. What we call trust is actually an outcome of strategic confidence. Once again this is not a static quality. As the dominant dynamic strategy unfolds, the nature of strategic confidence and, hence, trust changes in subtle ways.

Strategic confidence, which holds society together, lasts only as long as the success of its dynamic strategy. Once the strategy has been exhausted, strategic confidence declines and, in extreme cases, evaporates completely. And as strategic confidence declines, so too does trust and cooperation. In terms of figure 14.1, the economic distance between the self and the family will decline as the need for mutual support increases, while that between the family and other individuals and groups will increase. Some individuals and groups will even disappear from the concentric spheres. This occurred, for example, during the last "days" of the Roman Empire, following the exhaustion of the old conquest strategy, when families transferred their allegiances from the state to regional warlords. This is quite common in history, even in recent history as shown by the disintegration of the former Yugoslavia. In these circumstances the old society collapses, and in extreme cases involving total chaos many individuals even abandon their own families and attempt to survive by their own wits (another stumbling block for neo-Darwinism). This leads to the triumph of centrifugal over centripetal forces and, hence, to the total destruction of any form of society. It would have happened in the dinosaur dynasty just as it can be shown to have happened in the Roman Empire and all other ancient societies (Snooks 1997: ch. 6).

This model has been developed from the systematic observation of human society (Snooks 1997), but it can be simplified and applied to other animal societies. While the family is of central importance in human society it is of even more significance in the animal kingdom. Just as in human society, we can also explain the animal family in materialist terms. And, more significantly, we can explain relationships in animal society that extend beyond the family. The evidence is discussed in chapter 7.

A number of key issues are involved. The relationship between males and females is based, as demonstrated in chapter 7, on an exchange of food, shelter, protection, and status in return for sex, support, and companionship. Both partners contribute to a cooperative relationship by specializing according to comparative advantage. In other words, males and females seek out each other because, through this type of cooperative relationship, they are better able to maximize the probability of their individual survival and appetite satisfaction. As argued in chapter 12, they will select their mates according to the characteristics that best assist them in the particular dynamic strategy they are pursuing. This process, which I have called **strategic selection**, will lead either to genetic or nongenetic change. It generates a set of relationships and strategic outcomes that cannot be explained using the neo-Darwinian kin-selection model, because, in the main, sexual partners are not closely related genetically.

But what is the basis for those relationships in the extended family? I have already suggested (in chapter 7) that the genetic relatedness of kin-selection theory is not only mechanical and impossibly complex (and confused), but also incapable of explaining the evidence. The concentric-spheres model, however, shows that the strength of extended family relationships depends on the relative importance of other family members in helping a given individual achieve his/her objectives of maximizing the probability of survival and prosperity. This model can be applied not only to the extended family in animal society but also more widely. It can explain both the competition between individuals for dominance of their own social group and the cooperation between them when they are threatened by outsiders.

None of these relationships has anything to do with "genetic relatedness." The reason that in animal society the closest relationships are between family members is that, like us, they begin their lives in a close association with parents and siblings, and with a less close relationship with other members of their extended family. As they need to cooperate with others to achieve their individual aims, they end up investing most of their time and effort in individuals to whom they are genetically related. Owing to this early investment, these formative relationships tend to remain strong. The dynamics of life, therefore, is a matter of economics, not of genetics.

The Strategic Theory of Institutional Change

The dynamic-strategy model shows that institutions (societal rules) and organizations (societal groups) emerge and change primarily in response to changes in strategic demand generated by an unfolding dynamic strategy. Only the superficial, ephemeral forms are shaped by forces on the supply side—by what I have called relative institutional "prices" (Snooks 1997: 66–67). These include the relative abundance (which can be reflected in relative prices or exchange rates) of different natural and man-made resources. Hence, in a competitive environment, strategic demand for both institutional and organizational support will be met most efficiently by responding to the relative costs of various possible alternatives. But strategic demand creates its own supply, not the other way around.

The Nature of Institutions and Organizations

Societal rules, both formal and informal, are established and constantly altered to facilitate the dynamic strategies by which individuals attempt to achieve their materialist objectives. These rules are required, I argue, to economize on the scarcest resource in life and human society—the intellect. Similarly, societal organizations, from the simplest family relationships in the animal kingdom to the most complex society-wide relationships in human civilization, also respond largely to these dynamic strategies rather than to either the machinations of selfish genes or the dictates of mechanically "evolved" rules.

The differences between institutions and organizations can be outlined briefly. Institutions in human society cover the full spectrum of economic, political, and social activities and include: the rules by which business is con-

ducted; the way goods, services, and factors of production are bought and sold; the way business is financed; the rules of money supply; the way property rights are allocated; the way politics is conducted; and the way people interact at a social level. Organizations, on the other hand, are formal groups of individuals that employ these rules in their daily activities to achieve a range of short-run economic, political, and social objectives. Taken as a whole the guiding long-run objective is the maximization of survival and prosperity. A few examples of this distinction will help. In the political sphere a constitution is an institution and parliament is an organization; in the judicial sphere the formal laws are institutions and the courts are organizations; in the business sphere corporate laws and regulations are institutions and the corporations are organizations; and in the social sphere religious codes and laws are institutions and churches are organizations. This useful distinction was first proposed by the economic institutionalist J. R. Commons (1934). In less complex societies these formal arrangements are replaced by customs, taboos, and traditions, and in animal society by simple rules of behavior, shaped by their dynamic strategies, which are enforced by the leading adults in kinship groups and are passed on from generation to generation.

The incentives to which organizations respond are to be found in the strategic opportunities faced by organisms. Societal rules do not provide opportunities or incentives of their own volition as the economic institutionalists (for example, North 1990) claim; they merely communicate the opportunities generated by fundamental strategic forces. And they communicate them to organisms, not their genes as the neo-Darwinists would have us believe. There will, of course, be an interaction between demand and supply forces, but causality flows overwhelmingly from the former to the latter. Rules, therefore, do not "evolve" in a Darwinian manner as some economists have recently argued (for example, Hayek 1988). This is a deductivist fallacy.

It is in the process of **strategic imitation,** by which the vast majority of organisms emulate the action of successful strategic pioneers, that societal rules are employed.[1] Institutions are needed to economize not on benefit–cost information, as the economic institutionalists or rationalists argue, but on intelligence. This is true not only in animal society where intelligence is particularly scarce, but also in human society where most individuals find intellectual activity difficult and unhelpful. Intelligence is the scarcest resource on Earth, and probably nonexistent in the rest of the universe. The rulemakers, therefore, are the strategic followers, who demand guidance in their strategic pursuit, while the rulebreakers are the strategic pioneers, who attempt to break out of the restrictions of institutional conventions as they follow their new visions. As the "followers" constitute the vast majority of organisms in life, "rules" are essential to the dynamics of both animal and human society even though they are purely a response to it.

Strategic Institutions
Strategic institutions, therefore, are the rules of conduct required to support the emergence and development of the various dynamic strategies. In animal soci-

ety, even though the **strategic cerebrum** supervises the strategic pursuit (chapter 10), the demands made upon intellectual faculties are minimal.[2] Only *informal* "rules," in the form of customary behavior passed on from generation to generation through strategic imitation, are involved. As animal and human societies become increasingly complex, strategic costs rise with the increasing demands made upon intellectual resources. Owing to the need to reduce these costs by economizing on intelligence, these circumstances lead to the growing importance of *formal* institutions or rules. These rules can clearly be seen operating in the societies of higher animals as well as of humans (Hauser 2000: 249–53).

Strategic institutions differ with the type of dynamic strategy and the stage reached in the unfolding of that strategy. This can be seen most clearly in human society but it also exists in the animal world. A few examples will illustrate the point. As far as the economic and political system is concerned, a conquest strategy will lead to central control by a military strongman (king or dictator), because the society's wealth is concentrated in the hands of a small military elite who are determined to control the sources of their wealth; a commerce strategy, owing to the more extensive ownership of economic resources, will see the emergence of a regulated market system with a wider political franchise and a government led by either elected merchant princes or parliamentary representatives of the commercial elite; and a technological strategy will generate a free-market system that gives rise to widespread economic ownership and control, and to a parliamentary democracy based on universal franchise. The same is true of different systems of property rights, law, and social intercourse (Snooks 1997).

These ecosociopolitical relationships, which are systematic and predictable, are not the outcome of chance. Different strategies can best be facilitated by different economic and political systems. A society that switches from an exhausted commerce or technological strategy to a conquest strategy, for example, will witness the reemergence of a ruthless ruling elite that will dismantle former liberal institutions; it will experience considerably less economic and political freedom; control over its rules of exchange will pass from private (free markets) to public (forced labor, requisitioning, state distribution) hands; its property rights will revert from a widespread (even universal) to a restrictive and authoritarian basis; and its democratic rules of social intercourse will be replaced by totalitarian decree. Changes of this nature can be seen throughout human history, such as in Carthage (after 300 BC) and Greece (after 338 BC) as they turned from commerce to conquest; and in Germany and Japan (after the mid-1930s) as they turned irrationally from the technological to the conquest strategies (Snooks 1996: ch. 10; 1997: ch. 6). These institutional reversals—the outcome of changing dynamic strategies and, hence, of strategic demand—cannot be explained by Darwinian natural selection, neo-Darwinism, or evolutionary institutionalism. As Alfred Wallace (1871: 36) recognized more than a century ago, under natural selection "such a variety could not return to the original form; for that form is an inferior one, and could never compete with it for existence." But it can do so in ecosociopolitical reality, and it can be explained by the dynamic-strategy theory.

Strategic Organizations

The four dynamic strategies discussed in chapter 12 also call forth a set of characteristic and predictable organizations. In *The Ephemeral Civilization* (Snooks 1997: ch. 3) I divided strategic organizations into two main categories—major strategic organizations and support organizations. The major strategic organizations are those needed by decisionmakers to implement and expand society's dominant dynamic strategy. For those societies pursuing family multiplication these organizations include the family and the kinship teams required for hunting and gathering; for those pursuing conquest it covers military and imperial organizations; for those pursuing commerce it includes trading, financial, naval, and foreign-service organizations; and for those pursuing the technological strategy it encompasses industrial and commercial organizations together with comprehensive state bureaucracies.

Any society replacing its old and exhausted dynamic strategy will also need to gradually change its major strategic organizations. When switching from commerce to conquest, for example, a society—such as Athens during the fifth century BC or Venice in the sixteenth century—has to replace much of its commercial organizational structure with a military structure; while a society switching from commerce to technological change—such as Western Europe between the eighteenth and nineteenth centuries—has to replace much of its commercial organization with an industrial system. And the reverse of this was experienced when Germany and Japan transformed their ecosociopolitical structures in order to pursue the conquest strategy. Obvious? Try and explain it using the Darwinian concept of natural selection. And remember what Wallace had to say about the irreversibility of evolution driven by natural selection.

Support organizations also depend on the type of dynamic strategy pursued by any society. The nature of education and training, the type of manufacturing enterprises, the character of research and development, and even the forecasting methods employed (compare the nonscientific forms employed by conquest societies as far separated in time as Nazi Germany and ancient Rome) are determined by the dynamic strategy pursued and the stage reached in its unfolding. Once again this can best be seen in a society switching dynamic strategies as the old one is exhausted. With a shift from commerce to conquest—as in third-century BC Carthage or in fourth-century BC Greece or sixteenth-century Venice—the support organizations shift their focus from the skills and commodities required in trade to those needed for war. Of course, this was also experienced by Western democracies during the Second World War as they transformed themselves into centrally determined war machines to take on the fascist conquerors.

This model of strategic institutions and organizations, as suggested in chapter 7, can be adapted quite simply for analyzing the animal kingdom. In many animal families—particularly the big cats, primates, and the herbivores—there are power hierarchies consisting of dominant males, fertile females, immature males and females, and mature outcast males.[3] Each animal knows its place in the social structure and its behavior is regulated by unspoken rules—conventional behavior—that are transmitted from generation to generation through strategic imitation.

The nature of these "rules" and relationships, and the size and function of groupings beyond the family (herd, tribe, flock) are determined by the type of dynamic strategy being pursued by any given species. Under the family-multiplication strategy in many species the family group—which may not be permanent as it forms, dissolves, and re-forms as the seasons change—includes a sole dominant male with a variable number of fertile females. This group is constantly shadowed by younger outcast males who regularly challenge the dominant male, not to increase their genes in the gene pool as the neo-Darwinists claim, but to usurp the leadership of this group's family-multiplication strategy. The underlying reason is that strategic leadership is the best way to maximize survival and prosperity. Family groups also maintain associations with each other to form larger breeding populations, which provide potential partners with varying physical characteristics on which strategic selection can operate and which also provide protection against predators.

When the species in question exhausts the family-multiplication strategy, the family organization and its relationship to the wider population changes. By adopting the conquest strategy—as has always occurred in the past but which has been preempted in the modern era through man's excessive intervention in the lives of wild species—the family will display greater teamwork between mature males and, accordingly, a greater sharing of the returns from hunting and the sexual services of the females. At least until the conquest strategy is exhausted and their society descends into chaos, when it becomes a case of every individual for itself. By this time every sphere in the concentric-spheres model will have been stripped away, leaving the individual to its lonely fate.

Social Dynamics

The dynamic-strategy theory shows that institutions and organizations not only depend on strategic demand at a point in time but that they change over time in response to the way it changes. Shifts in strategic demand in turn are due to the changing fortunes—the unfolding—of the dominant dynamic strategy, and to the transition from one exhausted strategy to its replacement. Changes in institutions and organizations are achieved by individuals wanting to invest energy and resources in the unfolding dynamic strategy and needing to establish more effective rules and structures to do so.

The strategic demand for new organizations in human society is quite straightforward. Individuals require, form, and modify associations—such as trading, financial, shipping, insurance, industrial, and military organizations—to facilitate investment in, and the operation of, the prevailing dynamic strategy. Organizations are constantly being changed to enable strategists to adapt to new circumstances. But the demand for new institutions is less direct and, hence, more complex. To change existing formal rules it is necessary for the strategists to influence those who hold political power. This can be achieved either by tempting or pressuring political leaders. Temptation can be exercised by offering politicians a share of the strategic profits through either bribery or legal business arrangements. Political pressure is exercised by lobby groups that threaten to divert the political support of their strategic backers to opposition parties. In rare instances these days, governments may even provide, on their own initiative (!), strategic leadership to facilitate the perceived requirements of the society's strategists. I say "rare" because most governments in the Western world are captives of the neoliberal (economic rationalist) policies of orthodox economists (Snooks 2000: ch. 5).

A fascinating aspect of the dynamic-strategy theory is, as flagged earlier, that it can explain the *reverses* that occur from time to time in ecosociopolitical institutions. As institutions change in response to changing strategic demand generated by an unfolding dynamic strategy, the switch from an exhausted to an earlier strategy will cause a society's institutions/organizations to turn back on themselves. For example, all ancient and medieval commerce societies—such as Carthage, Greece, and Venice—went through the strategic sequence of conquest → commerce → conquest. This meant that the move to more democratic institutions—whether economic, political, or social—during the commerce phase was reversed during the conquest phase back toward autocracy. Similar reversals have taken place in the social relationships of the animal world. The waves of speciation during the dinosaur dynasty, for example, led on a number of occasions to the strategic sequence of genetic change → family multiplication → conquest → genetic change → family multiplication, etc. This would have generated the organizational sequence of nuclear family → extended family → roving packs of warlike males → nuclear family → extended family, etc.

This is a particularly important discovery. We have incorrectly come to think of ecosociopolitical institutions as "evolving" in some sort of Darwinian manner, largely because, over the past millennium, England, the pioneer in parliamentary democracy, underwent a gradual transition from autocratic monarchy to universal parliamentary democracy. But, as I show in *The Ephemeral Civilization* (Snooks 1997: ch. 10), this was only because of England/Britain's historically atypical strategic sequence of conquest → commerce → technological change. As I argue there, this radically different strategic sequence was a mid-eighteenth-century outcome of the fortuitous coincidence between Britain's exhausting commerce strategy and the exhausting global neolithic technological paradigm. If the Industrial Revolution—the new technological paradigm shift—had not taken place when it did (1780 to 1830), Britain in time-honored tradition would have turned to conquest on a global scale. There is even evidence that it was preparing to do so. And there would have been a reversal in the development of institutions/organizations from democracy to totalitarianism. The prospect and fact of institutional/organizational reversal explodes the myth of social evolution and of the role of natural selection in explaining the development of human culture.

While the dynamic-strategy mechanism generates a changing demand for institutions and organizations to facilitate the objectives of **materialist man** (or materialist organism), their design is influenced by supply-side forces. These forces constitute, as mentioned earlier, the costs associated with establishing a range of feasible rules and social structures that could be employed by any particular society to meet the change in strategic demand. I have called these costs relative institutional "prices." In the animal world this involves the metabolic energy that must be devoted to forming and supervising customary behavior and social relationships. The forms adopted depend upon which of the available alternatives meet the prevailing strategic demand most efficiently and, thereby, maximize material advantage. Of course, the most efficient institution or organization "available" to a particular society at a point in time is not necessarily the most efficient form available in a timeless sense. Time and place play an important role in this story.

Within the limits provided by the strategic-demand framework, changes in the *design* of institutions and organizations will also occur as relative institutional "prices" change. These relative "prices" change whenever there is a change in technology (or genetic structure) broadly conceived to include ideas relevant to the structure of human (or animal) society. But even these changes depend ultimately on the dynamic-strategy mechanism. In this analysis of the dynamics of institutional/organizational change, therefore, pride of place is assumed by strategic demand.

Conclusions

Social relationships in the animal and human worlds cannot be adequately explained either by Darwin's "social virtues" that are supposed to "evolve" as an outcome of natural selection or by the neo-Darwinists' genetic concept of "kin selection." Modern Darwinism—or sociobiology—will, therefore, never become the foundation for the social sciences and humanities as claimed by Edward Wilson. Indeed, it is now under challenge from the dynamic-strategy theory on its own home ground. Equally, the evolutionary institutionalists in the social sciences, who have taken their lead from either Darwin or the neo-Darwinists, will never be able to explain ecosociopolitical change.

In order to understand social dynamics in both the animal and human worlds we must reject the prevailing deductive fantasy and employ the historical method to reconstruct the real underlying mechanism. That mechanism can be modeled by the dynamic-strategy theory in which institutions and organizations are determined by a constantly changing strategic demand, which is an outcome of the unfolding dynamic strategy and the switching between dynamic strategies. The fact that backward-switching strategies will cause ecosociopolitical institutions to reverse themselves explodes the myth of social evolution and the role of natural selection in explaining social dynamics in both the animal and human worlds. Now that the dynamic-strategy model is complete, we are in a position to explore the laws of life.

Chapter 15

The Laws of Life

It is a remarkable fact that none of the sciences possesses a set of *general* laws that can explain the origins and dynamics of the real world. Science has only been able to develop laws that are restricted to either subsets of the dynamic whole or static relationships. Usually the latter. This is just as true of the physical and biological sciences as it is of the social sciences. While lawmaking is most advanced in the physical sciences, even in this less complex field of intellectual endeavor there are no general laws bridging dynamic relationships between the smallest and greatest levels. Why? Because there are no general dynamic theories that can embrace both the micro and macro worlds. Until now.

A recent breakthrough has been made in the social sciences. This may come as a surprise to some because the social sciences in general and history in particular are widely regarded as more backward in lawmaking than their more precocious siblings in the natural sciences. Indeed, physical scientists usually express thinly veiled contempt for even the most "scientific" of the social sciences—neoclassical economics; and biological scientists have long been threatening that they are on the verge of colonizing their less scientific brethren. It is somewhat ironical that they now find themselves preempted.

The social sciences possess a long-standing set of static microeconomic laws, but few in macroeconomics and none in history. What they have always lacked is a set of general dynamic laws. It was the attempt to rectify this deficiency that gave rise to my *Laws of History* (Snooks 1998a) and *Longrun Dynamics* (Snooks 1998b), which are based on the dynamic-strategy theory as applied to the development of human society over the past 2.4 million years (myrs) (Snooks 1996; 1997). These works provide the basis for the "social dynamics" that John Stuart Mill (1843) attempted to develop but was later forced to abandon.

This breakthrough in the social sciences, therefore, occurred from within rather than without, as had long been predicted. Why? Because the life sciences in general and sociobiology in particular were never in the race. Darwinists of both the original and sociobiological kind have failed to develop a general dynamic theory despite their long-standing claims to the contrary. It has been one of the objectives of this book to show why.

But how did the social sciences, like the unheralded athlete, come from nowhere in this race to breast the lawmakers tape well ahead of the fancied contenders? It is curious because the history of the social sciences consists of either a retreat into storytelling (a role with which most historians appear to be comfortable), a preoccupation with deductive fantasy by those employing physical-science methods (pursued by orthodox, neoliberal economists), or a slavish imitation of the flawed Darwinist theory (adopted by less orthodox institutional economists). These matters are discussed in detail elsewhere (Snooks 1993: ch. 1; 1997: ch. 4; 1998b: pt. II). The breakthrough in the social sciences came from a rejection of storytelling, of deductive gameplaying, and of neo-Darwinian mimicry. By embracing a new inductive approach—what I call **existential historicism**—it has been possible to reconstruct the general dynamic mechanism underlying the historical patterns not only of human society but also of all life. The laws of life (and history) are derived from modeling this dynamic mechanism.

A New Method of Lawmaking—Existential Historicism

Historicism has had a remarkably bad press. This is largely due to the politically motivated attack made in the middle decades of the twentieth century by a number of extreme deductivists under the intellectual influence of Karl Popper (Snooks 1998a: ch. 5). But it is also an outcome of the limited imagination of the old historicists as a group. The old historicists in the social sciences treated the patterns they detected in history as the dynamic mechanism itself rather than merely the outcome of the *underlying* process of change. They were, therefore, easy targets for extreme deductivists such as Karl Popper (1902–1994) and Friedrich Hayek (1899–1992). In the biological sciences the historicists (or paleontologists) are also afflicted with the same problem, which is exacerbated by their refusal to abandon the flawed Darwinian framework.

In chapter 3 I argue that while the problem of induction does exist—there are no mechanical rules for working from observation to lawmaking—philosophers have overlooked the problem of deduction—the failure to embrace the totality of reality—and that this problem is even more of a handicap in science. This is reflected quite clearly in the failure of the deductivists in either the biological or social sciences to construct realistic general dynamic theories. I have also argued at length elsewhere that the problem of induction can be reduced to manageable proportions through employing what I call the **existential quaternary method** (Snooks 1998a: ch. 7).

The existential quaternary method—the four steps of induction—includes the following steps:
- the identification of timescapes or historical patterns;
- the construction of a general dynamic model that explains these patterns;

- the derivation of the historical dynamic mechanisms underlying these timescapes;
- the construction of a model that explains the dynamics of institutions and organizations.

This is the realist method that underpins the modelbuilding in part III of this book. It is the method of existential historicism—a type of historicism that falsifies Popper's "poverty of historicism" hypothesis. As Popper failed to recognize this form of historicism, the poverty of the imagination that he ascribed to his methodological opponents must also be shared by him. Rather than exposing the poverty of historicism, Popper was responsible for its perversion.

Table 15.1   The Laws of Life—a checklist

The Primary Laws

1	The law of motivation in life
2	The law of competitive intensity
3	The law of strategic selection
4	The law of strategic optimization
5	The law of strategic imitation
6	The law of strategic struggle
7	The law of diminishing strategic returns
8	The law of strategic crisis
9	The law of societal collapse

The Secondary Laws

10	The law of cumulative biological/technological change
11	The law of genetic revolution
12	The law of (biological) eternal recurrence
13	The law of technological revolution

The Tertiary Laws

14	The fundamental law of institutional change
15	The law of social complexity
16	The law of social cohesion
17	The law of institutional economy

Source: See text.

The laws of life can be derived using the quaternary system of analysis. To do so we must focus not on the timescapes—the fallacy of the old historicists—but on the general dynamic model, the historical dynamic mechanisms, and the dynamic model of institutional change. The reason for excluding timescapes as a basis for lawmaking is that they merely show the pattern of *outcomes* generated by more fundamental dynamic processes and the pattern of outcomes can suddenly change (even go into reverse) as the dynamic mechanisms unfold. It was the old historicists' fatal mistake to view these patterns of outcomes as laws in themselves that could be universalized and extrapolated into the future.

It did not occur to them, or to antihistoricists like Popper, that dynamic mechanisms might underlie these patterns.

Just as there are three sources for the laws of life, so there are three categories of laws, each operating at a different level of biological activity. A checklist of these laws is provided in table 15.1. The laws derived from the general dynamic model are the **primary laws of life**. They are the unchanging laws that govern the behavior of organisms as they pursue their objectives of survival and prosperity by investing in the most effective of the four main dynamic strategies. They also govern the way species/societies/dynasties respond to the dynamic strategies (chapters 10 to 12). The **secondary laws of life** are derived from the dynamic mechanism underpinning the historical eras both before and since the exhaustion of the **genetic option** (chapter 13). As these historical mechanisms were reconstructed by applying the general model to the quantitative timescapes, the secondary laws can be thought of as being derived from the unchanging primary laws. Finally, the **tertiary laws of life** are derived from the dynamic-strategy model of institutional change (chapter 14). These laws can be thought of as being derived from both the primary and secondary laws.

The Primary Laws

The primary laws, as indicated above, have been derived from the general dynamic-strategy model developed in chapters 9 to 11 using the historical (or inductive) method. As we know, laws must be derived not from patterns in the historical record but from the underlying dynamic mechanisms. By constructing and employing this basic model—the general dynamic-strategy model—it is possible to explain the general nature of the dynamic process operating beneath the surface of life. It is self-evident that if the general model of life is valid, as I claim, then it must be based on a set of unchanging laws that can be isolated here and employed in other contexts. The easy part is deriving the laws. What is difficult—the outcome of decades of intensive effort—is constructing the general dynamic model. Indeed, as I show in *The Laws of History* (Snooks 1998a), scholars have struggled with this problem for the past two thousand seven hundred years—ever since the time of Hesiod.

It should also be possible to reverse this process of modelconstruction → laws, to become one of laws → modelconstruction. Starting with the laws of life it would have been possible, in consulting the timescapes, to reconstruct the dynamic-strategy theory. It is possible, therefore, to work both ways between laws and general dynamic theory once we have identified the historical patterns of life and human society. What is essential to both lawmaking/modelbuilding processes is an understanding of the real world of existential historicism rather than the virtual world of deductivism.

The primary laws are highly general and involve universal and necessary relationships that are strongly supported both empirically and logically. Derived from the general dynamic-strategy model, the fundamental laws can be set down in the following nine interrelated propositions.

1   THE LAW OF MOTIVATION IN LIFE states that the constant preoccupation of organisms throughout the history of life is the struggle to survive and prosper under varying degrees of scarcity.
     Discussion
     1.1   This law, which underpins the dynamic-strategy theory, was derived from a historical examination of life over the past 3,850 myrs and of human society over the past 2.4 myrs. As scarcity is demand-determined this condition will last as long as life itself.
     1.2   This law provides the driving force in life that is both self-starting and self-maintaining. It accounts for the origin and dynamics of life on any habitable planet in the universe.
     1.3   This law implies that life in general and human society in particular will never be based on altruism. This does not, however, involve a rejection of cooperation (very different from "altruism"), which is often necessary for the achievement of self-interest.
     1.4   The counterfactual conditional for this law is that the only alternative to the struggle for resources is poverty and, ultimately, extinction.

2   THE LAW OF COMPETITIVE INTENSITY states that the fundamental activity of organisms, involving the selection and pursuit of dynamic strategies, varies according to the intensity of competition for scarce resources at any given level of genetics/technology.
     Discussion
     2.1   In life, intense competition—the Darwinian scenario—leads not to the genetic strategy and speciation but to the conquest strategy and extinction. Ironically it is the non-Darwinian scenario of minimal competition that leads to the genetic strategy and speciation. Moderate competition is experienced during the family-multiplication and commerce strategies (chapters 9, 10, and 12).
     2.2   Complete isolation (the absence of external, but not internal, competition), a rare and temporary condition, leads not to genetic (or technological) change—as neo-Darwinists of the "periphera isolates" persuasion (Mayr 1963; Eldredge and Gould 1972) believe—but to stagnation and, on recontact with more open societies, extinction. This can be seen most clearly in human history, particularly, in the New World, Australasia, and the Pacific.

3   THE LAW OF STRATEGIC SELECTION states that organisms focus on and develop those specific physical and instinctual characteristics—a product of inheritance and mutation—that assist their strategic pursuit, while ignoring the rest.

Discussion

3.1   This law, which is central to the dynamics of life, governs genetic change that is an outcome not of "divine selection" or "natural selection" but of strategic selection—self-selection writ large. It covers both the emergence of new species in response to the genetic strategy and the development of add-on "technology" in support of nongenetic strategies. Genetic change, in other words, is a dynamic strategy that is deliberately pursued by organisms, but only when circumstances are favorable (see Law #2).

3.2   This law also governs the selection of physical and instinctual characteristics that support nongenetic strategies—such as fertility and mobility (together with sexual and adventurous drives) in the family-multiplication strategy and offensive and defensive weapons (together with aggressive drives) in the conquest strategy. Hence this law is far more general than Darwin's concept of natural selection purports to be.

3.3   Strategic selection is made possible in both intelligent and non-intelligent life through the mechanism of strategic imitation (chapters 10 and 12) in response to strategic demand.

3.4   This law accounts for the non-Darwinian genetic profile of most species, which reflects relatively rapid genetic change followed by much longer periods of minimal genetic change. While this pattern has recently become known as "punctuated equilibria" (Eldredge and Gould 1972)—although it was recognized by some of Darwin's peers—the causal implications of the name are totally misleading.

4   THE LAW OF STRATEGIC OPTIMIZATION states that a competitive "society" will adopt the dynamic strategy that promises to satisfy its materialist objectives most efficiently.

Discussion

4.1   Efficiency is measured in terms of the benefits and costs of metabolic energy and other resource use as between alternative dynamic strategies. This is worked out by organisms through trial and error, not by rational calculation. The optimum strategy is widely adopted within a society/species/dynasty through the strategic-imitation process.

4.2   Efficiency is a relative, not an absolute, concept. In other words organisms adopt the most efficient dynamic strategy in a particular environment, not the most efficient strategy in a timeless sense.

5   THE LAW OF STRATEGIC IMITATION states that successful strategic innovators who pioneer new dynamic strategies and who earn supernormal "profits" will be followed by a swarm of imitators, at first in a limited locality, but eventually on a widening front until the entire society/species/dynasty embraces the new strategy.

Discussion

5.1　　The strategic pioneers are the first to respond to changes in the environment—in relative factor "prices"—by investing in new strategic opportunities. This is a trial-and-error process.

5.2　　It is through strategic imitation that the "desire" of organisms to maximize their probability of survival and prosperity is achieved. In life, imitation is just as important as innovation.

5.3　　This is the mechanism by which successful individual "choice" is transformed into the dynamic strategies of life. Imitation is the real basis of animal and human "decisionmaking." It is the outcome of the remarkable survival instinct of organisms.

5.4　　It is also the way in which strategic selection becomes manifest in a society/species/dynasty (Law #3).

6　THE LAW OF STRATEGIC STRUGGLE states that organisms, in an attempt to achieve their materialist goals, struggle with each other for control of the dominant dynamic strategy and, hence, access to natural resources.

Discussion

6.1　　The strategic struggle in most species is played out by the males who, using the tactics of order and chaos, battle with each other for control of "family" and higher social groups. It is through these groups that the various dynamic strategies are pursued. The primary reason, therefore, for the struggle between males is not for sex to maximize the presence of their genes in the gene pool but for control of the sources (natural resources) of survival and prosperity—for control of the strategic pursuit.

6.2　　Strategic struggle is also the mechanism by which an unfolding dynamic strategy is translated into ecosociopolitical change (chapter 14).

7　THE LAW OF DIMINISHING STRATEGIC RETURNS states that the investment of energy/resources in a dominant dynamic strategy will ultimately experience diminishing returns, which will lead to a deceleration and eventual cessation of life/societal dynamics.

Discussion

7.1　　Stagnation will occur when the marginal return (generated by access to natural resources) and marginal expenditures (of energy/resources) of the dominant dynamic strategy are finally equated. At this stage the dominant strategy will have been exhausted and stasis will result, if only temporarily.

7.2　　This law also implies the existence of the countervailing law of increasing strategic returns experienced in the early stages of an unfolding dynamic strategy.

7.3　　This more general law replaces the classical (assuming a fixed supply of land in the long run) and neoclassical (assuming that all "factors of production" are fixed in the short run) economic laws of diminishing returns. Diminishing returns in reality arise not from the exhaustion of resources but from the exhaustion of dynamic strategies that provide access to resources that cannot be regarded as fixed. For a detailed discussion see Snooks 1998a: 202–3.

8   THE LAW OF STRATEGIC CRISIS states that the exhaustion of a dominant dynamic strategy in a competitive world leads not to the stationary state (long-run equilibrium) but to a strategic crisis that threatens the very existence of the society/species/dynasty.
Discussion
8.1   The reason that stasis turns into crisis is that the earlier exploitation of a viable dynamic strategy leads a society/species/dynasty to attain levels of population and consumption that can only be maintained by a continuous inflow of natural resources. Once the inflow dries up, the overnourished society collapses. And this inflow dries up completely with strategic exhaustion.
8.2   The more general dynamic law of strategic crisis displaces the Malthusian crisis (central to Darwin's natural selection concept), because strategic crisis is an outcome not of continued population pressure on resources but of the exhaustion of strategic opportunities.

9   THE LAW OF SOCIETAL COLLAPSE states that any species that exhausts its genetic style, any dynasty that exhausts its genetic paradigm, or any human society that exhausts its dominant dynamic strategy, and is unable to replace it, will collapse.
Discussion
9.1   In nature, crisis will lead to collapse only when the strategic potential of a species (its genetic style) or a dynasty (its genetic paradigm) has been entirely worked out. The only recourse for individuals faced with strategic exhaustion is the zero-sum dynamic strategy of conquest, which can lead only to the extinction of the species and, ultimately, of the dynasty.
9.2   In human civilization, collapse of industrial societies will occur whenever the old, exhausted strategy cannot be replaced with a new strategy. But under the technology option the adoption of the conquest strategy will not necessarily lead to the extinction of our species because we can, under normal circumstances, transcend the exhaustion of an old technological paradigm through the development of a new paradigm. Changing technological paradigms leaves a society changed but intact. This was not possible under the genetic option of the prehuman era, where a new genetic paradigm (a new dynasty) can only take place following the collapse of the old genetic paradigm (literally over the dead bodies of the old dynasty).
9.3   It is possible, however, that the collapse and extinction of our species could occur through a misguided global attempt to ban growth-inducing technological change in the mistaken belief that it will pre-

vent the destruction of the physical environment. This would occur if we were foolish enough to block the emergence of the next technological paradigm shift on environmental, or any other, grounds (Snooks 1996: 427–30).

The Secondary Laws

The primary laws of life operate at a very general level. Their purpose is to explain the general behavior of organisms, the strategies they employ to meet their objective of survival and prosperity, and the general consequences of these actions for the society, species, or dynasty. In order to explain the dynamic processes dominating specific historical eras, we need to apply these primary laws to our timescapes. This enables us to reconstruct the underlying historical dynamic mechanisms. The secondary laws, in turn, are arrived at inductively from these dynamic mechanisms, not from the timescapes (historical patterns) as the old historicists mistakenly thought. In effect, if not in practice, the secondary laws are derived from the primary laws.

By employing both the timescapes and the general model we can identify and explain the three great interlocking mechanisms that have been operating over the past 3,500 myrs. The general dynamic-strategy model, therefore, generates distinct but related processes of change under different historical circumstances, namely whether the genetic or technology options are operative. As shown in chapters 10 and 13 these historical mechanisms are the great genetic paradigm shifts, the great wheel of life, and the great technological paradigm shifts. The first leads to biological progress (in terms of output and "productivity"), the second to the eternal recurrence, and the third to economic progress. The secondary laws are derived from those mechanisms that are specific to either the genetic or technology options.

The four laws underlying the historical mechanisms that have dominated the changing fortunes of life over the past 3,850 myrs are as follows.

10 THE LAW OF CUMULATIVE BIOLOGICAL/TECHNOLOGICAL CHANGE states that the relationship between a series of genetic/technological paradigm shifts is geometric owing to the cumulative effect generated when the "technological" (either genetic or industrial) output of one paradigm becomes the input of the next.
Discussion
   10.1   Not only does the time lag between paradigm shifts (both genetic and technological) decline exponentially, so too does the period of both the transition and the global dispersion of organisms. The pace of the dynamics of both life and human society has been and is accelerating. A major implication is that the next technological paradigm shift in human history will appear soon and will take no more than a generation to unfold (Snooks 1996: ch. 13).

10.2   The geometric relationship for biological change appears to be quite regular and precise. Each great wave of biological activity (figure 9.3) is approximately one-third the duration of the wave preceding it. In other words, this relationship approximates a log-linear function.

11   THE LAW OF GENETIC REVOLUTION states that once the potential for an increase in biological output at the global level has been exhausted, a new genetic revolution (paradigm shift), which causes a quantum leap in *potential* access to natural resources, will occur *provided the genetic option has not already been exhausted.*

Discussion

11.1   The reason that an exhausted genetic paradigm is followed by a new one is that paradigmatic exhaustion leads to extinction and, thereby, to the removal of intense competition for natural resources. It is in this non-Darwinian world of minimal competition and abundant resources that speciation occurs under the driving influence of the genetic strategy (chapters 10 and 12). This is the law that delivers the coup de grâce to the Darwinian theory of "evolution."

11.2   The series of genetic revolutions will not occur indefinitely. Once the genetic option has been exhausted—once genetic change can no longer provide better access to natural resources owing to the saturation of all possible habitats in the most effective possible way—subsequent dynasties will merely invent different biological ways of achieving the same level of biological output and productivity (output per unit of input).

12   THE LAW OF (BIOLOGICAL) ETERNAL RECURRENCE states that, when the genetic option for life on any habitable planet has been exhausted, all future dynasties will merely repeat the biological achievement of the past in slightly different ways.

Discussion

12.1   This law operates the great wheel of life that turns unobserved in time without gaining any upward traction. The substitution of the great wheel for the great steps of life marks the end of biological progress, as measured in terms of total biomass and of output per unit of energy input.

12.2   This dynamic should be regarded as the normal outcome for life on habitable planets throughout the universe (chapter 13). The only way the eternal recurrence can be broken is if one species in any dynasty is able to pass the brain-size threshold required to replace the exhausted genetic option with the unanticipated technology option before the dynasty of which it is a part collapses and goes extinct. The probability (as we have seen) of life on any planet doing so is very low.

12.3   Life on Earth defied the odds through a long chain of unlikely and unsought events to generate the Intelligence Revolution (chapter 13). *As this Intelligence Revolution is an outcome of pure chance rather than of the desire of organisms to survive and prosper, there is no law governing its emergence, only its future course once under way.*

13   THE LAW OF TECHNOLOGICAL REVOLUTION (which only comes into operation when the technology option has replaced the exhausted genetic option) states that, once the potential for an increase in material living standards (real GDP per capita) at the global level has been exhausted, a new technological revolution (or paradigm shift) will occur, causing a quantum leap both in *potential* access to natural resources and in *potential* material living standards.

Discussion

13.1   With great good luck, life on Earth escaped the eternal recurrence—that biological black hole—and was able to substitute the technology option for the long (80 myrs) exhausted genetic option. This enabled one fortunate species to continually improve its access to natural resources through an endless sequence of technological revolutions (or paradigm shifts). In this way the human species was able to transcend the habitat limitations of genetic change and to generate ever increasing material living standards.

13.2   Unless we build barriers to this law—by banning growth-inducing technological change in favor solely of restorative technological change—human society will not collapse until the universe does, because there is no other limit to the technology option. Unless, that is, we damage the driving force in ourselves through genetic engineering.

13.3   With the introduction of the technology option, an analogous set of laws—the laws of history—were introduced. These new laws have governed the dynamics of human society ever since, and will continue to do so until the universe collapses. The laws of history are discussed in detail elsewhere (Snooks 1998a).

13.4   The introduction of the technology option limits the size and complexity of the brain of the leading species. Brain size and complexity will only increase to the threshold level that enables the leading species to successfully negotiate a series of technological·paradigm shifts.

The Tertiary Laws

As there is no independent mechanism underlying institutional/organizational change, there can be no independent laws to govern it. The modern quest by sociobiologists and evolutionary institutionalists to discover an independent general theory of societal rules and organizations, therefore, is similar to the medieval quest for the holy grail. Both are fruitless. While there certainly are regularities, or recurring patterns, in institutional/organizational structures throughout time and space, they are the outcome of the fundamental mechanism captured by the dynamic-strategy theory and not of any independent institutional/organizational mechanism. The laws governing institutional change, therefore, are merely derived from the fundamental laws of the dynamics of life. And the ephemeral nature of animal and human institutions can only be understood in terms of the eternal forces that drive life.

There are four laws governing institutional change. They constitute the tertiary laws of life.

14   THE FUNDAMENTAL LAW OF INSTITUTIONAL CHANGE states that all institutions and organizations, no matter how simple or complex, change in response to the unfolding and replacement of the dynamic strategies and substrategies of life.
     Discussion
     14.1   This is the law from which all other laws of institutional change follow. Its rationale is discussed in chapter 14.
     14.2   The nature of the strategic sequence is the key to understanding very long-run institutional change. If the strategic sequence is reversed (owing to changes in the underlying dynamic forces), then institutional change will be reversed. There is no such thing as institutional "evolution"—just as there is no such thing as biological "evolution"—in some sort of Darwinian sense. They are Victorian myths that were perpetuated into the twentieth century.
     14.3   The nature of all institutions/organizations changes according to the requirements of the strategists. It responds, in other words, to changes in strategic demand.

15   THE LAW OF SOCIAL COMPLEXITY has a Janus-like structure: the first form states that, as the dynamic strategy unfolds, the societal vehicle carrying it forward will increase in size and complexity; and the converse form states that, if the dynamic strategy is derailed, the vehicle of societal change will be transformed in the reverse direction.
     Discussion
     15.1   The explanation behind this law is that, as the dynamic strategy unfolds, both the requirements for and the capacity of new genetics/technologies make larger and more complex societal structures and rules necessary as well as feasible. As this can be reversed—the law is Janus-like—it is impossible to treat institutional change as "evolutionary."
     15.2   While this law applies to both nature and human society, it can be seen working most clearly in the latter case. For example, with the unfolding of the commerce strategy tribal kingdoms were transformed into city-states (ancient Greece) and nation-states were trans-

formed into empires (Britain); with the unfolding of the conquest strategy, tribal kingdoms (Rome and Tenochtitlan) and city-states (Greece and Venice) were transformed into empires; and as the technology strategy unfolds, nation-states are being transformed into mega-states (the European Union). And in the past, when ancient strategies were exhausted, empires (Rome and Venice) were transformed back into city-states and even tribal kingdoms (Western European monarchies). See Snooks 1997.

15.3 For complex modern societies there are other institutional laws—such as the "law of democratization"—that have no counterpart in simple prehuman society (Snooks 1998a: ch. 10).

16 THE LAW OF SOCIAL COHESION states that social structures, both simple and complex, will emerge and be viably maintained only while the dominant dynamic strategy is unfolding successfully.
Discussion

16.1 The explanation of this law (chapter 10) is that social cohesion is generated by strategic confidence, which is an outcome of a successful dynamic strategy. Strategic confidence generates the social cement that is commonly called "trust." Trust, therefore, is a product of neither human nature, nor intellect, nor institutional evolution as is usually claimed.

16.2 Strategic confidence is the "economic gravity" in the concentric-spheres model of behavior that, in a viable "society," balances the centrifugal force of the maximizing self (chapter 14).

16.3 When a dynamic strategy exhausts itself strategic confidence collapses and trust vanishes, thereby unleashing the centrifugal forces that generate social unrest and endanger the future of society.

17 THE LAW OF INSTITUTIONAL ECONOMY states that the form of the institution/organization chosen to facilitate strategic demand will be the one that does so most economically.
Discussion

17.1 The explanation of this law is that organisms—the dynamic strategists—attempt to maximize the probability of survival and prosperity by organizing their activities so as to get the best returns from energy/resource use.

17.2 This is not to say that institutions will take the most economical or efficient form in any timeless sense, only in a relative sense taking into account the backdrop of past "choices."

Conclusions

An integrated set of laws governing the dynamics of life has been developed in this chapter. This was only possible because of the prior construction of a general dynamic model—the dynamic-strategy theory. The difficult part in this process is not the derivation of the laws of life but rather the reconstruction of the dynamic mechanism underlying the fluctuations of life over the past 3,850 myrs. This is something that Darwin and the Darwinists have been unable to do. The natural selection hypothesis is not a realist dynamic theory and, not surprisingly, it is completely unable to explain the patterns in either the history of life or of human society. Hence, there are no operational or testable Darwinian laws, which is why Karl Popper regarded the theory of natural selection as unscientific. The ability to generate an integrated set of laws is a critical test for any theory of life. Darwinism fails that test.

The laws of life provide the opportunity to construct a new generation of models capable of analyzing detailed aspects of biological change, genetics, directed mutation, and of the "societies" of animals and humans. Some progress has already been made in this respect regarding the economic and political development of both First World and Third World societies (Snooks 1998b; 1999; 2000). It also enables us to make sensible predictions about the future not only of life on Earth (including that of human civilization) but of life on any habitable planet in the universe. These predictions are not about individual events but rather about dynamic processes and their outcomes. It is to the future that we look in the next and final chapter.

Chapter 16

The Future of Life on Earth and in the Universe

What does the future hold for life both on Earth and in the universe? It is a question often asked. Uncertainty and apprehension about what lies ahead has always been with us, but at the beginning of the twenty-first century it seems even more profound than in the past. Not only are we now concerned about the fate of our civilization—as we were throughout much of the turbulent twentieth century—we are also apprehensive about the health of our planet and the way in which we are genetically transforming our food and, soon, even ourselves through the new biotechnology. Soon we will face the disturbing prospect of being able, by genetic engineering, to transform even human nature.

We are right to be concerned. Where are the signposts that will guide us through this unknown and rapidly changing terrain? Darwinism, as we have come to realize, has failed to provide any directions, any certainties. Even neo-Darwinists like Edward Wilson and Richard Dawkins have acknowledged that "social choice" will replace natural selection as the agent of biological change. How then are we to understand what the future will bring? Where will our bold experiments with genetic engineering lead?

To answer these urgent questions we need to consult a general dynamic theory that can explain and predict changes in human civilization, the natural environment, biotechnological outcomes, and even the prospects for intelligent life on other habitable planets in the universe. Such an explanatory model—the dynamic-strategy theory—is presented in this book. It is the purpose of this chapter to briefly outline how this theory can provide the information about the future that we need to move confidently ahead.

The Future of Civilization and the Global Environment

As this is a subject dealt with in considerable detail elsewhere (Snooks 1996: ch. 13; 1997: ch. 13; 1998a: pt. II), only the briefest of outlines need be provided here. Our starting point is the collapse of Darwinism, which is unable to explain the past, let alone predict the future for human civilization and the global environment. This is, as we have seen, a critical test for any general theory of life.

History can best be explained by the dynamic-strategy theory. It is the outcome of a sequence of technological paradigm shifts (see figure 13.2) driven by **materialist man** pursuing his objective of maximizing the probability of survival and prosperity by investing in the most effective available dynamic strategy. These paradigm shifts include the paleolithic (hunting), neolithic (agricultural), and modern (industrial) revolutions. The paleolithic paradigm, which was initiated by the hunting revolution about 2 to 1.6 million years (myrs) BP, was exploited and finally exhausted by the pursuit of the family-multiplication strategy. This gave rise to the **great dispersion** of mankind during which *Homo erectus* spread slowly but steadily around the world. The neolithic paradigm, which was initiated by the agricultural revolution around 10,600 BP in the Old World and 7,000 BP in the New World, was exploited and finally exhausted (during the mid-eighteenth century) by the pursuit of the conquest and commerce strategies. And the modern paradigm, which began with the Industrial Revolution at the end of the eighteenth century, has been exploited ever since by the technological strategy.

The industrial technological paradigm has been spreading around the globe for the past two centuries, initially throughout Western Europe, North America, Japan, Australasia, and now throughout other parts of Europe, Asia, and South America. Only sub-Saharan Africa remains outside its embrace. But the expanding strategic core will gradually draw even this final continent into its sphere of influence (Snooks 1999). Once it does so, the resource-accessing potential of the modern paradigm will be exhausted and a new technological revolution—what I have elsewhere called the Solar Revolution—will take place and human society will begin to exploit a new technological paradigm. As these paradigm shifts are occurring at an exponential rate, we can expect the new revolution to appear soon, possibly by the middle of the twenty-first century, and that it will be transmitted around the world in a generation or so. Thereafter it is possible that technological revolution will become continuous (Snooks 1996: ch. 13).

This dynamic-strategy mechanism has important implications for the future of the natural environment and for our reactions to it. In the past, each time a technological paradigm faced exhaustion, the pressure on natural resources and the subsequent environmental damage occurring at the center of global strategic activity increased dramatically. But once the revolution had taken place, the pressure on natural resources was released, and damage to the environment abated, even reversed itself. This pressure on the natural environment can be seen in the reduction in hunting and gathering resources in the Middle East (fertile crescent) and in the Mesoamerican isthmus just prior to the neolithic revolutions in the Old and New Worlds; and it was clearly evident in the large-scale reduction of forests and the exploitation of marginal lands in Western Europe in the eighteenth century. Pressure on these resources was subsequently released as the new technological paradigm was established.

Today we are again witnessing a growing pressure on natural resources, together with associated environmental damage. While it is important to reduce the adverse impacts of this damage through the deployment of *restorative* technology, we must not attempt to eliminate *growth-inducing* technological change as some radical ecologists have been suggesting. We should, in other words, be careful not to confuse the signs of a coming technological revolution with the signs of environmental collapse. This confusion, which is common among writers on the environment—including Edward Wilson (1998)—is due to their failure to develop a realist general dynamic theory of life. Darwin and the Darwinists are of no assistance whatsoever in this, as in other matters.

Once the Solar Revolution occurs—and it will occur soon—the current strategic pressure on the environment will be released (Snooks 1997: ch. 13). Of course the environment will change as it has always done over the past 3,850 myrs of life on Earth, but it will not collapse and it will provide a higher material living standard for all, especially in the Third World (Snooks 1999). No doubt some life forms 2,500 myrs ago were unhappy about the pollution generated by their toxic waste product, oxygen. But, as we now know, oxygen was essential to the rise of new and more complex life forms such as those that occupy the Earth today.

The real danger to the global environment is that we will be tempted to outlaw growth-inducing technological change in the mistaken belief that economic growth is the enemy of the planet. Biological/economic growth has been the essence of the dynamics of life for the past 4,000 myrs. If economic growth was ever outlawed the Solar Revolution would stall, which in turn would force the dynamic strategists to seek other ways of achieving their objective of survival and prosperity, as life forms have always done. If we wish to survive and prosper it is not feasible to engineer stasis or equilibrium. It is just not possible to stand still. We must continue to change or be swept away.

In the event that we were foolish enough to ban growth-inducing technological change, the options available to the dynamic strategists would be severely constrained. Of the four universal dynamic strategies only conquest would provide returns comparable to the outlawed technological strategy. In effect we would be responsible for artificially exhausting the **technology option**, just as the dinosaurs exhausted the **genetic option**. And, in this context, conquest would ultimately lead, as it has always done, to the extinction of the existing dynasty—in this case mankind—and to the total destruction of the environment. It is essential, therefore, that we understand the options before us and that we do not lose our nerve as the next technological paradigm shift approaches. Darwinism can tell us nothing about these options or the choices we must make.

The Biological Future of Humanity

Natural scientists tell us, with undisguised pride, that the recent "breakthrough" in biotechnology will usher in a new age—"the age of biological control" (Wilmut, Campbell, and Tudge 2000: 290). Yet they express the need for caution, because there are no signposts to guide us in the use of this new biotechnology. Natural selection, which they mistakenly claim can explain genetic change in the past, will be of no use in the future because, they tell us, for the first time in the history of life on Earth a single species will be able to manipulate the genetic structure not only of itself but of the rest of nature. They even claim that we might be able to bring back to life those species, such as mammoths, that have left behind frozen DNA.

Edward Wilson, as we have already noted (chapter 8), has acknowledged that the genetic control promised by biotechnology will lead to "hereditary change" depending "less on natural selection than on social choice," and that "volitional evolution" will replace Darwinian evolution. Clearly this leaves a vacuum. Further, Ian Wilmut, one of the pioneers of this new biotechnology, stresses that in order to prevent this new knowledge getting out of control "we need to know far more" about the dynamics of life, and suggests that we look not to Darwinism but to the methodology of biotechnology itself for guidance (Wilmut, Campbell, and Tudge: 22). It is rather bizarre, although understandable in the present intellectual vacuum in the biological sciences, that the biotechnologists have asked us to place our faith in their research methods and ethics. This is a little like Dracula asking to be placed in charge of the blood bank.

The fundamental problem here is that the biotechnologists have little idea about the dynamics of life, whether past, present, or future. Darwinism has failed its most faithful followers as well as the rest of us who are placing our collective future in their hands. This failure is reflected in the way that biologists employ Darwinism. Serious biologists begin any new work by making a few references to the importance of Darwinism—such as "nothing makes sense in biology except in the light of evolution" (Theodosius Dobzhansky)—and then get on with their detailed biological studies. Essentially they have always placed their faith in their own research methodology rather than in the teachings of Darwin. It is similar to the way that serious Soviet economists acknowledged their dependence on Marx in the opening sentences of a new work and then proceeded with their essentially non-Marxist analysis. Of course, less serious biologists/economists merely told Darwinist/Marxist fairy stories about nature/society.

This can be seen in the fascinating book by Wilmut, Campbell, and Tudge on the methods and wider implications of cloning ("nuclear transfer technology") that arise from their "creation" of Dolly the sheep. In their book they mention Charles Darwin only once (in relation not to natural selection but, ironically, to his farmyard experiments with hybrid vigor) and neo-Darwinism only once (in relation not to kin selection, the selfish gene, or the extended phenotype, but to Richard Dawkins's support for cloning). The reason for this scant reference to Darwinism is that it has nothing to tell biotechnologists about the dynamics of life. It is just a faith to which most biologists feel they must express token allegiance.

While most biologists in the past were able to get away with this apparently comforting sleight of hand, today we desperately need to understand the dynamics of life. In the past this confusion did not matter because it only shaped the larger story that biologists told about life on Earth, not their detailed biological studies that owed nothing to this fable. Natural selection is a story that most biologists preferred to the creation story, because it seemed more "scientific" and it granted greater professional status.

But today everything has changed. Today biotechnology is providing us with the ability to make major and rapid changes to the genetic structure of life. In the future it will even be possible to change human nature. Today we need to understand exactly what it is we are changing and where this is likely to lead us. This requires a knowledge not only of the techniques of biotechnology but also of the dynamics of life. Fairy stories are no longer sufficient. Today we urgently need to understand reality.

If we decide to change human nature in the future, it will be absolutely essential that we clearly understand the impact that this is likely to have on life and civilization. To change human nature would be to change the driving force in life. If any society decided to do this—and all biotechnologists agree that eventually it will be possible—it would condemn itself to extinction. Quite clearly it is better to understand this in advance rather than through trial and grievous error.

There are many groups who currently want to change human nature. These include religious groups that think we all should have loving natures, ecologist/socialist groups that wish to eliminate the acquisitive spirit of capitalism, feminist groups that want to modify the dominating male drive, unworldly philosophers (such as Peter Sloterdijk in Germany—Laubichler 1999: 1859) who want to tame the "bestial dimension" of human nature through biotechnology now that humanist education has failed, and so on. These ideals are based on wishful thinking rather than an understanding of the basic conditions of life. Any attempt to use biotechnology to create an "ideal" human nature or "ideal" society would destroy the driving force in life, which in turn would lead to the extinction of those societies—even our entire species—that sanctioned such radical genetic change. We must, therefore, trade in our fairy stories for a realistic theory of life while there is still time. This book is dedicated to developing such a theory.

The biotechnologists and their fellow travelers claim to be heralds of a "new age of biological control." But is this true? The dynamic-strategy theory demonstrates that *life forms have always exercised biological control*. Under the genetic option that prevailed for 4,000 myrs, life forms were able, through **selective sexual reproduction**—an instrument of **strategic selection**—to employ the dynamic strategy of genetic change to improve their access to natural resources. The reason that this ability has not been exercised over the past one hundred and fifty thousand years is that the technology option provides a more direct, precise, and quick-acting alternative to achieving the same end. The technological strategy has now developed to the stage that it can give us more sensitive control of genetics. Genetic change, therefore, is back on the strategic agenda after one hundred and fifty thousand years of retrenchment, this time not as the liberator of technology but as its slave.

Hence, the only difference between now and then is that biotechnology is a more efficient instrument of genetic control than is selective sexual reproduction. Yet although biotechnology bypasses selective sexual reproduction through the newly discovered cloning techniques, strategic selection will be just as relevant in the future as it has been in the past. Strategic selection can explain the new strategic instrument of biotechnology just as it can explain the old instrument of selective sexual reproduction. The main characteristic of the "new age," therefore, is not our ability to shape genetic change but the more efficient way that we are able to do so.

Unfortunately it is this greater efficiency of genetic control that threatens to extinguish life on Earth. In the past, genetic change shaped by selective sexual reproduction was an extremely slow process, taking hundreds of thousands or millions of years. Hence, if a handful of families over a few generations made some grievous mistakes in their use of selective sexual reproduction—such as excessive inbreeding—these could be corrected fairly easily by eliminating the "deviants" and developing mating rules to avoid these mistakes in the future. This is the most likely source of incest taboos in all human societies.

Today, through biotechnology, the new strategic instrument of genetic change, entire societies could make changes to the human genome that are mistakenly thought to be advantageous—and may be in the short run—only to discover that in the long run it is a path that leads to extinction because it undermines the "desire" to survive and prosper. If we do manage to cripple ourselves genetically, together with much of our mammalian kin, and go extinct, life on Earth will resume where the dinosaurs left off by riding the great wheel—by enduring the eternal recurrence.

It is ironical that the "new age" will lead not only to the abandonment of Darwinism but also to the improved status of the social sciences. For the first time in human history it really matters that we understand the nature of human society. It is just as important as the breakthrough in biotechnology itself. Without the new biological methods we would not have been able to achieve direct and precise control over the genetic structure of life, but equally without a realistic understanding of the dynamics of life and human society this breakthrough could destroy us. From the discussion in earlier chapters it is clear that the revolution in the social sciences must come from within, because sociobiology—that much heralded reformer of the humanities—is not even able to explain the society of other animals. Darwinism is in a state of terminal collapse.

Are We Alone in the Universe?

Dynamic-strategy theory, unlike Darwinism, can also provide answers to some of the central questions about life in the universe. The most interesting question is not whether life exists in some of the billions of solar systems that comprise the universe—it seems highly likely that it does—but whether other life systems have produced intelligent species. What we need to know is: Are we alone in the universe? And we need to know it because our ultimate survival as a species could well depend on the answer.

Is there intelligent life in other solar systems? Our dynamic-strategy theory suggests that the answer is more complex than usually realized. The conventional wisdom is that intelligent life is the normal outcome of an evolutionary process. Hence, if life begins somewhere in the universe, provided the physical environment is appropriate and there are no major catastrophes, it is only a matter of time before intelligent forms emerge. If Earth is any guide, this should take about 4,000 myrs.

Of particular interest is whether there are forms of life elsewhere in the universe that are more advanced than our own. This issue is usually considered in relation to the period of time between the Big Bang, about 12,000–15,000 myrs ago, and the formation of our own planet. As this is about 7,500–10,500 myrs, it is usually concluded that life probably began much earlier in other parts of the universe that emerged before our own solar system. On the assumption that the "evolution" of life is, as Darwin (*Origin*: 448) thought, a rough function of time, it is usually deduced that those life systems that began significantly before ours must be more advanced—but clearly not sufficiently advanced to have made their presence known to us.

To test this simple hypothesis an increasing amount of expert time, financial resources, and sophisticated equipment is being devoted to listening to the skies in order to locate any signals that might have been produced by more advanced life forms. Earth has been advertising its own presence to the rest of the universe for more than a century (since the radio-wave experiments of Heinrich Hertz in 1888) through the generation of radio and, later, television signals. Obviously more advanced life forms would have been doing the same for much longer.

If we do establish the presence of more advanced life forms in the universe, then we will have good cause to be apprehensive. As we have seen, the desire on the part of all life on Earth to survive and prosper—and life elsewhere in the Universe will not differ in this respect—has, in competitive situations, always led to the more advanced forms eliminating the less advanced. The fossil record is quite clear about this. Accordingly, while it may be expedient to listen to the skies, we would do well not to advertise our own presence any more than is necessary. But, some readers might object, surely more advanced life forms will also be enlightened. Only, we should reply, if they had eliminated from their own species the universal driving force in life through the use of biotechnology. And had they successfully achieved this we will never hear from them because they would have destroyed themselves.

There are two important reasons, which arise from the dynamic-strategy theory, for thinking that either we are alone in the universe or if not, other life forms are no more intelligent than we are. In either case we are not likely to encounter them in our own lifetimes.

In the first place the dynamic-strategy theory shows that the emergence of life on Earth was not a steady progression and that it was not the outcome of Darwinian "evolution." Rather it has been, and still is, the outcome of the strategic pursuit and it has taken place via a series of genetic paradigm shifts, at least until the genetic option was exhausted. Thereafter life on Earth became enmeshed in the eternal recurrence, which would have continued, with plant and animal dynasties rising and falling until the end of time, had not the **Intelligence Revolution** occurred as the outcome of a long sequence of highly unlikely and unsought events.

The point is that life elsewhere in the universe may not have been so lucky in negotiating a similar long sequence of highly unlikely events. If this is true, then life forms elsewhere are probably still caught up in their planets' own eternal recurrence. The elapse of time since the first emergence of life, therefore, is not an appropriate measure of the progression of life. Instead it depends on how lucky an alien dynasty might be in escaping the black hole of eternal recurrence. My view is that, as the materialist organism neither knows nor cares that it is caught up in the eternal recurrence, the great wheel of life will be the normal dynamic mechanism for biological systems throughout the universe. Those able to break free, like us on Earth, will be the exception rather than the rule, because it is purely a matter of chance.

But, if a life system in an older part of the universe did manage to break out of the eternal recurrence, would it be more advanced intellectually than our own? The dynamic-strategy theory suggests that this is unlikely. If we reran the history of life on Earth it is highly unlikely that we would escape again in a mere 65 myrs (since the fall of the dinosaur dynasty), if at all. In which case we would not be we! The same would be true of other life systems in the universe. An early start in the race of life, therefore, is no guarantee of an early intellectual revolution. It all depends on the length of time biological systems are carried around and around by the great wheel of life. We made our escape quite quickly in these difficult circumstances.

Even in those cases in which alien life systems did manage to make a relatively early escape from this biological black hole, the level of intelligence they achieved would not be significantly different from our own. The reason, as shown by the dynamic-strategy theory, is that brain size and complexity will grow only until a threshold level is reached that enables the life form in question to substitute the technology option for the genetic option. Once this substitution has been achieved the average level of intelligence of this species will remain static. I can see no compelling reason why that intellectual threshold level would differ greatly between advanced life systems throughout the universe. Only when they gain direct control over their genome through biotechnology will it be possible to increase intelligence levels further as the outcome of a strategic act. But even then it may be less risky and more effective to employ technological rather than genetic change in their strategic pursuit. In other words, the issue of how advanced are life forms that have escaped the eternal recurrence in other parts of the universe will depend more on their technological achievements than on their intellectual capability. And these technological achievements will depend upon their social dynamics. The ball once more is in the social sciences' court.

Epilogue

A Modern Theory for the Modern World

From this study it has become clear that Darwinism was merely a stopgap theory of the Victorian age. Owing to its totally unrealistic assumptions and its failure to accurately determine the conditions under which genetic change has occurred in the past, Darwinism is fatally flawed. This is not a surprising conclusion because the natural selection hypothesis was the product of a dubious methodology—the use of the farmyard analogy for nature—by a brilliant naturalist who unfortunately had no idea about the properties of dynamic processes. Even the focus of Darwinism is wrong, as it treats directional genetic change as if it were the dynamics of life. Yet despite these obvious flaws, Darwinism was adopted and adapted by the relatively new discipline of genetics during the course of the twentieth century.

Why? Largely because it provided a sense of identity for the new geneticists, together with a rationale and status for their upstart discipline. They employed Darwinism as a cloak thrown over the shoulders of their discipline, but a cloak they retailored to suit an outlook and purpose very different to that of its original owner. In the process the neo-Darwinists changed natural selection from an economic to a sociological concept: from the struggle of organisms for existence to the struggle of genes to increase their influence in the gene pool.

In essence the neo-Darwinists have attempted to reconstruct Darwin's macrobiological theory by employing the micro building blocks of genetics. But they have failed. Indeed the absurdity of neo-Darwinism was inadvertently revealed by their most enthusiastic advocates including Bill Hamilton, John Maynard Smith, Edward Wilson and, particularly, Richard Dawkins. They did this by taking neo-Darwinism to its logical extreme: to a world ruled by genes. This is just as absurd as the parallel suggestion that human society is ruled by technological ideas. By removing the cloak of Darwinism they have revealed that the emperor is indeed without clothes.

True, this extreme form of Darwinism is not accepted by paleontologists such as Niles Eldredge, Stephen Gould, and others, but even they have not been able to escape the Darwinian net. While these historicists are aware that Darwin's original theory of natural selection is unable to explain the patterns they detect in the fossil record—namely "punctuated equilibria"—they have maintained their faith in Darwinian natural selection and have attempted to buttress it by developing a host of ad hoc supporting arguments. These are the discipline's flying buttresses, which merely add to the mazelike confusion of evolutionary theory. In the end they have, like those at the cathedral of Concepción that so intrigued Charles Darwin, been unable to prevent the collapse of Darwinism. It has all ended in a rather grand pile of ruins.

Darwinism fails many tests of an empirical and methodological nature, but probably the most critical are its inability either to predict conditions under which genetic change occurs and/or to explain the dynamics of human society. Any theory of life that expects to survive and prosper in the long run must be able to explain the dynamics of both nature and human society. Charles Darwin and Darwinists of all descriptions have been at a loss to do so. The neo-Darwinist attempt to brush these critical tests aside as being irrelevant is unscientific and totally unconvincing.

But the difficulties for Darwinism do not end here. This discipline has no way of predicting what is likely to happen to the dynamics of life in the future. Natural selection is unable to explain not only the past (particularly the past one hundred and fifty thousand years when genetic change became redundant), but also the rise of biotechnology; and it is unable to predict where this new technology might take us in the future. Even if Darwinism could explain the dynamics of life before the emergence of modern man one hundred and fifty thousand years ago, it could not be regarded as either a general or a universal theory. Even the sociobiologists indirectly recognize this difficulty with their antiquated theory from the Victorian age.

This book attempts not only to expose the fatal flaws and absurdities in Darwinian theories of all types, but also to present an entirely new theory of life that places the organism back at the center of a strategic pursuit for survival and prosperity. It has been argued that the dynamic-strategy theory can explain the strategic activities of the organism, society, species, and dynasty, together with all the main historical patterns in life and human civilization; that it can be used to make sensible and nontrivial predictions about the future of life not only on Earth but also throughout the universe; and that the laws of life that it generates can be employed to develop a range of detailed models relevant to nature and human society. The dynamic-strategy theory, therefore, is everything that the natural selection model is not: it is general, universal, operates at both the micro and macro levels, and its predictions are borne out by historical fact. It is a modern theory for the modern world.

Of course, this new vision of life could be conveniently ignored. This is the usual response when the existing intellectual paradigm is challenged, particularly by an outsider. Biologists, as we have seen, happily ignored the fatal flaws in Darwinism throughout the twentieth century. Why should the twenty-first century be any different? Why not just shut the doors of the cathedral of neo-Darwinism and continue to pretend that the "selfish gene" rules life by using us as its "survival machines"—its "lumbering robotic" slaves?

The answer is quite clear. For the first time in history the story we tell about our past, including our deep past, really matters. If we persist with these fairy tales—this science fiction—about the "selfish gene," or even the earlier farmyard analogy, we could well destroy the very forces that have enabled us to get this far. Unless we come to grips with the dynamics of life we will surely end as did the dinosaurs 65 myrs ago.

Notes

Chapter 1 The Rise and Fall of Darwinism

1. A recent prominent example is the Princeton philosopher Peter Singer (2000).

2. Despite the need to replace outdated Victorian ideas with those relevant to the twenty-first century, one high-profile geneticist, Steve Jones, in a book curiously entitled *Almost Like a Whale: The Origin of Species Updated* (1999), has recently served up a rehash of Darwin's ideas, using the same chapter titles, major headings, and even summaries from the 1859 version of *The Origin of Species*. He has merely attempted to spice up these Victorian leftovers with some new anecdotes about domesticated and wild species, together with a sprinkling of humor. While this limited reworking of Darwin has none of the excitement, seriousness, or imagination of the original, it is just as flawed. The subtitle could be more accurately rendered "The Origin of Species *Outdated*."

3. In parallel fashion, neoclassical economists—who have always attempted to construct growth theories from the building blocks of production theory—have made the same mistake (Snooks 1998b: ch. 3).

Chapter 2 The Chapel of Evolution

1. This is similar to the modern discipline of economics where the dominant neoclassical economists—or "productionists" as I prefer to call them—may be experts in production theory but they can tell us next to nothing about the dynamics of human society (Snooks 1998b; 1999; 2000).

2. Self-regulation of population densities by most types of animals is widely discussed in the literature. This involves regulating population densities without destroying natural resources and before predators and climatic changes force them to do so. Regulatory mechanisms include the estrous cycle—involving a limited fertility period—and the reproductive diapause—whereby fertilized females delay the birth of their young until the return of more favorable climatic conditions. Well-known examples include mammals such as kangaroos, and insects such as grasshoppers. Less well-known is that reproduction is also controlled by viruses and plants. For a discussion of these issues, see Chitty 1971; Frith and Calaby 1969; Hickman, Roberts, and Larson 1993; Hodek 1983; Krebs 1994; Lee and Cockburn 1985; Pener and Orsham 1983; Raven and Johnson 1992; Sinclair 1989; Tauber et al. 1983; Wynne-Edwards 1971. My explanation for this self-regulation—an outcome of strategic selection (see ch. 12)—differs from those provided by the above authors, who invoke, among others, the group-selection thesis despised by neo-Darwinists.

Chapter 3 Flaws in the Foundations

1. Extracts of all Darwin's notebooks can be found in Gruber and Barrett 1974.

2. A very different argument about Darwin's "discovery" of natural selection is presented in Gruber and Barrett 1974. It involves a psychological rather than a historical analysis. While it draws on a detailed study of Darwin's notebooks, it is shaped by Gruber's belief in a particular theory of knowledge. He argues that "the developed whole [of natural selection] *must* be foreshadowed or prefigured at earlier and more primitive steps in the process of creative development" (Gruber and Barrett 1974: 174; emphasis in original). And again:

> Theory construction is a matter of rumination and schematization; it is primarily an internal effort, involving play with ideas—information and generalizations—already incorporated in existing schemes of thought; its end product is the alteration of those schemes. (Gruber and Barrett 1974: 108)

In other words, Gruber argues that the idea of natural selection "must" have been implicit in Darwin's thinking before he began his domestic experiments and before he read Malthus's *Essay on the Principle of Population* in September 1838. Hence, by applying a pre-existing psychological theory of knowledge Gruber overlooks the critical role of analogy in Darwin's thinking. To do so he is forced to downplay Darwin's own references to the use of "induction" (Gruber means analogy) in both *Origin* and his *Autobiography* (Gruber and Barrett 1974: 173). The psychologist claims to understand more about the method of genius than does the genius himself!

3. There has been a growing body of literature over the past decade suggesting that beneficial mutations in at least some organisms respond to their need to survive. This is called "adaptive mutation" or even "directed mutation." For recent reviews of this literature, see Foster 1998; 1999. It is also discussed in chapter 12 above as evidence for my strategic selection theory.

4. See endnote 2, chapter 2.

Chapter 4 Stress in the Structure

1. As I show in *Ephemeral Civilization* (Snooks 1997), the emergence and development of human society is not an "evolutionary" process. Rather it is characterized by regular and major reversals in societal forms that can only be explained in terms of reversals in the sequence of dynamic strategies, such as conquest → commerce → conquest.

2. This is closer to my concept of biological transformations, which occur in species as a result of employing the dynamic strategy of genetic change, as opposed to those changes that are the response to strategic demand generated by nongenetic strategies. Darwin appears to have intuitively felt that there must be a more general dynamic theory, but he was held back by his commitment to the farmyard theory of natural selection. He remained a prisoner of the farmyard analogy.

Chapter 5 The Buttress-Builders

1. I have independently made a similar attack on neo-Darwinism (in its sociobiological form) in *The Ephemeral Civilization* (Snooks 1997: 97–101). Yet at the same time I find myself in agreement with the neo-Darwinists that the macroevolutionary theories of the paleontologists, which propose an active role for species as well as individuals, are fanciful, even metaphysical.

2. This is very similar in nature to the growth models of neoclassical economists, who occupy a similar deductive niche to the neo-Darwinians. Because of the limitations of neoclassical microeconomic "productionist" building blocks employed to construct macroeconomic growth theory, they postulate a linear equilibrium growth path. The realist solution is to develop a general dynamic model using the inductive method (Snooks 1998b).

3. In the social sciences the historical economists (naive historicists) of the late nineteenth century in both Germany and Britain fell into the same trap. Unable to scramble out of this trap, their discipline perished soon afterward. Their memory was later revived and desecrated by Karl Popper (1957; 1971). See Snooks 1998a: chs. 4 and 5.

Chapter 6 The Cathedral-Reconstructionists

1. Dawkins makes the point in one of his more recent books that while *The Selfish Gene* received immediate acceptance and acclaim, he had noticed a growing resistance some decades later. In my opinion the rapid acceptance was due to a combination of persuasive rhetoric—Mayr (1999: 1856) has referred to him and Wilson, with a hint of irony, as one of the "media stars" of the profession—and the repackaging of an old idea—it was Hamilton who had early difficulties in promoting kin-selection. The more recent criticism of neo-Darwinists is the outcome of the way Dawkins and Wilson have exposed the absurdity of their collective position.

2. This was one of Dawkins's better predictions.

3. One has the feeling that Michael Rodgers, at the time at Oxford University Press, was interested in developing the field of "science speculation" at Oxford when he accepted Richard Dawkins's first book, *The Selfish Gene*, a project that Oxford discontinued after he moved to W. A. Freeman where he started again by publishing Dawkins's second book, *The Extended Phenotype*!

Chapter 7 The Image-Makers and Evangelists I

1. See endnote 2, chapter 2.

2. Although John Maynard Smith, through his familiarity with economic game theory (which is to be expected from someone with these two first names), appears less confused than most sociobiologists about the altruism/cooperation distinction, he is still committed to a genetic explanation of animal society.

Chapter 9 A New Story of Life

1. The statistical, or biometric, "models" employed by Sepkoski and his colleagues are devoid of real analytical content. While Sepkoski (1978: 245) admits they are only "a description of the fundamental patterns" rather than "a complete causative account," these logistic/equilibrium models are often treated as explanations rather than technical stories. They are, in fact, comparative-static models that focus on convergence to equilibrium, dealing with the manner in which the convergence path might be affected by a large exogenous shock such as an extinction. No explanation is given for the forces driving the system (which are assumed to result from "initial conditions" and "constant probabilities of branching") or those that determine the "apparent" equilibrium point or equilibrium path. It is enough for these authors that the convergence path can be "described" by a logistic curve—a mathematical relationship with absolutely no analytical content. These are the empty statistical boxes of paleontology. Strikingly similar are neoclassical economic "growth" models, which are also notorious for their sterility and lack of relevance

to the real world (Snooks 1998b: 34–48; 1999: 107–15). Despite these analytical reservations, it should be recognized that Sepkoski has brought together in graphical form much useful data on marine animals.

2. As one might expect, the dating of the appearance of life on Earth is controversial. The earliest fossil evidence is contained in rocks in southern Greenland, which have been recently dated at 3,850 myrs (Nutman et al. 2000: 3052).

3. In any case, if the analysis of Michael Benton (1985) is correct—that the extinction of land animals around 65 myrs BP was only 14 percent and not much greater than the "background" extinction rate—there is not much scope for the contribution of catastrophes to the collapse of the dinosaurs.

4. But at the beginning of the twenty-first century political leaders in the First World—and, through their control of the IMF, World Bank, and WTO, also in the Third World—take their advice from deductive economists (neoliberals or economic rationalists) whose impact may be just as disastrous. They are the global crisis makers of the twenty-first century (Snooks 2000).

5. Others have nominated *Equatorius africanus* as the first ape to leave the treetops occasionally at about 15 myrs BP (Zimmer 1999: 1335).

Chapter 10 The Dynamic-Strategy Theory of Life

1. Most trained biologists today are unaware that there is any fundamental difference between modern Darwinism and Charles Darwin's original model of natural selection. Find one and ask him or her.

2. When I first employed the term "economic distance" to measure the strength of economic relationships between the self and all other individuals and groups (Snooks 1994), I was not aware of Dawkins's (and, later, other sociobiologists') use of the term "generation distance" to measure the strength of the genetic relationship between an individual and his/her kin. This is an interesting coincidence of terms for diametrically opposed concepts.

Chapter 11 The Driving Force

1. The term "dynamic materialism" was coined in the first volume of my global history trilogy, *The Dynamic Society* (Snooks 1996: 170).

Chapter 12 Strategic Selection

1. When I first developed the dynamic-strategy theory—in *The Dynamic Society* (Snooks 1996)—I employed a slightly different terminology for these four strategies in nature and in human society. They were genetic/technological change, procreation/family multiplication, symbiosis/commerce, and predation/conquest. But because the strategies are the same whether in nature or in human society, and in order to simplify the analysis, I now use the same set of names for both human society and nature.

2. Throughout this book the statement that a species/society pursues a dynamic strategy means merely that the majority of individuals in that species/society are pursuing that strategy. Elsewhere (chapter 11 and Snooks 1996) I explain that successful individual strategies of the few are adopted by the many through the process of strategic imitation.

Chapter 13 Life's Dynamic Mechanisms

1. Of course, serious large-scale wars can still break out because of the rise of irrational forces in powerful countries, as occurred twice during the twentieth century, but these wars will last for only a few years, not the hundreds of thousands of years as in the era of the dinosaurs. The reason is that, under the technological strategy, conquest is not as profitable as industrial production in the longer term. Conquest as a dynamic strategy is irrational in the modern age and if it breaks out will soon be contained. See Snooks 1996: ch. 10; and Snooks 2000.

Chapter 14 The Dynamics of Social Organization

1. Studies of wild animals—including birds, dolphins, monkeys, and apes—demonstrate the important role of imitation in their behavior. See, for example, Byme 1995; Dugatkin 1992, 1997; Russon 1997; Tomasello and Call 1997; Whiten and Custance 1996. There is a large literature on this subject. Only those scientists who restrict their observations to the laboratory—where imitation is defined, somewhat curiously, as the response of animals to *human* actions—seem to doubt this.

2. While intelligence is scarce in the animal world, it is quite clear that many species possess the ability to recognize and quantify objects and other organisms and to predict their actions and reactions. They are also capable of learning by trial and error and imitating the actions of others. See Hauser 2000.

3. For a discussion of "rules" in animal society, see Grinnell and McComb 1996; Grinnell, Packer, and Pusey 1995; Harcourt and de Waal 1992; Packer and Ruttan 1988; and Trivers 1985.

Glossary of New Terms and Concepts

This glossary includes the new terms and concepts (arising from my dynamic-strategy theory) dealt with in this book. When a new term or concept is first mentioned in a chapter it appears in bold type. Italics in the glossary are used to indicate that additional concepts are also defined here.

Biotransition is the term used in this work in place of the misleading term "evolution," which has been irreparably contaminated by Darwinian natural selection. It involves the transmutation of species arising from genetic change and is part of the wider process of the dynamics of life.

Commerce strategy. The dynamic strategy of commerce, or symbiosis, is a more specialized means of achieving prosperity, practiced by a variety of plants, bacteria, viruses, insects, and other animals, and, most spectacularly, by some human societies. It can be thought of in nature as an alternative *dynamic strategy* to *family multiplication* and in human society as an alternative to *conquest*. Under this strategy an organism/society that manages to gain a monopoly over an important resource or location may exchange it for a resource or locational access held by a different organism/society. This exchange is to their mutual advantage.

Concentric-spheres model of behavior. In the concentric-spheres model represented in figure 14.1—originally developed to explain human society—the self is at the center of a set of concentric spheres that define the varying strength of cooperative relationships between it and all other individuals and groups in society. The strength of the social relationships between the self and others—which can be measured by the *economic distance* between them—will depend on how essential they are to maximizing the probability of the self's survival and prosperity. As the economic distance—a measure of materialistic rather than genetic relatedness—between the center and each sphere increases, the degree of cooperation (*not* "altruism") between them diminishes. It is because economic forces transcend biological forces that the concentric-spheres model can explain what the kin-selection model cannot—the varying cooperative relationships between any given individual and all others in society. The concentric-spheres model, therefore, involves two sets of forces, one centrifugal and the other centripetal in nature. The centrifugal force is the incessant *strategic desire* of the self to survive and prosper—a desire that leads, on average, to the individual placing itself above all others. It is the driving force in life. The centripetal force—the economic gravity holding society together—is the need to cooperate with other individuals and groups in order to pursue the dominant *dynamic strategy* more effectively. It is through both competition and cooperation that the *materialist organism* maximizes the probability of its survival and prosperity.

Conquest strategy. This has played a major role in the dynamics of both life and human society. In life, conquest has been responsible for the elimination—the extinction—of those species and dynasties that had exhausted their *genetic styles* and/or *genetic paradigms*. By doing so they provided other species and potential dynasties with the opportunity to show what they could do. If the dinosaurs, for example, had not eliminated themselves through their world wars, the mammals would never have had the chance to show what their passion for the intellect could do on life's stage. And in human society, conquest was responsible for taking the agricultural revolution around the world—at first the Old World and then the New World—thereby ultimately exhausting the neolithic technological paradigm (by the mid-eighteenth century in the Old World) and making way for the Industrial Revolution (1780 to 1830). This transforming role could only be identified in retrospect. In prospect, conquest has always been adopted because, once a species' *family-multiplication strategy* had been exhausted, it was the only way that organisms could hope to survive and prosper.

Deduction, the problem of. See *problem of deduction.*

Dynamic materialism is the term coined in *The Dynamic Society* (Snooks 1996: 169–207) to encapsulate the process by which organisms—effectively economic decisionmakers—attempt to achieve their materialist objectives in a competitive environment by adopting the most cost-effective *dynamic strategies* and *dynamic tactics.* The essential components of dynamic materialism are a typically competitive environment in which organisms struggle with each other for scarce natural resources in order to survive and prosper. Those circumstances bring out a heightened ambition, together with an aggressive and creative energy in what I have called the *materialist organism* in life and *materialist man* in human society.

Dynamic strategies. To achieve its objective of survival and prosperity, the *materialist organism* must find a way to gain consistent, long-term access to the resources required for generating metabolic energy and for providing shelter. It must pursue a dynamic strategy that will deliver a reliable return on the energy it expends in its participation in life. The most appropriate dynamic strategy will depend on the physical environment—on the availability of natural resources—and on the degree of competition with other organisms in their own and neighboring species. The dynamic strategies of life are fourfold: *genetic/technological change, family multiplication, commerce* (symbiosis), and *conquest.* These strategies are "dynamic" because they are employed by organisms (including man) to maximize the probability of survival and prosperity over the lifetime of the individual, and because, when successful, they lead to the transmutation of species, to population increase, to geographic expansion of the species or dynasty, and to biological or economic growth (an increase in "output" per unit of input).

Dynamic-strategy theory is a general dynamic theory that transcends the specialist preoccupations of both the biologist and the social scientist. It can explain the dynamics of both life and human society.

Dynamic tactics of order and chaos are employed in the *strategic struggle* between various organisms attempting to gain control of the dominant *dynamic strategy* and to influence the distribution of natural resources. The dominant individuals will employ the tactics of order to maintain compliance, and the dissenters will employ the tactics of chaos to overthrow the existing power structure. It is this struggle that leads to a change in the structure of institutions and organizations in animal (including human) society.

Dynasty of life is that group of species employing a series of related *genetic styles* that in aggregate constitute the prevailing *genetic paradigm*. A dynasty comprises both the participants in and the outcome of a genetic revolution. They generate the *genetic paradigm shift*.

Economic distance is the measure of the strength of material relationships between the self and all other individuals and groups in society (figure 14.1). Economic distance is inversely related to the importance of other individuals and groups in maximizing the material objectives of the self. Hence family and close associates occupy the spheres nearest to the self, and strangers are to be found on the most distant spheres. It is an important component of the *concentric-spheres model of behavior*. This model replaces the flawed neo-Darwinist theory of kin selection.

Eternal recurrence is the ultimate outcome of the exhaustion of the resource-accessing capacity of life's *genetic option*. Subsequent biological activity involves the rise and fall of life's *dynasties* without any improvement in their access to the Earth's resources. This is called in my model the *great wheel of life* (figure 13.4). The only way to break out of this eternal recurrence is through the replacement of the genetic option with the *technology option*.

Existential historicism constitutes a new approach to deriving the laws of life and human history. Rather than looking for laws in the regularity of events, the existentialist seeks laws in the regularity of dynamic processes or mechanisms that underlie the superficial patterns. The existentialist draws a distinction between the ephemeral events and the eternal processes.

Existential quaternary method (or the four steps of induction). This is an attempt to answer the critics who claim that there are no rules of induction, by establishing a procedure by which the laws of life and history can be derived from empirical data. The four steps of induction involve the identification of historical patterns, both quantitative and qualitative; the construction of a general dynamic model; the derivation of specific historical mechanisms; and the use of the general and specific dynamic models to explain institutional change. As it turns out the "problem of induction" is less debilitating than the *problem of deduction*.

Family-multiplication strategy involves the exploitation of unused resources through procreation and migration to new regions. By increasing the size of the extended family and gaining greater control over natural resources, the family head is able to achieve the universal objectives of survival and the maximization of material advantage. This is the force that drove the primitive dynamic which has been called here the *great dispersion*. Family multiplication is a strategy pursued by plants as well as animals and humans.

Genetic change, dynamic strategy of. See *genetic strategy*.

Genetic competence is the effectiveness with which a *dynasty* is able to exploit life's *genetic option* in order to gain access to the Earth's resources.

Genetic option. In the history of life, organisms have been able to improve their access to the Earth's resources through the genetic option—the use of the *dynamic strategy of genetic change*—and, when it was totally exhausted, the *technology option*. With each *genetic paradigm shift* the capacity of the genetic option was progressively exploited until it was finally exhausted by the time the dinosaurs were in their prime (80 myrs ago). The technology option, however, only displaced the defunct genetic option when the intellectual capability of the hominids enabled them to employ the technology strategy (2.4 myrs ago). During the almost 80 myrs between these events, nature was dominated by the *great wheel of life*. If life had relied only on the genetic option it would never have escaped the *eternal recurrence*.

Genetic paradigm. This is the genetic basis on which an entire *dynasty of life* gains access to the Earth's resources. Each paradigm consists of a series of *genetic styles*—more commonly known as species—which employ specialized genetic techniques to survive and prosper.

Genetic paradigm shift. This is a genetic revolution that enables a quantum leap in *potential* access to natural resources owing to the occurrence of a major genetic innovation or cluster of innovations (figures 13.1 and 13.2). It occurs when the former *genetic paradigm* has been totally exhausted and the old *dynasty of life* has collapsed. With this genetic revolution the new dynasty of life is able to achieve greater access to—or productivity in the use of—natural resources and to generate a greater biomass than the old extinct dynasty. The development of life has taken place within a dynamic structure defined by the great genetic and *technological paradigm shifts*. The genetic paradigm shifts have been based on the Prokaryote, Eukaryote, and Endothermic Revolutions.

Genetic strategy. The most controversial feature of my *dynamic-strategy theory* is its treatment of genetic change. Rather than viewing the emergence of species as the outcome of either "divine selection" or "natural selection," I see it as the result of *strategic selection*, or self-selection writ large. Genetic change in other words is a *dynamic strategy* that is deliberately pursued by organisms—similar to the role of technological change in human society—when the circumstances are favorable. Organisms select those mutations that assist their dominant dynamic strategy and they ignore the rest.

They do this by cooperating and mating with those individuals who have an edge in accessing natural resources as demonstrated by their material success. This occurs not just when pursuing genetic change as a dominant dynamic strategy, but also sometimes under the nongenetic strategies in response to the strategic demand that they generate for military weapons (*conquest*) and fertility and mobility aids (*family multiplication*). Accordingly, the concept of strategic selection is far more general in its application than Charles Darwin's concept of natural selection, to which he was forced to add "sexual selection" as a separate and puzzling process. This is why the dynamic-strategy theory, in contrast to Darwinian evolution, is a *general* dynamic theory.

Genetic styles. Within any *genetic paradigm* there is a series of genetic styles that are the outcome of species pursuing the *dynamic strategy of genetic change* in order to gain access to the Earth's resources and hence to survive and prosper.

Grandest pile of ruins I ever saw. A morality play about the fallibility of our modern gods. See prologue.

Great dispersion. This is the mechanism by which species pursuing the dynamic strategy of *family multiplication*—procreation and migration—spread from their region of origin to the rest of the globe.

Great steps of life. The dynamic structure of life is defined by a series of *genetic* and *technological paradigm shifts*, which I have called the great steps of life (figures 13.1 and 13.2). These great steps, which outline the genetic/technological potential for biological/economic development, are driven by the major genetic/technological revolutions. Owing to the nature of the dynamics of life, these great steps have been changing exponentially in two dimensions: increasing in height while reducing in depth. What this implies is that the dynamics of both life and human society have been accelerating over the past 3,500 myrs. There have been three genetic paradigm shifts and three technological paradigm shifts over that vast period of time, and they are involved in a systematic geometric relationship.

Great waves of life (of geometrically declining duration) are generated by the rise and fall of dynasties and encompass the long waves (of about 30 myrs) generated by the rise and fall of groups of species, which in turn contain the shorter waves (of up to 6 myrs or so) generated by the rise and fall of individual species (figure 9.3). At all levels these waves are generated by the expansion, exhaustion, and replacement of *dynamic strategies, genetic/technological styles*, or *genetic/technological paradigms*. This pattern of waves within waves, therefore, constitutes the dynamic form of our model.

Great wheel of civilization. This was the mechanism driving the *technological paradigm shift* between the neolithic and modern eras (Snooks 1996: figure 12.10). It was the dynamic of ancient society. Each rotation of the great wheel brought the Dynamic Society closer to the limit of the old neolithic paradigm through population expansion and the transmission of ideas. It is a process of economic change in which each cycle of ancient society begins again anew. It is the mechanism underlying the rise and fall of ancient civilizations. It is the eternal recurrence of the ancient world driven by the *dynamic strategy of conquest* which, for a time, enables individual societies to exceed the global living standards determined by the ancient *technological paradigm*. The conqueror must rebuild his empire anew on each and every occasion. The only escape is through the *technological strategy* via the great linear waves of economic change.

Great wheel of life. This is the mechanism of the *eternal recurrence* (figure 13.4). It is the process of biological batonpassing whereby a dynasty emerges to exploit a *genetic paradigm*, exhausts their version of it, collapses and is followed in the same way by a new dynasty operating within the same paradigm. The great wheel of life rotates without gaining traction, without being able to generate a new paradigm shift to replace the old exhausted paradigm, without gaining more intensive access to the Earth's natural resources. The great wheel of life comes into operation in a mature life system when the *genetic option* has been exhausted. At this point in time it is no longer possible through genetic change to increase the global level of the biomass of life. The only way to break out of this eternal recurrence is through the replacement of the genetic option with the *technology option*.

Imitative information, in contrast to the information required by rationalist theory, is readily available and easily evaluated. It is gained from direct observation of individuals in their "societies." In human society it is garnered from professional advice, from the media, and from books, magazines, and electronic sources; and in animal society it is obtained from interaction within kinship groups. What we all need to know is: who is successful and why, not how to calculate accurate rate of return calculations. See *strategic imitation*.

Intelligence Revolution. This revolution was the outcome of mammals, in contrast to dinosaurs, following the strategic pursuit with finesse rather than force. It enabled one line of mammals—*Homo sapiens*—to escape the *eternal recurrence*: the *great wheel of life*.

Materialist man, a subset of the *materialist organism*, is a central concept in long-run dynamics. Materialist man is related, yet very different, to the neoclassical concept of *homo economicus*. Rational economic man is not a dynamic force in society, but rather an abstract collection of preferences and rational choices concerning consumption and production—a set of optimizing conditions. Economic theorists have divorced these behavioral outcomes from more fundamental human motivational impulses. Materialist man on the other hand is a real-world decisionmaker who attempts to survive and, with survival, to maximize material advantage over his lifetime. This does not require perfect knowledge or sophisticated abilities to rapidly calculate the costs and benefits of a variety of possible decisionmaking alternatives, just an ability to recognize and imitate success.

Materialist organism. A systematic examination of the history of life undertaken in this book and elsewhere (Snooks 1996; 1997; 1998a) suggests that organisms attempt at all cost to survive *and*, having survived, to prosper—to maximize their consumption subject to the prevailing physical, social, and genetic/technological constraints. This is the dynamic concept of the materialist organism. It is the all-important driving force in the dynamic system, striving at all times, irrespective of the degree of competition, to increase its access to natural resources. It is this most basic force in life—which I have called *strategic desire*—that accounts for the dynamic-strategy theory's self-starting and self-maintaining nature. More intense competition merely raises the stakes of the strategic pursuit. Also see *materialist man*.

Metabolic demand is the demand generated by an organism's metabolic system for organic resources (fuel) in order to sustain life. It is the source of *strategic desire*, which is the driving force in life.

Nongenetic strategies. The reason that the genetic profile of a species consists of a series of long steps is that, once it has emerged, the component organisms turn from the pursuit of genetic to nongenetic strategies. While speciation is an outcome of the dynamic strategy of *genetic change*, the success of the species depends largely on the pursuit of nongenetic strategies by individuals. The nongenetic strategies include *family multiplication, commerce,* and *conquest*.

Primary laws of life are the unchanging laws that govern the behavior of organisms as they pursue their objectives of survival and prosperity by investing time, energy, and resources in the most effective of the four main *dynamic strategies*. They also govern the way species/societies/dynasties respond to the dynamic strategies. They are the laws derived from the general dynamic-strategy model.

Problem of deduction. We are told repeatedly of the "problem of induction"—the lack of mechanical rules for drawing inferences from empirical data—but the more serious "problem of deduction" is ignored. Only by the systematic examination of empirical data is it possible to derive realistic premises. It is not sufficient to argue that those logical models based on unrealistic assumptions will be weeded out through a process of empirical falsification—a process Karl Popper calls "error elimination"—because in reality very few theories are rejected in this way. Old theories just fade into the background when new unverified theories come along, often to be revived at a later date. Also, it is highly likely that, without the assistance of systematic observation, the deductivist will completely overlook some of the main explanatory variables.

Quaternary inductive method. See *existential quaternary method.*

Secondary laws of life are derived from the dynamic mechanism underpinning the historical eras both before and since the exhaustion of the *genetic option*. As these historical mechanisms were reconstructed by applying the general model to the quantitative *timescapes*, the secondary laws can be thought of as being derived from the unchanging *primary laws*.

Selective sexual reproduction is based on those perceived characteristics in other individuals that will improve the prospect of survival and prosperity of the selecting individual. If this perception turns out to be correct, those physical and instinctual characteristics will also be passed on to the individual's offspring. Mate choice is an important technique for implementing all four *dynamic strategies*, not just of *genetic change*. Selective sexual reproduction provides individuals with greater control over their dynamic strategies.

Strategic cerebrum. Organisms in pursuit of survival and prosperity control their dynamic strategies through what I call *strategic instruments*. These strategic instruments include brains in organisms that possess them and special genes in those that do not. These instruments are the strategic cerebrum and the *strategic gene*. A watershed occurred in life when some organisms were able to replace the strategic gene with the strategic cerebrum.

Strategic confidence, which is the outcome of a successful *dynamic strategy*, is the force that keeps animal (including human) society together. A successful dynamic strategy leads to an effective network of competitive/cooperative relationships, together with all the necessary "rules" and "organizations." In societal transactions, individuals relate directly to the successful strategy and only indirectly to each other. It is not a matter of mutual "trust" as such—of having confidence in the nature of other individuals—but rather having confidence in the wider dynamic strategy in which they are all involved and on which they all depend. What we know as "trust" is derived from strategic confidence. Once the dynamic strategy has been exhausted and cannot be replaced, strategic confidence declines and, in extreme cases, disappears. And as strategic confidence declines, so too does "trust" and cooperation. Strategic confidence is communicated directly to individuals in both animal and human society by the rise and fall in material standards of living.

Strategic demand is the central concept in the dynamic-strategy model of both life and human society. It is an outcome of the unfolding dynamic strategy and exerts a long-run influence over the employment of resources, the institutional and organizational structure of animal and human society, and the genetic and technological structure of organisms and human societies. Shifts in strategic demand occur as the dominant dynamic strategy unfolds and as one dynamic strategy replaces another. These shifts elicit changes in the way organisms employ resources and interact with each other.

Strategic desire. The attribute that first separated living from nonliving cells some 4,000 myrs ago was not the ability to reproduce systematically but the intense need to obtain organic material to fuel their internal metabolic systems. This need to meet *metabolic demand* is called strategic desire. It is the driving force in life. Replication (or *family multiplication*), which is merely one of the *dynamic strategies* employed by organisms to survive and prosper, is secondary to strategic desire. Without strategic desire, life would not exist.

Strategic gene. Organisms in pursuit of survival and prosperity control their *dynamic strategies* through what I call *strategic instruments*. These strategic instruments include brains in organisms that possess them and special genes in those that do not. These instruments are the *strategic cerebrum* and the strategic gene. Strategic genes are used by organisms to respond to the changing physical and social environments, which determine the availability of nutrients and presence of competitors, and to activate the most appropriate dynamic strategy. When nutrients are abundant this will be the *family-multiplication strategy*; when competition is intense it will be the *conquest strategy*. In lower life forms, therefore, these strategic genes play the same role as central nervous systems in higher life forms—they select and supervise the most appropriate dynamic strategy to fulfill the *strategic desires* of the *materialist organism*. As such they do not drive life, they merely facilitate it. It is essential to realize that these strategic genes emerge in response to the driving ambition of their organisms. They do not have an independent existence or any driving ambitions of their own as claimed by the neo-Darwinists. It is significant that, in some forms of life, organisms replaced the strategic gene with the strategic cerebrum.

Strategic imitation. It is in the process of strategic imitation, by which the vast majority of organisms emulate the action of successful strategic pioneers, that societal "rules" are created and employed. Institutions are needed to economize not on benefit–cost information, as the economic institutionalists or rationalists argue, but on intelligence. This is true not only in animal society where intelligence is particularly scarce, but also in human society where most individuals find intellectual activity difficult and unhelpful. Intelligence is the scarcest resource on Earth, and probably nonexistent in the rest of the universe. The rulemakers, therefore, are the strategic followers, who demand guidance in their strategic pursuit; while the rulebreakers are the strategic pioneers, who attempt to break out of the restrictions of institutional conventions as they follow their new visions. As the "followers" constitute the vast majority of organisms in life, "rules" are essential to the dynamics of both animal and human society even though they are purely a response to it. See Snooks 1996; 1997.

Strategic instruments are the agents employed by organisms to supervise their strategic pursuit. They include the *strategic gene* in primitive life forms and the *strategic cerebrum* in more advanced life forms.

Strategic leader. In animal societies males battle with each other to gain control over the sources of their *dynamic strategy*, which are territories that provide access to food and shelter. They battle, in other words, to become the leading *strategist* in their group. The conflict between them is part of the struggle for "political" control of their "society." Having maintained or achieved this strategic control, which ensures his survival, the strategic leader is in a good position to maximize his prosperity, which involves the consumption of food, sex and, for the time being, leisure. While procreation assumes greater significance when it is part of the dynamic strategy of *family multiplication*, even then it is a means to a more important end.

Strategic pursuit. All life is dedicated to the strategic pursuit, in which the pioneering *strategists* explore the economic potential of the most effective *dynamic strategy* and its substrategies. It is important to focus on the strategic pursuit rather than the means by which this driving force is translated into a material surplus. In the life sciences both the neo-Darwinists and the neo-liberal economists have focused on static physical structures rather than the dynamics of the strategic pursuit. In doing so they have failed to develop general dynamic theories.

Strategic selection is a central concept of this work. It is a dynamic process in which organisms are themselves responsible for selecting or rejecting benign (nonlethal) mutations. It operates through the *strategic imitation* process. If an individual experiences a beneficial mutation that enables better access to natural resources—within the context of the *dynamic strategy* it is pursuing—that individual will increase its prospects of survival and prosperity. This success will attract the attention of others. Those with similar abilities will cooperate with each other to improve their joint prospects. The point of strategic selection is that individual organisms—rather than gods, genes, or blind chance—are responsible for selecting comrades, mates, and siblings that have the necessary physical and instinctual characteristics to successfully pursue the prevailing dynamic strategy. It is important to realize that strategic selection operates under varying degrees of competition, not just under intense Darwinian competition, and that it responds to each of the four dynamic strategies, not just the *genetic strategy*. Also it is associated with the welfare of the self and not that of future generations. Strategic selection is a form of self-selection at the "societal" level which replaces the "divine selection" of the creationists and the "natural selection" of the Darwinists.

Strategic sequence. Typically organisms in a species will pursue a sequence of strategies from the time they begin to diverge from the parent species until they finally go extinct. The sequence prior to the emergence of human society 2 myrs ago was typically *genetic change, family multiplication* or *commerce*, and *conquest*. Each *dynamic strategy* is exploited until it is exhausted, which leads to a temporary crisis until a new strategy can be employed in their *strategic pursuit*. If, in a normally competitive environment, a new strategy is not adopted by a species following the exhaustion of an old strategy, that species will collapse and go extinct prematurely. Accordingly this strategic sequence leads not to a linear development path but to a series of waves consisting of phases of expansion, stagnation, crisis, decline, and renewed expansion.

Strategic struggle is the main "political" instrument by which established individuals/species (old *strategists*) attempt to maintain their control over the sources of their prosperity, and by which emerging individuals/species (new strategists) attempt to usurp such control. Although it employs "political" instruments it is fundamentally an "economic" struggle—a struggle for survival and prosperity in the face of scarce resources. In the process these individuals/species employ the *dynamic tactics of order and chaos*. See *strategic leader*.

Strategic substitution. The *Intelligence Revolution* differed from the three earlier genetic revolutions (or *genetic paradigm shifts*) because it enabled the substitution of the *technology option* for the long-exhausted *genetic option*. It was because of this strategic substitution that the fourth and final genetic revolution generated a *technological* rather than a *genetic paradigm shift*. While the enabling condition for this strategic substitution was the achievement of a threshold level of brain size (in the range 700 cc to 1,000 cc in man), the driving force was provided by the *strategic desire* of one small branch of the mammal dynasty—the hominids—to acquire more precise and precipitate control over the means of intensifying their access to natural resources.

Strategists comprise the dynamic group in animal and human society that invests time and resources in pursuing and profiting from one of the four *dynamic strategies*. The strategists are a diverse group. We must distinguish between the strategic pioneers (the more ambitious and less riskaverse) and the strategic followers; between the old strategists (supporters of the traditional strategy) and the new strategists (supporters of the emerging strategy); and between the surplus-creating strategists and the surplus-consuming strategists. While there is synergy between the pioneers and the followers and the surpluscreators and surplusconsumers, the old and new strategists are generally involved in a struggle against each other for control of society's dominant dynamic strategy. See *strategic struggle*.

Stratologists. The new vision of animal and human societies as organizations dedicated to the pursuit of strategic objectives is the focus of interest in the study of stratology. The stratologist is contrasted with the neo-Darwinist in biology and the neoliberal in economics, with their more restricted vision of animal and human society—a vision that fails the test of dynamic reality.

Technological change, dynamic strategy of. See *technological strategy*.

Technological paradigm shifts. The progress of human society takes place within a dynamic structure defined by the great technological paradigm shifts in which growing resource scarcity is transcended by mankind breaking through into an entirely new technological era, thereby opening up extended possibilities for further economic growth. This involves the introduction of an entirely new set of techniques, skills, institutions, and outcomes. There have been three great technological paradigm shifts in human history: the paleolithic paradigm shift when hunting displaced scavenging; the neolithic paradigm shift when agriculture displaced hunting; and the industrial paradigm shift when urban centers displaced rural areas as the major source of growth (figure 13.2 and 13.3).

Technological strategy. This strategy is the dominant dynamic of modern society and, in the past, has been employed by economic decisionmakers to transcend exhausted technological paradigms. It was the outcome of the Intelligence Revolution and enabled life to escape the *eternal recurrence*. The technological strategy was at the very center of the Paleolithic (hunting), Neolithic (agricultural), and Industrial (modern) Revolutions or *technological paradigm shifts*. And it will be the dominant dynamic strategy of the future. Unlike the *conquest* and *commerce strategies* it leads to an increase in material living standards not only for its host civilization but for human society as a whole.

Technological styles. Within the modern technological paradigm, the dynamic *strategists* of competing nation-states attempt to secure a comparative advantage in their pursuit of extraordinary profits by developing new technological substrategies or technological styles. These technological styles—which historically have included steam-powered iron machinery using coke (1780s–1830s); steel, synthetic chemicals, and complex machinery (1840s–1890s); electricity and the internal combustion engine (1900s–1950s); automated processes, microelectronics, lasers, new construction materials, and biotechnology (since the 1950s)—emerge within the existing industrial paradigm as it unfolds at the global level. In dynamic-strategy theory technological styles are analogous to *genetic styles*.

Technology option. The emergence of intelligence enabled the most intellectually advanced line of the dynasty of mammals to replace the *genetic option* with the technology option, which was to spawn a series of *technological paradigm shifts* and, within each of these, a series of *technological styles*. This released life from the *eternal recurrence*.

Tertiary laws of life. These are the laws governing institutional change in animal and human society. They are constructed from the dynamic-strategy model of institutional change and can be thought of as being derived from both the *primary* and *secondary laws*.

Timescapes are those portraits of reality provided by a visual representation of long-run quantitative and qualitative data. These portraits emerge from the statistical record of the course taken by life and human society over vast expanses of time. They show us the nature of real-world relationships, such as the great waves of life and of economic change. They provide a glimpse of dynamic processes operating in life and in society, and constitute the building blocks of existential models. As fact is stranger than fiction they provide a breadth of vision required to build realist dynamic models that is missing in the deductive approach. See *existential historicism* and *existential quaternary method*.

References

Abbate, E., et al. 1998. "A One-Million-Year-Old *Homo* Cranium from the Danakil (Afar) Depression of Eritrea." *Nature* 393, no. 6684 (4 June): 458–60.

Alexander, R. D., and G. Borgia. 1978. "Group Selection, Altruism, and the Levels of Organization of Life." *Annual Review of Ecology and Systematics* 9: 449–74.

Alvarez, L. W., W. Alvarez, F. Asaro, and H. V. Michel. 1980. "Extraterrestrial Cause for the Cretaceous–Tertiary Extinction." *Science* 208, no. 4448 (6 June): 1095–1108.

Asfaw, B., T. White, O. Lovejoy, B. Latimer, S. Simpson, and G. Suwa. 1999. "*Australopithecus garhi*: A New Species of Early Hominid from Ethiopia." *Science* 284, no. 5414 (23 April): 629–35.

Bahn, P. G. 1993. "50,000-Year-Old Americans of Pedra Furada." *Nature* 362, no. 6416 (11 March): 114–15.

Bakker, R. T. 1978. "Dinosaur Renaissance." Pp. 125–41 in *Evolution and the Fossil Record: Readings from Scientific American* (with introductions by Léo F. Laporte). San Francisco: W. H. Freeman and Company.

———. 1986. *The Dinosaur Heresies: New Theories Unlocking the Mystery of the Dinosaurs and Their Extinction.* New York: Morrow.

Balter, M., and A. Gibbons. 2000. "A Glimpse of Human's First Journey Out of Africa." *Science* 288, no. 5468 (12 May): 948–50.

Benton, M. J. 1985. "Mass Extinction among Non-marine Tetrapods." *Nature* 316, no. 6031 (29 August): 811–14.

Black, R. M. 1988. *The Elements of Palaeontology.* Cambridge: Cambridge University Press.

Byme, R. 1995. *The Thinking Ape.* Oxford: Oxford University Press.

Cairns, J., J. Overbaugh, and S. Miller 1988. "The Origin of Mutants." *Nature* 335, no. 6186 (8 September): 142–45.

Calder, N. 1984. *Timescale: An Atlas of the Fourth Dimension.* London: Chatto & Windus, The Hogarth Press.

Campbell, I. H., G. K. Czamanske, V. A. Fedorenko, R. I. Hill, and V. Stepanov. 1992. "Synchronism of the Siberian Traps and the Permian–Triassic Boundary." *Science* 258, no. 5089 (11 December): 1760–62.

Carneiro, R. L. 1974. "Editor's Introduction." Pp. ix–lvii in *The Evolution of Society: Selections from Herbert Spencer's Principles of Sociology.* Edited and with an introduction by Robert L. Carneiro. Chicago: University of Chicago Press.

Carrier, D. R. 1987. "The Evolution of Locomotor Stamina in Tetrapods: Circumventing a Mechanical Restraint." *Paleobiology* 13: 326–41.

Chitty, D. 1971. "The Natural Selection of Self-Regulatory Behavior in Animal Populations." Pp. 136–70 in *Natural Regulation of Animal Populations,* ed. I. A. McLaren. New York: Atherton Press.

Commons, J. R. 1934. *Institutional Economics: Its Place in Political Economy.* New York: Macmillan.

Cowen, R. 1990. *History of Life.* Boston: Blackwell Scientific Publications.

———. 2000. *History of Life.* 3rd ed. Malden, MA: Blackwell Science.

Crawford, M., and D. Marsh. 1989. *The Driving Force: Food in Evolution and the Future.* London: Mandarin.

Cullotta, E. 1999. "A New Human Ancestor." *Science* 284, no. 5414 (23 April): 572–73.

Darwin, C. 1839. *Narrative of the Surveying Voyages of His Majesty's Ships Adventure and Beagle, Between the Years 1826 and 1836, Describing Their Examination of the Southern Shores of South America, and the Beagle's Circumnavigation of the Globe, Volume III: Journal and Remarks, 1832–1836.* London: Henry Colburn.

———. 1859. *On the Origin of Species by Means of Natural Selection; or, the Preservation of Favoured Races in the Struggle for Life.* 1st ed. London: John Murray.

———. 1868. *The Variation of Animals and Plants Under Domestication.* London: John Murray.

———. 1869. *On the Origin of Species by Means of Natural Selection; or, the Preservation of Favoured Races in the Struggle for Life.* 5th ed. London: John Murray.

———. 1871. *The Descent of Man and Selection in Relation to Sex.* London: John Murray.

———. 1872. *On the Origin of Species by Means of Natural Selection; or, the Preservation of Favoured Races in the Struggle for Life.* 6th ed. London: John Murray.

———. 1948. *The Origin of Species by Means of Natural Selection, or, The Preservation of Favored Races in the Struggle for Life;* and *The Descent of Man and Selection in Relation to Sex.* New York: Random House.

———. 1979. *The Origin of Species by Means of Natural Selection or the Preservation of Favoured Races in the Struggle for Life.* New York: Avenel Books.

———. 1988. *Charles Darwin's* Beagle *Diary.* Edited by R. D. Keynes. Cambridge: Cambridge University Press.

Davis, J. M. 1973. "Imitation: A Review and Critique." Pp. 43–72 in *Perspectives in Ethology,* ed. P. P. G. Bateson and P. H. Klopfer. New York: Plenum.

Dawkins, R. 1976. *The Selfish Gene.* Oxford: Oxford University Press.

———. 1982. *The Extended Phenotype: The Gene as the Unit of Selection.* Oxford: Freeman.

———. 1989. *The Selfish Gene.* New ed. Oxford: Oxford University Press.

———. 1998. *Unweaving the Rainbow: Science, Delusion and the Appetite for Wonder.* Harmondsworth, England: Penguin Books.

Dennett, D. C. 1995. *Darwin's Dangerous Idea: Evolution and the Meanings of Life.* New York: Simon & Schuster.

Desmond, A. J., and J. Moore. 1991. *Darwin.* London: Michael Joseph.

Dobzhansky, T. G. 1937. *Genetics and the Origin of Species.* New York: Columbia University Press.

———. 1970. *Genetics of the Evolutionary Process.* New York: Columbia University Press.

Dugatkin, L. A. 1992. "Sexual Selection and Imitation: Females Copy the Mate Choice of Others." *American Naturalist* 139: 1384–89.

———. 1997. *Cooperation among Animals: An Evolutionary Perspective.* New York: Oxford University Press.

Duve, C. de. 1996. "The Birth of Complex Cells." *Scientific American* 274, no. 4 (April): 38–45.

Easteal, S., C. Collet, and D. Betty. 1995. *The Mammalian Molecular Clock.* Austin and New York: R. G. Landes/Springer-Verlag.

Eccles, J. C. 1989. *The Evolution of the Brain: Creation of the Conscious Self.* London: Routledge.

Eldredge, N. 1985. *Time Frames: The Rethinking of Darwinian Evolution and the Theory of Punctuated Equilibria.* New York: Simon & Schuster.

———. 1989. *Macroevolutionary Dynamics: Species, Niches and Adaptive Peaks.* New York: McGraw-Hill.

———. 1995. *Reinventing Darwin: The Great Debate at the High Table of Evolutionary Theory.* New York: Wiley.

Eldredge, N., and S. J. Gould. 1972. "Punctuated Equilibria: An Alternative to Phyletic Gradualism." Pp. 82–115 in *Models of Paleobiology,* ed. T. J. M. Schopt. San Francisco: Freeman, Cooper.

Falconer, H., assisted by H. Walker. 1859. *Descriptive Catalogue of the Fossil Remains of Vertebrata from the Sewalik Hills, the Nerbudda, Perim Island, &C. in the Museum of the Asiatic Society of Bengal.* Calcutta: printed by C. B. Lewis.

Feynman, R. P. 1967. *The Character of Physical Law.* Cambridge, MA: MIT Press.

Fisher, P. E., D. A. Russel, M. K. Stoskopf, R. E. Barrick, M. Hammer, and A. A. Kuzmitz. 2000. "Cariovascular Evidence for an Intermediate or Higher Metabolic Rate in an Ornithischian Dinosaur." *Science* 288, no. 5465 (21 April): 503–5.

Fisher, R. A. 1930. *The Genetical Theory of Natural Selection.* Oxford: Clarendon Press; New York: Oxford University Press.

Fortey, R. 2000. "Olenid Trilobites: The Oldest Known Chemoautotrophic Symbionts?" *Proceedings of the National Academy of Sciences of the United States of America* 97, no. 12 (6 June): 6574–78.

Foster, P. L. 1998. "Adaptive Mutation: Has the Unicorn Landed?" *Genetics* 148 (April): 1453–59.

———. 1999. "Mechanisms of Stationary Phase Mutation: A Decade of Adaptive Mutation." *Annual Review of Genetics* 33: 57–88.

Frith, H. J., and J. H. Calaby. 1969. *Kangaroos.* Melbourne: F. W. Cheshire.

Gibbons, A. 1998. "Solving the Brain's Energy Crisis." *Science* 280, no. 5368 (29 May): 1345–47.

Gould, S. J. 1980. "Is a New and General Theory of Evolution Emerging?" *Paleobiology* 6, no. 1 (Winter): 119–30.

———. 1982. "Darwinism and the Expansion of Evolutionary Theory." *Science* 216, no. 4544 (23 April): 380–87.

Gould, S. J., and R. C. Lewontin. 1979. "The Spandrels of San Marco and the Panglossian Paradigm: A Critique of the Adaptationist Programme." *Proceedings of the Royal Society of London* B205: 281–88.

Grinnell, J., and K. McComb. 1996. "Maternal Groupings as a Defense against Infanticide by Males: Evidence from Field Playback Experiments on African Lions." *Behavioral Ecology* 7: 55–59.

Grinnell, J., C. Packer and A. E. Pusey. 1995. "Cooperation in Male Lions: Kinship, Reciprocity or Mutualism?" *Animal Behaviour* 49: 95–105.

Groves, C. P. 1989. *A Theory of Human and Primate Evolution.* Oxford: Clarendon Press; New York: Oxford University Press.

Gruber, H. E., and Paul H. Barrett. 1974. *Darwin on Man: A Psychological Study of Scientific Creativity by Howard E. Gruber Together with Darwin's Early and Unpublished Notebooks Transcribed and Annotated by Paul H. Barrett.* Foreword by Jean Piaget. New York: E. P. Dutton.

Haldane, J. B. S. 1932. *The Causes of Evolution.* London: Longmans, Green; New York: Harper & Brothers.

Hall, B. G. 1988. "Adaptive Evolution That Requires Multiple Spontaneous Mutations. I. Mutations Involving an Insertion Sequence." *Genetics* 120 (December): 887–97.

Hamilton, W. D. 1964. "The Genetical Evolution of Social Behaviour." *Journal of Theoretical Biology*, no. 7: 1–52.

Harcourt, A. H., and F. B. M. de Waal. 1992. *Coalitions and Alliances in Humans and Other Animals*. Oxford: Oxford University Press.

Hauser, M. D. 2000. *Wild Minds: What Animals Really Think*. London: Allen Lane, The Penguin Press.

Hawkesworth, D. L. 1988. "Coevolution of Fungi with Algae and Cyanobacteria in Lichen Symbioses." Pp. 125–48 in *Coevolution of Fungi and Plants and Animals*, ed. K. A. Pirozynski and D. L. Hawkesworth. London: Academic Press.

Hayek, F. A. von. 1988. *The Fatal Conceit: The Errors of Socialism*. Edited by William Warren Bartley. London: Routledge; New York: Routledge, Chapman & Hall.

Hempel, C. G. 1966. *Philosophy of Natural Science*. Englewood Cliffs, NJ: Prentice-Hall.

Hickman, Jr, C. P., L. S. Roberts, and A. Larson. 1993. *Integrated Principles of Zoology*. St. Louis: Mosby Year Book.

Hodek, I. 1983. "Role of Environmental Factors and Exogenous Mechanisms in the Seasonality of Reproduction in Insects Diapausing as Adults." Pp. 9–33 in *Diapause and Life Cycle Strategies in Insects*, ed. V. K. Brown and I. Hodek. The Hague: Dr W. Junk Publishers.

Howe, M. J. A. 2001. *Genius Explained*. Cambridge: Cambridge University Press, "Canto" series.

Hull, D. L. 1973. *Darwin and His Critics: The Reception of Darwin's Theory of Evolution by the Scientific Community*. Cambridge, MA: Harvard University Press.

Johns, H. M. 1983. *Collins Guide to the Ferns, Mosses and Lichens of Britain and Northern and Central Europe*. London: Collins Books.

Jones, S. 1999. *Almost Like a Whale: The Origin of Species Updated*. London: Doubleday.

Kershaw, A. P. 1986. "Climatic Change and Aboriginal Burning in North-East Australia During the Last Two Glacial/Interglacial Cycles." *Nature* 322, no. 6074 (3 July): 47–49.

Keynes, J. M. 1936. *The General Theory of Employment, Interest, and Money*. London: Macmillan.

Knoll, A. H. 1991. "End of the Proterozoic Eon." *Scientific American* 265, no. 4 (October): 42–49.

Krebs, C. J. 1994. *Ecology: The Experimental Analysis of Distribution and Abundance*. New York: HarperCollins College Publishers.

Kurland, J. A. 1977. *Kin Selection in the Japanese Monkey*. Basel: S. Karger.

Laubichler, M. D. 1999. "Frankenstein in the Land of *Dichter* and *Denker*." *Science* 286, no. 5446 (3 December): 1859–60.

Leakey, R. E., and R. Lewin. 1992. *Origins Reconsidered: In Search of What Makes Us Human*. London: Little Brown.

Lee, A. K., and A. Cockburn. 1985. *Evolutionary Ecology of Marsupials*. Cambridge: Cambridge University Press.

Lumsden, C. J., and E. O. Wilson. 1981. *Genes, Mind, and Culture: The Coevolutionary Process*. Cambridge, MA: Harvard University Press.

Lyell, C. 1830–1833. *Principles of Geology: Being an Attempt to Explain the Former Changes of the Earth's Surface, by Reference to Causes Now in Operation*. London: J. Murray.

McCollum, M. A. 1999. "The Robust Australopithecine Face: A Morphogenetic Perspective." *Science* 284, no. 5412 (9 April): 301–5.

Malthus, T. R. 1798. *An Essay on the Principle of Population, As It Affects the Future Improvement of Society. With Remarks on the Speculations of Mr. Godwin, M. Condorcet, and Other Writers.* London: J. Johnson.

Maynard Smith, J. 1958. *The Theory of Evolution.* Harmondsworth, England: Penguin Books.

————. 1972. "Game Theory and the Evolution of Fighting." Pp. 8–28 in *On Evolution,* ed. J. Maynard Smith. Edinburgh: Edinburgh University Press.

————. 1974. *Models in Ecology.* Cambridge: Cambridge University Press.

————. 1976. "Evolution and the Theory of Games." *American Scientist* 64: 41–45.

————. 1993. *The Theory of Evolution.* Cambridge: Cambridge University Press.

Maynard Smith, J., and E. Szathmary. 1995. *The Major Transitions in Evolution.* Oxford: W. H. Freeman Spektrum.

Mayr, E. 1942. *Systematics and the Origin of Species from the Viewpoint of a Zoologist.* New York: Columbia University Press.

————. 1963. *Animal Species and Evolution.* Boston: Belknap Press/Harvard University Press.

————. 1999. "Structure of Theories of Biology." *Science* 285, no. 5435 (17 September): 1856–57.

Mendel, G. 1866. *Versuche über Pflanzenhybriden.* Brunn: Verlage des Vereines.

Mill, J. S. 1843. *A System of Logic, Ratiocinative and Inductive: Being a Connected View of the Principles of Evidence and the Methods of Scientific Investigation.* London: J. W. Parker.

Nature. 1990–2001.

Newell, N. D. 1978. "Crises in the History of Life." Pp. 179–92 in *Evolution and the Fossil Record: Readings from* Scientific American (with introductions by Léo F. Laporte). San Francisco: W. H. Freeman and Company.

North, D. C. 1990. *Institutions, Institutional Change, and Economic Performance.* Cambridge: Cambridge University Press.

Nutman, A. P., V. C. Bennett, C. R. L. Friend, and V. R. Mcgregor. 2000. "The Early Archaean Itsaq Gneiss Complex of Southern West Greenland: The Importance of Field Observations in Interpreting Age and Isotopic Constraints for Early Terrestrial Evolution." *Geochimica et Cosmochimica Acta* 64, no. 17 (September): 3035–60.

Packer, C., and L. Ruttan. 1988. "The Evolution of Cooperative Hunting." *American Naturalist* 132: 159–98.

Paterson, H. E. H. 1985. "The Recognition Concept of Species." Pp. 21–29 in *Species and Speciation,* ed. E. S. Vrba. Pretoria: Transvaal Museum.

Pener, M. P., and L. Orsham 1983. "The Reversibility and Flexibility of the Reproductive Diapause in Males of a 'Short Day' Grasshopper, *Oedipoda minata.*" Pp. 67–85 in *Diapause and Life Cycle Strategies in Insects,* ed. V. K. Brown and I. Hodek. The Hague: Dr W. Junk Publishers.

Pennisi, E. 1999. "Did Cooked Tubers Spur the Evolution of Big Brains?" *Science* 283, no. 5410 (26 March): 2004–5.

Popper, K. R. 1957. *The Poverty of Historicism.* London: Routledge & Kegan Paul.

————. 1965. *Conjectures and Refutations: The Growth of Scientific Knowledge.* New York: Basic Books.

————. 1971. *The Open Society and Its Enemies* 1: *The Spell of Plato*; 2: *The High Tide of Prophecy: Hegel, Marx and the Aftermath.* 5th, revised ed. Princeton, NJ: Princeton University Press.

Rampino, M. R., and R. B. Stothers. 1988. "Flood Basalt Volcanism during the Past 250 Million Years." *Science* 241, no. 4866 (5 August): 663–68.

Raup, D. M., and J. J. Sepkoski. 1984. "Periodicity of Extinctions in the Geologic Past." *Proceedings of the National Academy of Sciences of the United States of America* 81: 801–5.

———. 1986. "Periodic Extinction of Families and Genera." *Science* 231, no. 4740 (19 March): 833–36.

Raven, P. H., and G. B. Johnson. 1992. *Biology*. St. Louis: Mosby Year Book.

Redecker, D., R. Krodner, and L. E. Graham. 2000. "Glomalean Fungi from the Ordovician." *Science* 289, no. 5486 (15 September): 1920–21.

Reed, K. E. 1997. "Early Hominid Evolution and Ecological Change through the African Plio-Pleistocene." *Journal of Human Evolution* 32, no. 2/3 (March): 289–322.

Ricqles, A. de. 1974. "Evolution of Endothermy: Histological Evidence." *Evolutionary Theory* 1: 51–80.

Rothenbuhler, W. C. 1964. "Behavior Genetics of Nest Cleaning in Honey Bees. IV. Responses to F_1 and Backcross Generations to Disease-Killed Brood." *American Zoolologist* 4, no. 2 (May): 111–23.

Rowlett, Ralph M. 1999. "Fire Use." *Science* 284, no. 5415 (30 April): 741.

Ruben, J. A., C. Dal Sasso, N. R. Geist, W. J. Hillenius, T. D. Jones, and M. Signore. 1999. "Pulmonary Function and Metabolic Physiology of Theropod Dinosaurs." *Science* 283, no. 5401 (22 January): 514–16.

Russell, D. A. 1969. "A New Specimen of *Stenonychosaurus* from the Oldman Formation (Cretaceous) of Alberta." *Canadian Journal of Earth Sciences* 6, no. 4, part 1 (August): 595–612.

Russon, A. E. 1997. "Exploiting the Expertise of Others." Pp. ß174–206 in *Machiavellian Intelligence 2*, ed. A. Whiten and R. W. Byrne. Cambridge: Cambridge University Press.

Schumpeter, J. A. 1912. *Theorie der wirtschaftlichen Entwicklung*. Leipzig: Duncker & Humblot.

Science. 1990–2001.

Science. 1999. "Earlier Emergence of Eukaryotes." *Science* 285, no. 5430 (13 August): 981, 983.

Sepkoski Jr, J. J. 1978. "A Kinetic Model of Phanerozoic Taxonomic Diversity, I: Analysis of Marine Orders." *Paleobiology* 4, no. 3: 223–51.

———. 1979. "A Kinetic Model of Phanerozoic Taxonomic Diversity, II: Early Phanerozoic Families and Multiple Equilibria." *Paleobiology* 5, no. 3: 222–51.

———. 1984. "A Kinetic Model of Phanerozoic Taxonomic Diversity, III: Post-Paleozoic Families and Mass Extinctions." *Paleobiology* 10, no. 2: 246–67.

Service, R. F. 1999. "Exploring the Systems of Life." *Science* 284, no. 5411 (2 April): 80–83.

Sinclair, A. R. E. 1989. "Population Regulation in Animals." Pp. 197–241 in *Ecological Concepts*, ed. J. M. Cherrett. Oxford: Blackwell.

Singer, P. 2000. *A Darwinian Left: Politics, Evolution, and Cooperation*. New Haven: Yale University Press.

Smith, A. 1776. *An Inquiry into the Nature and Causes of the Wealth of Nations*. London: printed for W. Strahan and T. Cadell.

Snooks, G. D. 1993. *Economics without Time: A Science Blind to the Forces of Historical Change*. London: Macmillan; Ann Arbor: University of Michigan Press.

———. 1994. *Portrait of the Family within the Total Economy: A Study in Longrun Dynamics: Australia, 1788–1990*. Cambridge: Cambridge University Press.

———. 1996. *The Dynamic Society. Exploring the Sources of Global Change*. London: Routledge.

———. 1997. *The Ephemeral Civilization: Exploding the Myth of Social Evolution*. London: Routledge.

———. 1998a. *The Laws of History.* London: Routledge.

———. 1998b. *Longrun Dynamics. A General Economic and Political Theory.* London: Macmillan; New York: St. Martin's Press.

———. 1999. *Global Transition. A General Theory of Economic Development.* London: Macmillan; New York: St. Martin's Press.

———. 2000. *The Global Crisis Makers. An End to Progress and Liberty?* London: Macmillan; New York: St. Martin's Press.

Spencer, H. 1851. *Social Statics: Or, the Conditions Essential to Human Happiness Specified, and the First of Them Developed.* London: John Chapman.

———. 1860–62. *First Principles.* New York: D. Appleton and Co.

Steele, E. J., R. A. Lindley, and R. V. Blanden. 1998. *Lamarck's Signature: How Retrogenes Are Changing Darwin's Natural Selection Paradigm.* St. Leonards, NSW: Allen & Unwin.

Stenseth, N. C. 1999. "The Evolutionary Synthesis." *Science* 286, no. 5444 (19 November): 1490.

Tauber, M. J., C. A. Tauber, J. R. Nechols, and J. J. Obrycki. 1983. "Season Activity of Parasitoids: Control by External, Internal and Genetic Factors." Pp. 87–108 in *Diapause and Life Cycle Strategies in Insects*, ed. V. K. Brown and I. Hodek. The Hague: Dr W. Junk Publishers.

Tomasello, M., and J. Call. 1997. *Primate Cognition.* New York: Oxford University Press.

Trivers, R. L. 1971. "The Evolution of Reciprocal Altruism." *Quarterly Review of Biology* 46: 35–57.

———. 1972. "Parental Investment and Sexual Selection." Pp. 136–79 in *Sexual Selection and the Descent of Man*, ed. B. Campbell. Chicago: Aldine.

———. 1974. "Parent–Offspring Conflict." *American Zoologist* 14: 249–64.

———. 1985. *Social Evolution.* Menlo Park, CA: Benjamin/Cummings.

Vrba, E. S. 1985. "Environment and Evolution: Alternative Causes of the Temporal Distribution of Evolutionary Events." *South African Journal of Science* 81: 229–36.

Wallace, A. R. 1871. *The Action of Natural Selection on Man.* New Haven, CT: C. C. Chatfield & Co.

Weismann, A. 1885. *Die Continuität des Keimplasma's als Grundlage einer Theorie der Vererbung.* Jena: Verlag von Gustav Fischer.

White, T. D., G. Suwa, and B. Asfaw. 1994. "*Austraopithecus Ramidus*, A New Species of Early Hominid from Aramis, Ethiopia." *Nature* 371, no. 6495 (22 September): 306–12.

Whiten, A., and D. Custance. 1996. "Studies of Imitation in Chimpanzees and Children." Pp. 291–318 in *Social Learning in Animals: The Roots of Culture*, ed. C. M. Heyes and B. G. Galef. San Diego: Academic Press.

Williams, G. C. 1966. *Adaptation and Natural Selection: A Critique of Some Current Evolutionary Thought.* Princeton, NJ: Princeton University Press.

Wilmut, I. , K. Campbell, and C. Tudge. 2000. *The Second Creation: Dolly and the Age of Biological Control.* London: Headline.

Wilson, E. O. 1975. *Sociobiology: The New Synthesis.* Cambridge, MA: Belknap Press of Harvard University Press.

———. 1978. *On Human Nature.* Cambridge, MA: Harvard University Press.

———. 1998. *Consilience: The Unity of Knowledge.* New York: Knopf.

Wright, S. 1931. "Evolution in Mendelian Populations." *Genetics* 16: 97–159.

———. 1932. "The Roles of Mutation, Inbreeding, Crossbreeding, and Selection in Evolution." *Proceedings of the Sixth International Congress of Genetics* 1: 356–66.

Wynne-Edwards, V. C. 1962. *Animal Dispersion in Relation to Social Behaviour.* Edinburgh: Oliver & Boyd.

———. 1971. "Self-Regulating Systems in Populations of Animals." Pp. 99–135 in *Natural Regulation of Animal Populations*, ed. I. A. McLaren. New York: Atherton Press.

———. 1986. *Evolution through Group Selection*. Oxford: Blackwell Scientific.

Zimmer, C. 1999. "Kenyan Skeleton Shakes Ape Family Tree." *Science* 285, no. 5432 (27 August): 1335, 1337.

Index

About the Author

Graeme Donald Snooks is Coghlan Research Professor in the Institute of Advanced Studies at the Australian National University. Almost two decades ago, he embarked on a research program to develop a realist dynamic theory of the changing fortunes of human society and life from their beginnings. This has given rise to the widely acclaimed dynamic-strategy theory, which Professor Snooks is employing to rethink all aspects of the life sciences.

The results of this research have been published in a number of well-received trilogies, including the global history trilogy (*The Dynamic Society*, *The Ephemeral Civilization*, and *The Laws of History*) and the social dynamics trilogy (*Longrun Dynamics*, *Global Transition*, and *The Global Crisis Makers*). *The Collapse of Darwinism* is the beginning of a new trilogy in which Professor Snooks is exploring the cognitive sciences.